Advanced Research in VLSI and Parallel Systems

Proceedings of Previous Conferences

Proceedings of Caltech Conference on Very Large Scale Integration, 1979, Charles L. Seitz, editor. Computer Science Department, California Institute of Technology, 256-80, Pasadena, CA 91125.

Proceedings of Second Caltech Conference on Very Large Scale Integrations, 1981, Charles L. Seitz, editor. Computer Science Department, California Institute of Technology, 256-80, Pasadena, CA 91125.

Proceedings, Conference on Advanced Research in VLSI, MIT, January 1982, Paul Penfield, Jr., editor. Artech House, Inc., 610 Washington St., Dedham, MA 02026.

Third Caltech Conference on Very Large Scale Integration, 1983, Randal Bryant, editor. Computer Science Press, Inc., 1803 Research Boulevard, Rockville, MD 20850.

Proceedings, Conference on Advanced Research in VLSI 1984, Paul Penfield Jr., editor. Artech House, Inc., 610 Washington St., Dedham, MA 02026.

1985 Chapel Hill Conference on Very Large Scale Integration, Henry Fuchs, editor. Computer Science Press, Inc., 1803 Research Boulevard, Rockville, MD 20850.

Advanced Research in VLSI: Proceedings of the Fourth MIT Conference, Charles Leiserson, editor. MIT Press, 55 Hayward Street, Cambridge, MA 02142.

Advanced Research in VLSI: Proceedings of the 1987 Stanford Conference, Paul Losleben, editor. MIT Press, 55 Hayward Street, Cambridge, MA 02142.

Advanced Research in VLSI: Proceedings of the Fifth MIT Conference, March 1988, Jonathan Allen and F. Thomas Leighton, editors. MIT Press, 55 Hayward Street, Cambridge, MA 02142.

Advanced Research in VLSI: Proceedings of the 1989 Decennial Caltech Conference, Charles L. Seitz, editor. The MIT Press, 55 Hayward Street, Cambridge, MA 02142.

Advanced Research in VLSI: Proceedings of the Sixth MIT Conference, April 1990, William J. Dally, editor. The MIT Press, 55 Hayward Street, Cambridge, MA 02142.

Advanced Research in VLSI: Proceedings of the 1991 University of California / Santa Cruz Conference, Carlo H. Séquin, editor. The MIT Press, 55 Hayward Street, Cambridge, MA 02142.

All of the above volumes are available directly from the publishers.

Advanced Research in VLSI and Parallel Systems

Proceedings of the 1992 Brown/MIT Conference

edited by
Thomas Knight and John Savage

The MIT Press
Cambridge, Massachusetts
London, England

This book was printed and bound in the United States of America.

Library of Congress Cataloging-in-Publication Data

Advanced research in VLSI and parallel systems : proceedings of the 1992 Brown/MIT conference / edited by Thomas Knight and John Savage.
p. cm.
Includes bibliographical references and index.
ISBN 0-262-11166-7
1. Microprocessors—Congresses. 2. Integrated circuits—Very large scale—Congresses. 3. Computer architecture—Congresses. 4. Multiprocessors—Congresses. I. Knight, Thomas. II. Savage, John E., 1939– .
TK7895.M5A38 1992
621.39'2—dc20 92-3855
CIP

Contents

Circuit Technology

Verification and Fault Tolerance

Timing

Network Communication

Preface

This volume constitutes the proceedings of the Brown/MIT Conference on Advanced Research in VLSI and Parallel Systems, the fourteenth in an annual series of conferences begun in 1979. While the field has grown and changed significantly over the past fourteen years, this conference has always sought to bring together researchers from a wide variety of disciplines. VLSI research is a broad area with may subspecialities which often do not interact. This year, in recognition of the increasing importance of VLSI in parallel computation and large systems, the conference title this year has been expanded to explicitly include these areas.

The focus of this conference in the early years was on devices, simple design tools, and the education of a community of users to a radically new and potent technology. We learned from each other, designed and fabricated simple circuits, and created necessary tools.

The success of this early effort allowed us to expand the community to enbrace research on complexity issues. No longer were VLSI circuits simple. A whole new set of issues arose surrounding verification, cooperative design, and multi-level simulation tools. Thousand transistor designs gave way to hundred thousand and million transistor designs within the community. The practical side of the research community developed architectures and tools to control complexity while the theoretical side developed general architectural frameworks and performed analysis to show the utility of these general frameworks as well as limits on the performance of VLSI chips.

The focal point of the conference continues to shift. While traditional areas, such as circuit technology, architecture, CAD, timing, and fault tolerance, continue to be important, new areas and interests are emerging. Sensible design now requires consideration of system level design issues. For processors, this means attention not only to how we construct parallel systems, but also how we program them. Compilers are constructed before or during the chip design. Power, which was a secondary issue in most single chip designs, takes center stage in the design of system level solutions. Testing and reliability, subjects which were once thought of reluctantly, at a late stage in the design cycle, are now some of the most important issues facing system designers. Systems with hundreds or thousands of components are now being actively pursued in the community. Theory and analysis now play a central role in design activities. They are used to confirm and predict the performance of designs and to verify their correctness. All of these changes are reflected in this conference.

Many people have made this conference possible. We express our thanks to the authors who have submitted their most recent research results to this volume. The members of the program committee, and in many cases unnamed co-workers, worked many hours in reading and selecting the twenty-one contributed papers for the conference from the sixty-three submitted. Thanks go to Bernard Chern and Bob Grafton of the Computer Engineering Division of the National Science Foundation for continued NSF support of this conference. Special thanks also go to Daniel Lopresti for handling local arrangements at Brown University, and to Denise Sergent, who excelled in handling the administration of the conference. Finally, we express our sincere appreciation to our general chairman Charles Leiserson, without whom, quite literally, this conference would not have happened.

Tom Knight
John Savage

Program Committee

Tom Knight, Massachusetts Institute of Technology, Cambridge (co-chair)
John Savage, Brown University, Providence (co-chair)

Sandeep Bhatt, Yale University, New Haven
Gaetano Borriello, University of Washington, Seattle
Rich Brown, University of Michigan, Ann Arbor
Randy Bryant, Carnegie Mellon University, Pittsburgh
Srinivas Devadas, Massachusetts Institute of Technology, Cambridge
Gershon Kedem, Duke University, Durham
Monica Lam, Stanford University, Stanford
Jack Raffel, Lincoln Labs, Lexington
Sartaj Sahni, Univeristy of Florida, Gainsville
Carlo Sequin, University of California at Berkeley
Bob Sproull, Sun Microsystems, Inc., Billerica
Jeff Vitter, Brown University, Providence

Advanced Research in VLSI and Parallel Systems

Keynote Address:

Balancing the New Supercomputing Equation

Steve Nelson
Cray Research

Abstract

Significant changes in the supercomputing environment are in motion. These changes are the result of forces which are larger than any of the individual participants.

First, there is the application pull as expressed by the "grand challenge" problems. These problems will easily consume TeraFLOPs of performance. And then there is the technological push by which "super- chip" technology makes possible inexpensive processors with superior price-performance. True supercomputers are always defined as the best-of-class at any given time. The danger is that users will underestimate the ultimate cost of systems using massive numbers of these inexpensive processors and the significant amounts of associated memory. This technology will, however, be much more scalable than in the past and will result in an ever wider range of price and performance choices.

Balance, the efficient utilization of all system resources, will determine which systems provide the best value. There are significant challenges in maintaining this balance with regard to data movement, given the ability to dramatically increase the rate of computation at each node. Each computation has its I/O cost. Those who concentrate on "cheap FLOPs" and neglect the requirement of delivering and retiring data around compute-intensive nodes will limit their success. Likewise, software innovations through new algorithms, compilers, operating systems, and development tools will turn out to be much more important than precisely how many calculations can be performed per ALU-second.

Architecture

Design and Implementation of the Message-Driven Processor

William J. Dally, *Salim Ahmed, *Paul Carrick, Andrew Chien, *Roy Davison, J.A. Stuart Fiske, *Greg Fyler, Waldemar Horwat, John S. Keen, *Steve Lear, Richard A. Lethin, *Mark Vestrich, *Ted Nguyen, Michael Noakes, Peter R. Nuth, and D. Scott Wills

Artificial Intelligence Laboratory
Massachusetts Institute of Technology
545 Technology Square
Cambridge, MA 02139

* Platform Architecture Group
Intel Corporation
3065 Bowers Avenue
Santa Clara, CA 95052-8131

Abstract

The Message-Driven Processor (MDP) is an integrated multicom puter node. It incorporates a 36-bit integer processor, a memory management unit, a router for a 3-D mesh network, a network interface, and a 4K-word × 36-bit SRAM, and an ECC DRAM controller on a single 1.1M transistor VLSI chip. The MDP is not specialized for a single model of computation. Instead, it incorporates efficient primitive mechanisms for communication, synchronization and naming. These mechanisms efficiently support most proposed parallel programming models. Each processing node of the MIT J-Machine consists of an MDP with 1 MByte of DRAM. MDPs have been operational since June 1991 and J-Machines built from them have been on-line since July 1991. This paper describes the design and implementation of the MDP component.

1 Introduction

The Message-Driven Processor (MDP) is a multicomputer node incorporating a 36-bit integer processor, a memory management unit, a 3-D router, a communications controller, a 4K × 36-bit SRAM, and a DRAM controller with ECC on a single 1.1M transistor chip. An MDP, with optional external DRAM, forms a processing node of the MIT J-Machine, a distributed-memory, MIMD, concurrent computer [15]. A J-Machine is an ensemble of up to 65,536 message-driven processors (MDPs). This limit is set by the addressability of the router and the bandwidth of the network. The nodes are connected by a high-speed three-dimensional mesh network. Each J-Machine network channel has a bandwidth of 288Mbits/s.

The MDP supports a broad range of parallel programming models (including shared-memory[7], data parallel[22], dataflow[35], actors[21], and explicit message-passing[32]) by providing low-overhead primitive mechanisms for communication, synchronization, and translation. Communication mechanisms are provided that permit a user-level task on one node to send a message to any other node in a 4K-node machine in $< 2\mu s$. No processing resources on intermediate nodes are consumed and buffer memory is automatically allocated on the receiving node. On message arrival, a task is created and dispatched in $< 1\mu s$. Synchronization is provided by presence tags on all storage locations. Three separate register sets allow fast task switching. The translation mechanism maintains bindings between arbitrary names and values; it may be used to support a global virtual address space. These mechanisms have been selected to be both general and amenable to efficient hardware implementation. To support fine-grain concurrent programming systems, the mechanisms are designed to efficiently handle small objects (8 words) and small tasks (20 instructions).

This design was chosen to make the most efficient use of available chip and board area. Packaging a small amount of memory on each node gives us an extremely high memory bandwidth (3Gbits/s per chip or 200Tbits/s in a fully populated system). Memory consumes most of the chip area; from one point of view, the system is a memory with processors added to each node to improve bandwidth for local operations. The fast communication and global address space prevent the small local memories from limiting programmability or performance. The 3-D network gives the highest throughput and lowest latency for a given wire density[11, 13]. It allows the processing nodes to be packed densely and results in uniformly short wires.

The MDP builds on previous work on multicomputer design. Like the Caltech Cosmic Cube [33, 2], the Intel iPSC [31], the N-CUBE [30], and the Ametek [1], each MDP in the J-Machine has a local memory and communicates with other nodes by passing messages. Because of its low overhead, the J-Machine can exploit concurrency at a much finer grain than these early message passing computers. Delivering a message and dispatching a task in response to the message's arrival takes $< 3\mu s$ on the J-Machine, as opposed to 5ms on an iPSC-1 or $300\mu s$ on an iPSC-2. Like the BBN Butterfly [7] and the IBM RP3 [5] the J-Machine provides a global virtual address space. The same IDs (virtual addresses) are used to reference local (on the same node) and remote (on another node) objects. Like the InMOS transputer [3], the Caltech MOSAIC [27], and the Intel iWARP [4], a J-Machine node is a single chip processing element integrating a processor, memory, and a communication unit. The J-Machine is unique in that it extends these previous efforts with efficient primitive mechanisms for communication, synchronization and naming [12]. The J-Machine uses a direct communication network based on the work reported in [10, 16, 13, 34, 17, 20].

2 Instruction Set Architecture

The MDP extends a conventional microprocessor instruction-set architecture (ISA) with instructions to support parallel processing. Specifically, the MDP provides efficient hardware mechanisms for communication, synchronization and naming. This section describes the MDP ISA with particular emphasis on these mechanisms. Further details of the MDP ISA are described in [14, 15, 9, 19].

Register Set: The MDP provides separate register sets to support rapid switching between three execution levels: background, priority 0 (*P0*), and priority 1 (*P1*). The MDP executes at the background level when there are no pending messages. Each arriving message creates a task and initiates execution at *P0* or *P1* depending on the priority of the message. The MDP executes the highest priority task at any point in time. Thus, the arrival of a *P1* message while the MDP is executing a *P0* task will cause the MDP to switch execution levels (and thus register sets). When the *P1* task completes, execution at *P0* is resumed by switching to the *P0* register set which holds the register state of the suspended task.

The register set at each priority level includes:

4 general-purpose data registers:	`R0-R3`
4 address registers:	`A0-A3`
4 ID registers:[1]	`ID0-ID3`
1 instruction pointer:	`IP`

In addition to the registers listed above, there are several status and control registers not discussed here. Quick task switching at the same priority level is facilitated by the relatively small size of each register set.

The MDP operates on tagged data; every register and every word of memory holds a 36-bit word which consists of a 32 bit data value and a 4 bit tag. Twelve of the tags are pre-defined for data types that are commonly used (such as SYMBOLS, INTEGERS, BOOLEANS, etc.) and the remaining four are left for the user to define. These tags are used by the processor to check the type of operands to any numeric or logical operation. Attempting an operation on an operand of the wrong type results in a trap to fault-handling code. Tags are also used to provide fast trapping for two primitive synchronization types supported by the hardware.

In addition to the type checking performed by the hardware, structure references are normally bounds checked by the processor. Each address register `A0-A3` contains a physical address pointer consisting of a base and a length field. This pointer defines a *segment* in physical memory. Memory is read or written by indexing off one of the address registers. References

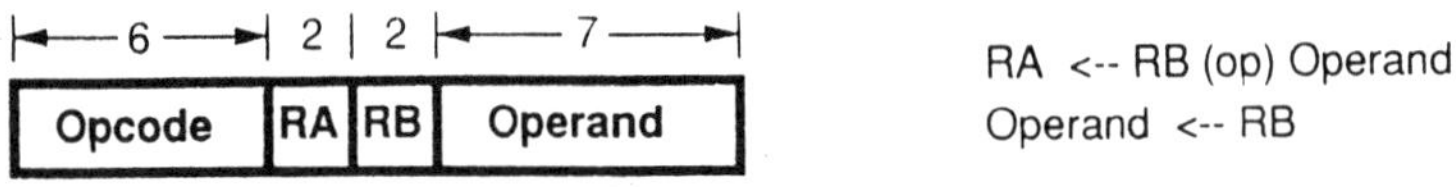

Figure 1: MDP Instruction Format: Two 17-bit three-address instructions are packed in each 36-bit word.

beyond the bounds of the segment trap to an exception handler.

The ID registers hold names of objects. An object's name never changes, but its location in physical memory may vary over time. An object may move around within the memory space of a particular node, or it may leave one node and move to another. The ID registers enable an executing program to keep track of objects in spite of such physical movements. The MDP provides instructions which efficiently translate an object's ID to a physical location. This allows a J-Machine to support a global name space on a distributed memory machine.

Instructions: The MDP executes 17-bit, fixed format, three-address instructions including the usual complement of arithmetic, logical, and control operations. The instruction format is shown in Figure 1. Each instruction specifies an operation, two register operands, and a third operand that may be a register, a memory location, or a constant. Two 17-bit instructions are packed in each 36-bit word. Any instruction-stream word that is not tagged as an instruction is considered a constant and loaded into general purpose register `R0`. This provides an efficient means for loading arbtrary 36-bit constants. The MDP instruction set is summarized by category in Figure 2.

Communication: To achieve low-overhead communication for fine-grain messages, the MDP provides hardware support for end-to-end message delivery. Message formatting, injection, delivery, buffer allocation, message buffering, and task scheduling are managed in hardware to reduce overhead.

An MDP transmits a message using a series of `SEND` instructions each of which injects one or two words into the network. A typical message send is shown in Figure 3. The first `SEND` instruction specifies the absolute address of the destination node in $< x, y, z >$ format for the network hardware. The second `SEND` reads the first two words of the message out of registers `R1` and `R2` and enqueues them for transmission. The final instruction enqueues two additional words of data, one from `R3`, and one from memory. The

Data Movement:

READ	WRITE	READR	WRITER	RTAG	WTAG	LDIP
LDIPR	CHECK					

Arithmetic and Logic:

CARRY	ADD	SUB	MULH	MUL	ASH	LSH
ROT	AND	OR	XOR	FFB	NOT	NEG
LT	LE	GE	GT	EQUAL	NEQUAL	EQ
NEQ						

Communication:

SEND	SENDE	SEND2	SEND2E

Translation:

XLATE	ENTER	PROBE

Special Instructions:

NOP	INVAL	SUSPENDCALL

Branches:

BR	BNIL	BNNIL	BF	BT	BZ	BNZ

Figure 2: The MDP instruction set includes communication and translation instructions.

```
SEND    R0           ; send net address
SEND2   R1,R2        ; header and receiver
SEND2E  R3,[3,A3]    ; selector and continuation - end msg.
```

Figure 3: MDP assembly code to send a 4 word message uses three variants of the SEND instruction.

use of the SENDE instruction marks the end of the message and causes it to be transmitted into the network. This sequence executes in 4 clock cycles (250ns).

An injected message is delivered to the destination node by the network (Section 3). At the destination, it is buffered in a hardware-managed FIFO queue in the internal RAM of the MDP. Separate queues are maintained for *P0* and *P1* messages.

Naming is supported in the MDP via translation instructions and segmented addressing. Addressing memory through segment descriptors permits arbitrarily sized objects to be relocated and protected. The **ENTER** instruction enters an arbitrary translation from a 36-bit key to a 36-bit data value in a set-associative cache (translation table) mapped into the on-chip memory. The **XLATE** instruction looks up the data value (if any) associated with a key. These instructions may be used to translate virtual object-IDs into physical segment descriptors or node numbers to support a global virtual address space. A segmented memory management unit may be used with the translation mechanism to provide protected addressing.

Synchronization: The 4 tag bits of the MDP word may specify 16 data types. Two of these types represent *futures.* A *future* tag indicates that the 32 bit data portion of the word does not yet hold the intended value. The *future* is used to synchronize between tasks. An attempt to reference values tagged *future* causes an exception. The exception handler typically suspends the task until the data is available.

Task Scheduling: When a message reaches the head of the highest priority non-empty queue, a task is created to handle it by changing the thread of control and creating a new addressing environment as shown in Figure 4. Every message header contains a message *opcode* and the message *length.* The message opcode is loaded into the IP to start a new thread of control. The length field is used along with the queue head to create a message segment descriptor (automatically written to A3) that represents the initial addressing environment for the task. The message handler code may open additional segments by translating object IDs in the message into segment descriptors. Creating a task to handle a message takes 3 cycles.

The dispatch mechanism is used directly to process messages requiring low latency (e.g., combining and forwarding). Other messages (e.g., remote procedure call) specify a handler that locates the required method (using the translation mechanism described above) and then transfers control to it.

For example, a remote procedure call message is handled by the call handler code as shown in Figure 5. The execution of this handler is depicted in Figure 6. The first instruction gets the method ID (offset 1 into the message segment referenced by A3). The next instruction translates this method ID into a segment descriptor for the method and places this descriptor in A0. In one of its operating modes, the MDP can use A0 as a pointer to a segment of code, and IP as an index into that segment. This allows code to be easily relocated at runtime. The final instruction of the CALL handler transfers control to the method by loading the IP with a short integer offset. Thereafter, the MDP will fetch instructions from the called method.

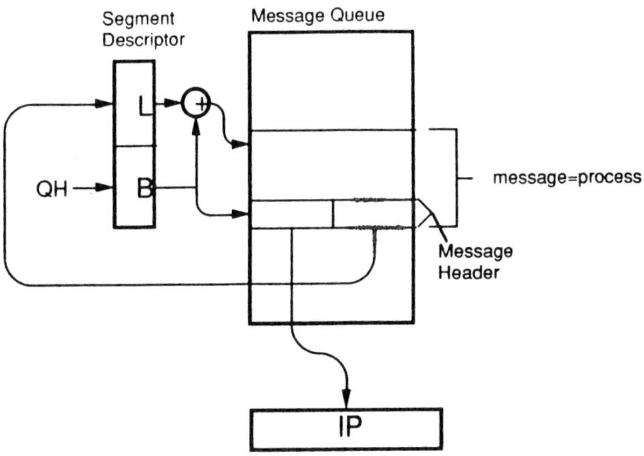

Figure 4: Message dispatch. In three clock cycles, a new task is created by (1) setting the IP to change the thread of control and (2) creating a message segment to provide the initial addressing environment.

```
MOVE    [1,A3],R0   ; get method id
XLATE   R0,A0       ; translate to segment descriptor
LDIP    INITIAL_IP  ; load instruction pointer to transfer control
                    ; to method
```

Figure 5: MDP assembly code for the CALL message handler.

The method code may then read in arguments from the message queue. The argument object identifiers are translated to physical memory base/length pairs using the XLATE instruction. If the method needs space to store local state, it may create a context object. When the method has finished execution, or when it needs to wait for a reply, it executes a SUSPEND instruction which dequeues its message and passes control to the next message in the queue.

Further details of the MDP instruction set architecture are described in [14, 15, 9, 19].

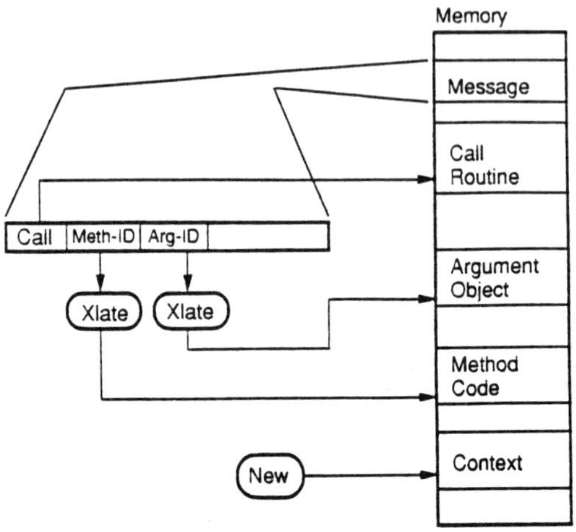

Figure 6: The CALL message handler invokes a method by translating the method identifier to find the code, creating a context (if necessary) to hold local state, and translating argument identifiers to locate arguments.

3 Network Architecture

The MDP contains a router and network interface that provide fast end-to-end message handling in a 3-D mesh network. Figure 7 shows the functions performed by the MDP network. The origin MDP executes a number of SEND instructions to enqueue words of the message in a FIFO for transmission through the network. The message is then transmitted to the destination through a number of intervening routers. The routers operate autonomously from the MDP integer core, allowing both communication and processing to occur simultaneously. Finally, at the destination node, the message is enqueued in a circular buffer in the processor's internal memory.

Each node's router communicates with the routers on the six neighboring nodes via bidirectional 9-bit wide channels that are cycled at 32MHz, twice the processor frequency. Unlike previous asynchronous routers[16, 28, 18], the MDP router is synchronous. All routers in the system are controlled by the same distributed clock. A novel pad design helps tolerate clock skew between routers and permits the channels to be turned around every clock without danger of conduction overlap. Messages are routed deterministically in dimension order: in the X direction first, then Y, then Z. Dimension order routing is provably deadlock-free[17] and permits the router to be partitioned by dimension into three identical, simple modules. The simple design results in a one clock cycle per node latency for message routing. The network

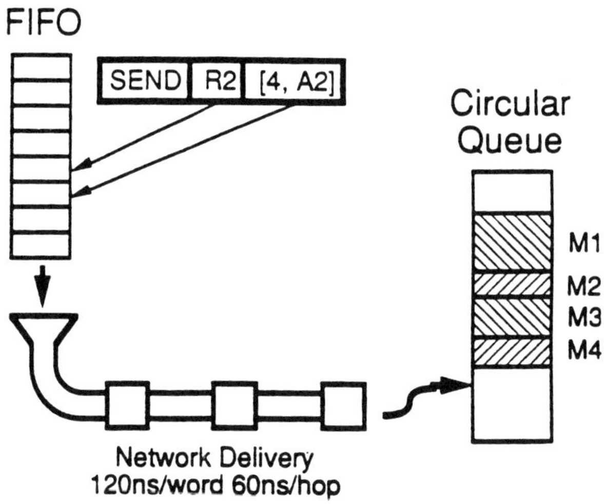

Figure 7: MDP Hardware provides end-to-end support for communication. A message is formatted by SEND instructions, delivered by the network, and buffered in an automatically allocated circular queue at the receiving node.

supports two message priorities on logically separate networks that share the same physical wires. This permits high priority messages to be used to handle faults involving the low priority network.

Network I/O interfaces provide hardware message handling for messages at their terminal nodes. A *network output* interface contains an 8-word FIFO buffer to match the rate of message generation (via the SEND instruction described above) to the network throughput. This interface also processes the destination address to compute the direction the message must travel in along each dimension. This permits the use of absolute addressing with a simple equality comparison in the routers. A *network input* interface buffers words arriving from the network until they are enqueued in circular message queues in the MDP's internal memory.

4 Implementation

The MDP was implemented conservatively to minimize risk. Among other things, this led to a simple microarchitecture, a conservative synchronous router, the extensive use of standard cells, and a generous non-overlap period for the two-phase clock.

Microarchitecture: The microarchitecture of the MDP consists of the eleven modules described below.

1. **Control and Prefetch:** Fetches and decodes an instruction sequence. Generates state sequences that control the operation of the remaining blocks.

2. **Diagnostic Port:** Permits control of the processor (e.g., reset, halt, single step, run,...) and access to both the on-chip and off-chip RAMs. Commands, addresses and data are scanned in and out serially.

3. **RALU:** The general registers and ALU. The standard arithmetic, logical and comparison instructions are performed in this block. The RALU in combination with (1) has most of the functionality of a RISC processor with tags.

4. **AAU:** The address arithmetic unit generates all memory addresses for data read and write, instruction fetches, network enqueuing, and task dispatch. All memory management and message queue buffer management functions are performed in this block.

5. **Memory:** The internal memory is a 4K $\times$ 36-bit 6-transistor static read/write memory. It includes two 144-bit wide input row buffers for enqueuing network data, one output row buffer for reading instructions, and comparators for implementing set associative access. This subsystem is actually two modules, the full-custom RAM array, and the standard-cell RAM interface. Both the prefetcher and the network-input access memory 144-bits at a time in order to reduce memory contention.

   Four instruction words (8 instructions) at a time are read from the RAM into instruction row buffers in the IMEM. Similarly, incoming message words are accumulated in queue row buffers and written four at a time into the RAM. This buffering of instructions and incoming messages reduces memory contention.

6. **External Memory Interface:** This subsystem generates timing signals, multiplexes addresses and data, and performs error detection and correction to interface standard dynamic memory components to the MDP.

7. **Network Input:** The network input module accepts data from the network, formats the 9-bit flits into 36-bit words, assembles words into the queue row buffers, and signals the control module to write the queue row buffers to memory.

8. **Network Output:** The network output contains a FIFO send buffer and logic to sequence messages into the network. This unit also sub-

tracts the node address from the message header to generate direction routing bits for the network.

9. **Router:** The router handles routing and flow control to deliver messages across the network. It is a synchronous 3-dimensional mesh router that provides two logically separate networks on one set of wires. It routes messages obliviously using absolute addresses and direction routing bits generated by the network output.

10. **Clock:** A two-phase non-overlapping clock generator matched to the clock load of the MDP.

11. **Pads:** Full-custom signal, power, and ground pads.

Two phase clocking is used throughout the chip. Instructions execute in from one to three cycles under hardwired control. Prefetch and decode of the next instruction are overlapped with the last cycle of an instruction's execution; however, instruction execution is not pipelined.

The *DIAG* module contains a serial scan path that is used to control the MDP and to access its internal state. All memory locations (internal and external) can be read and written via this path. The processor can also be reset, halted, interrupted, or single-stepped via the DIAG port.

Design Methodology: The MDP was implemented using Intel standard cells except for the on-chip RAM, clock generator, and pads. Using standard cells sacrificed a factor of 3-4 in area and 2-3 in performance over what would be possible with full-custom design. The advantage was a factor of 10 increase in productivity which was essential to complete the chip with our small design team.

About $35K$ standard cells containing $\approx 210K$ transistors[2] were used in the ≈ 700 sheets of schematics drafted at MIT. These schematics were transferred to Intel for layout. Many of the datapaths were laid out by hand to exploit the regularity of the design. Less regular collections of logic were laid out using automatic place and route CAD tools.

We wrote many of our own tools from scratch (netlist translators, logic simulators, timing verifiers) because we found no tools that were simultaneously efficient working with large designs, bug-free, and available in source form.

Three simulators were constructed representing the MDP at different levels: an instruction-level model, an RTL model, and a gate-level model.

Instruction Level: The instruction-level simulator [23] runs MDP machine code quickly, and has been used primarily for compiler development and architectural verification. It can be directed to produce assertions that specify the sequence of values in registers when running a program. These assertions are accepted by the RTL model of the MDP, to check its operation against the instruction-level simulator.

Register Transfer Level (RTL): The design of the modules described in section 4 was originally captured in an RTL simulator, a 34,000 line "C" program that contains structures that maintain the state of each register in the MDP and statements that simulate the function units and interconnect. The RTL simulator is accurate to the clock-phase and models all inter-module signals. Its results can be automatically compared to both instruction-level and logic-level simulations.

Gate/Switch Level: Each RTL module was reduced to a gate-level design – a composition of Intel standard cells. At the gate-level, the design consists of about 700 sheets of schematics. A gate-level simulator built into the RTL simulator is used to simulate the gate-level design and to compare its behavior with the RTL model. The gate-level simulator includes RC timing models and simulates the design's behavior in 100ps time increments.

The use of a standard cell library that did not include discrete transistors and transfer gates allowed the design to be simulated at the gate level, which is significantly faster and requires less memory than simulating the design at the switch level. (All simulations were performed using SPARC workstations.) We wrote our own fast logic simulator[26], which proved useful: we were able to customize the modeling, fix bugs, and add new features as necessary. This also permitted us to link the logic model and the RTL model into the same program. By running the two simulations simultaneously, they could be checked against one another with powerful semantics similar to test-vectors, avoiding the need to maintain huge files of vectors on disk.

The clock-generator and pad drivers were simulated both using a circuit simulator, and at the switch-level connected to the rest of the MDP. As layout progressed, parasitic capacitances were extracted periodically and back-annotated into the schematics to adjust and test their effect on timing. Small critical sections of the RAM were verified using a circuit simulator.

Layout was verified against schematics using a netlist comparison program. In many cases, this required repartitioning of the schematic hierarchy in order to be isomorphic to the layout hierarchy. As a final check, we extracted switch-level schematics from the layout and ran the regression suite on them. This also served to verify that our standard-cell models were correct.

To meet performance goals, almost all of the automatically synthesized logic (generated with the Berkeley MIS-II system) was replaced by hand-

Module	Dimensions $mm \times mm$	Area mm^2	Transistors $\times 10^3$
AAU	3.7×7.0	25.9	75.0
RALU	3.7×2.9	10.7	39.0
DIAG	0.9×1.1	1.0	3.7
PREF	0.9×1.1	1.0	3.2
CNTL	1.1×2.6	2.9	8.7
IMEM	7.8×0.5	3.9	13.0
EMI	1.6×1.8	2.9	9.0
NETIN	1.8×0.7	1.3	4.4
NETOUT	2.1×1.8	3.8	18.0
ROUTERS	8.4×1.3	10.9	29.0
RAM	8.8×4.9	43.1	880.0
CLOCK	0.7×0.8	0.6	0.1
PADS	50.5×0.2	8.4	2.6
Full Chip	10.2×15.0	153.0	1087.0

Figure 8: MDP area and transistor count by module

generated schematics. Significant repartitioning of critical functions was performed to reduce critical paths. Over 200 additional schematics were added during the sizing process and almost all of the original 500 sheets of schematics were modified.

Layout: The layout of the MDP chip was performed by Intel Corporation with the assistance of four MIT graduate students. The layout for most of the modules (all but the RAM, clock, and I/O pads) consisted of Intel standard cells. The large data paths in the AAU, RALU, Network Output, and Memory (RAM Interface) were laid out by hand-placing and hand-connecting standard cells in a graphics editor. All other standard-cell modules were laid out using an Intel-proprietary standard-cell place and route program.

The area and transistor counts for the different modules is shown in Figure 8. Figure 9 shows a die photograph of the MDP chip along with a floorplan showing each of these modules.

Schedule and Status: Architecture studies leading to the MDP were started in October 1986. Work on the RTL model of the microarchitecture began in June 1988. Schematic entry at MIT started in November 1988. The task of translating schematics into layout commenced in June 1989. Layout was completed in December 1990. These project phases overlapped.

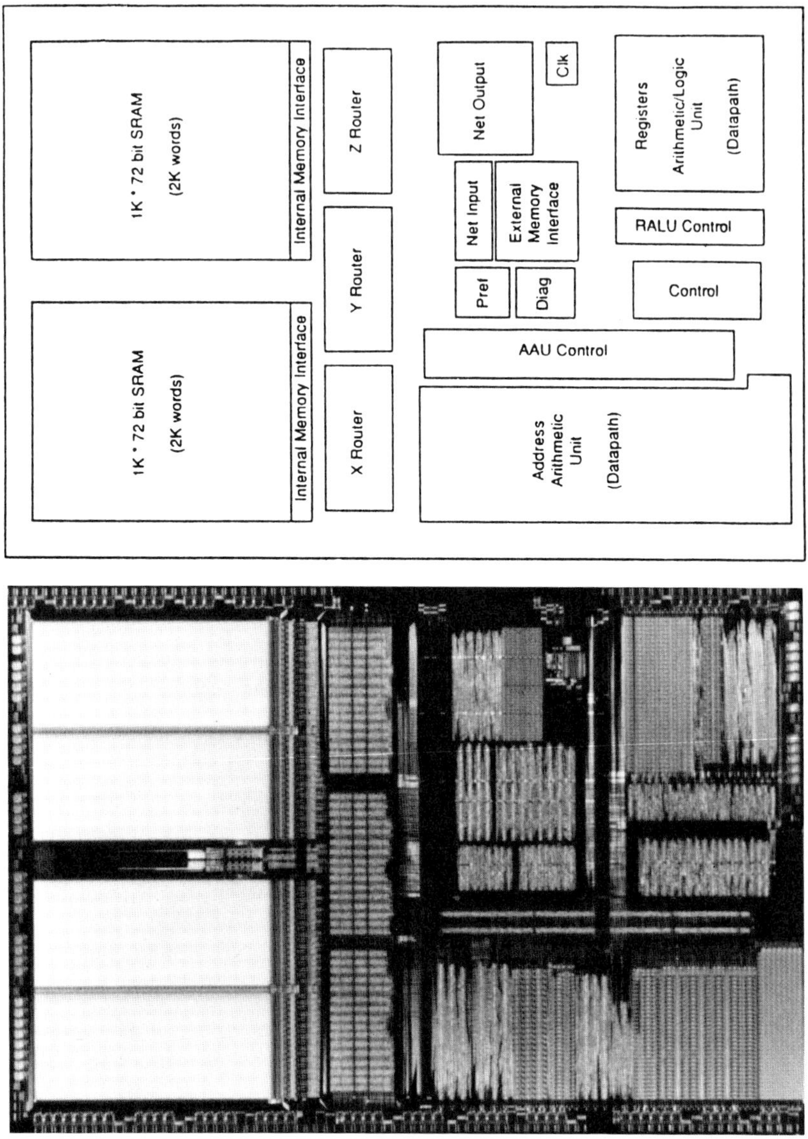

Figure 9: MDP chip die photograph and floorplan

Figure 10: J-Machine Processor Board: This board contains 64-nodes each comprising an MDP and three 1M×4 DRAMs.

Several design errors in the RTL were discovered rather late. The flexibility provided by the CAD tools permitted corrections to be propagated down to the schematic and layout levels fairly easily.

First silicon was received in June 1991 and was running programs within a few hours. Although we thoroughly simulated the logic design, running our validation tests and applications on the hardware has uncovered 8 bugs. These bugs were not caught during our RTL and logic simulations because our tests did not cover the very specific conditions under which they occur. All of the bugs involve interactions between instructions, asynchronous events, and/or diagnostic mode. While we simulated all of these features in isolation, the interactions that exposed the bugs were not simulated. Most of these bugs have simple software work-arounds, but for performance reasons a second revision of the layout with modified control logic and some metal fixes was sent for fabrication in December 1991. We plan to use several thousand of these chips to build research multicomputers at MIT.

System Design: A photograph of a J-Machine processor board is shown in Figure 10. The board measures 20.5 by 24 inches and contains 64 processing nodes. Each node consists of an MDP chip (in a 168-pin PGA package)

Figure 11: J-Machine Chassis: 1024 nodes are packaged on 16 64-node boards. Elastomer connectors make 2240 connections between each pair of boards.

and three 4Mbit DRAMs. Each pair of nodes shares a set of elastomeric connectors to communicate with the corresponding nodes on the boards above or below the board in a stack. A total of 32 elastomeric connectors held in 4 connector holders provide 2240 electrical connections between adjacent boards. Of these connections, 960 are used for signalling and the remaining are ground returns. No power is supplied through the elastomers. Power and ground are supplied to each board directly from bus bars. The center area of the board contains the final stage of the clock distribution network along with diagnostic fanout and multiplexing logic and temperature and airflow monitors.

A photograph of our chassis for a 1024-node system is shown in Figure 11. The chassis contains a *stack* of 16 processor boards, power supplies, and distribution bus bars. The boards are held together by a set of 20 tie-rods to compress the elastomer connectors. Larger systems are built by combining several chassis with each stack is connected to its neighboring stacks by 128 (16×8) short 60-pin ribbon cables (one for each pair of nodes on the periphery).

In addition to the processor board and chassis, we have also designed a diagnostic interface board, and are in the process of designing a SCSI disk interface, a distributed graphics frame buffer, and an S-bus interface. More details of the J-Machine system design are given in [29].

5 Software

The J-machine is intended as a platform for software experiments in fine-grained parallel programming. To this end, we have implemented and are studying software systems for different fine-grain programming models. Fine-grained programs typically execute from 10-100 instructions between communication and synchronization actions. Reducing the grain size of a program increases both the potential speedup due to parallel execution and the potential overhead associated with parallelism. Special hardware mechanisms to reduce the overhead due to communication, process switching, synchronization, and multithreading are therefore central to the design of the MDP.

A Concurrent Smalltalk (CST) programming system consisting of an operating system kernel (COSMOS) and an optimizing CST compiler has been developed for the J-Machine[23]. This system is currently operational on the prototype hardware. Concurrent Smalltalk is a parallel object-oriented programming language that includes an asynchronous message send and distributed objects. COSMOS is an operating system for the J-Machine that provides a global virtual namespace, object-based memory management, support for distributed objects, and low-overhead context switching. Its memory management system provides fast, transparent access to storage distributed across the machine. COSMOS is designed to efficiently support fine-grain concurrent computation where tasks are very short (40 user instructions) and data objects are very small (8 words). The CST compiler and the COSMOS runtime system also provide floating-point arithmetic, simple arrays, and garbage collection for CST programs.

We have experimented with several methods of executing dataflow programs on the J-machine[36, 35]. The simplest of the systems translates each node of the dataflow graph into a sequence of MDP instructions. A dataflow node with two inputs takes 20 MDP instructions to simulate it, i.e., to store the first data value, match it with the second value when it arrives, perform the dataflow operation, and send the resulting value to two destinations. This makes use of the `CFUT` tag and fault handler.

An obvious optimization is increasing the granularity of tasks in order to reduce scheduling overhead, retaining only enough exposed parallelism to keep processors from sitting idle. [36, 35] describe a system that simulates a hybrid dataflow/von Neumann architecture[25], which, while having some advantages, turned out to require excessive emulation overhead. We are currently building a system on top of the Berkeley TAM project[8] that addresses the inefficiencies of our earlier systems.

6 Conclusion

The MDP has been built to demonstrate the utility of general-purpose communication and synchronization mechanisms in a multicomputer building block. Its mechanisms have been shown to efficiently support dataflow [35] and object-oriented programming [24, 6] models using a shared address space. Our studies have shown that the use of a few simple mechanisms provides orders of magnitude lower communication and synchronization overhead than is possible with multicomputers built from off-the-shelf microprocessors. Communication and synchronization performance is competitive with processing nodes specialized to a single model of computation such as iWARP[4] (systolic) or the Transputer [3] (CSP).

Computers built from fine-grain processing nodes, such as the MDP, consisting of a small but powerful processor and a small amount of memory, are more cost-effective than those built from fewer coarse-grain nodes. Fine-grain nodes have a larger fraction of their silicon area devoted to processing and have higher arithmetic, memory, and communication bandwidth per unit cost. Large scale parallel machines built from fine-grained processors have a larger total amount of memory within a given latency of a processor. An efficient network design gives global memory latency and bandwidth competitive with coarse grain machines.

Our software studies have suggested improvements that could be made to the MDP. More registers, floating-point hardware, and better mapping mechanisms would be useful. The current implementation was very conservative, so there are certainly opportunities for streamlining the implementation, decreasing the cycle time and number of clocks per instruction. A commercial custom VLSI product based on the architectural mechanisms in the MDP is very plausible.

The MDP project has demonstrated the feasibility of building experimental systems with limited resources. By concentrating on the novel mechanisms of the MDP and keeping the design simple and modest in other respects we were able to complete the design of the chip, its system level hardware, and several programming systems with a handful (< 8) of graduate students and engineers in 2.5 years.

Acknowledgment

The research described in this paper was supported in part by the Defense Advanced Research Projects Agency under contracts N00014-88K-0738 and N00014-91-J-1698, in part by a National Science Foundation Presidential Young Investigator Award, grant MIP-8657531, with matching funds from General Electric Corporation, IBM Corporation and AT&T Corporation, and in part by assistance from Intel Corporation. Many members of the MIT Concurrent VLSI Architecture Group, in particular Scott Furman, Shaun Kaneshiro, Ellen Spertus, and Deborah Wallach, contributed to the design and programming of the J-Machine. Discussions with Chuck Seitz, Steve Ward, Anant Agarwal, and Tom Knight were helpful during the early stages of the J-Machine project. We thank Lisa Sardegna for her help preparing this manuscript.

Footnotes

[1] The background register set does not include ID registers. They only exist at *P0* and *P1*.

[2] The remaining $\approx$ 890K devices are contained in the full custom portions of the chip; mostly in the RAM.

References

[1] Ametek Computer Research Division. *Series 2010 Product Description*, 1987.

[2] William C. Athas and Charles L. Seitz. Multicomputers: Message-Passing Concurrent Computers. *Computer*, 21(8):9 – 24, August 1988.

[3] Iann Barron, Peter Cavill, David May, and Pete Wilson. Transputer Does Five or More MIPS Even When Not Used in Parallel. *Electronics*, pages 109–115, November 1983.

[4] Shekhar Borkar et al. iWARP: An Integrated Solution to High-Speed Parallel Computing. In *Proceedings of the Supercomputing Conference*, pages 330–338. IEEE, November 1988.

[5] W.C. Brantley, K.P. McAuliffe, and J. Weiss. RP3 Processor-Memory Element. *IEEE Transactions on Computers*, pages 782–789, 1985.

[6] Andrew A. Chien and William J. Dally. Experience with Concurrent Aggregates (CA): Implementation and Programming. In *Proceedings of the Fifth Distributed Memory Computers Conference*, Charleston, South Carolina, April 8-12 1990. SIAM.

[7] W. Crowther, J. Goodhue, E. Starr, R. Thomas, W. Milliken, and T. Blackadar. Performance Measurements On A 128-Node Butterfly Parallel Processor. 1985.

[8] David E. Culler, Anurag Sah, Klaus Erik Shauser, Thorsten von Eicken, and John Wawrzynek. Fine-Grain Parallelism with Minimal Hardware Support: A

Compiler-Controlled Threaded Abstract Machine. In *Proceedings of the Fourth International Conference on Architectural Support for Programming Languages and Operating Systems*, pages 164–175, 1991.

[9] William Dally et al. Message-Driven Processor Architecture. Version 11 AI Memo No. 1069, MIT, 545 Technology Sq., Cambridge, MA, August 1988.

[10] William J. Dally. *A VLSI Architecture for Concurrent Data Structures*. Kluwer Academic Publishers, 1987.

[11] William J. Dally. Wire-Efficient VLSI Multiprocessor Communication Networks. In Paul Losleban, editor, *Proceedings of Stanford Conference on Advanced Research in VLSI*, pages 391–415. MIT Press, 1987.

[12] William J. Dally. Mechanisms for Concurrent Computing. In *Proceedings of the International Conference on Fifth Generation Computer Systems*, pages 154–156, December 1988.

[13] William J. Dally. Performance Analysis of k-ary n-cube Interconnection Networks. *IEEE Transactions on Computers*, 39(6), June 1990.

[14] William J. Dally et al. Architecture of a Message-Driven Processor. In *Proceedings of the 14th International Symposium on Computer Architecture*, pages 189–205. IEEE, Computer Society Press, June 1987.

[15] William J. Dally et al. The J-Machine: A Fine-Grain Concurrent Computer. In G.X. Ritter, editor, *Proceedings of the IFIP Congress*, pages 1147–1153. North-Holland, August 1989.

[16] William J. Dally and Charles L. Seitz. The Torus Routing Chip. *Distributed Computing*, 1:187–196, 1986.

[17] William J. Dally and Charles L. Seitz. Deadlock Free Message Routing in Multiprocessor Interconnection Networks. *IEEE Transactions on Computers*, C-36(5):547–53, May 1987.

[18] William J. Dally and Paul Song. Design of a Self-Timed VLSI Multicomputer Communication Controller. In *Proceedings of the International Conference on Computer Design*, pages 230–234. IEEE, Computer Society Press, October 1987.

[19] William J. Dally and D. Scott Wills. Universal Mechanisms for Concurrency. In G. Goos and J. Hartmanis, editors, *Proceedings of PARLE-89*, pages 19–33, Eindhoven, The Netherlands, June 1989. Springer-Verlag.

[20] Charles M. Flaig. VLSI Mesh Routing Systems. Dept. of Computer Science Technical Report 5241:TR:87, California Institute of Technology, 1987.

[21] Carl E. Hewitt. Representing Organizations in Actor Systems. *MPSG Memo 88:1*, 1988.

[22] W. Daniel Hillis. The Connection Machine. Technical Report AI Memo 646, Massachusetts Institute of Technology Artificial Intelligence Laboratory, Cambridge, MA 02139, September 1981.

[23] Waldemar Horwat. Concurrent Smalltalk on the Message-Driven Processor. Master's thesis, MIT, May 1989.

[24] Waldemar Horwat, Andrew Chien, and William J. Dally. Experience with CST:Programming and Implementation. In *Proceedings of the ACM SIGPLAN 89 Conference on Programming Language Design and Implementation*, 1989.

[25] Robert Alan Ianucci. A Dataflow / von Neumann Hybrid Architecture. Technical Report MIT/LCS/TR 228, MIT Laboratory for Computer Science, Cambridge, MA, 1988.

[26] Richard A. Lethin. A Simulator for the Message-Driven Processor. Master's thesis, MIT, February 1991.

[27] C. Lutz et al. Design of the Mosaic Element. In *Proc. MIT Conference on Advanced Research in VLSI*, pages 1–10. Artech Books, 1984.

[28] John Y. Ngai and Charles L. Seitz. A Framework for Adaptive Routing in Multicomputer Networks. In *ACM Symposium on Parallel Algorithms and Architectures*, 1989.

[29] Michael Noakes and William J. Dally. System Design of the J-Machine. In William J. Dally, editor, *Sixth MIT Conference of Advanced Research in VLSI*, pages 179–194, Cambridge, MA 02139, 1990. The MIT Press.

[30] John F. Palmer. The NCUBE Family of Parallel Supercomputers. In *Proc. IEEE International Conference on Computer Design*, page 107. IEEE, 1986.

[31] David C. Rudolph and Daniel A. Reed. CRYSTAL: Operating System Instrumentation for the Intel iPSC/2. Dept. of Computer Science, Univ. of Illinois, Urbana, Ill. 61801, 1989.

[32] Charles L. Seitz. Concurrent VLSI Architectures. *IEEE Transactions on Computers*, C-33(12):1247–1265, December 1984.

[33] Charles L. Seitz. The Cosmic Cube. *Communications of the ACM*, 28(1):22–33, January 1985.

[34] Paul Y. Song. Design of a Network for Concurrent Message Passing Systems. Master's thesis, MIT, 545 Technology Square, Cambridge, MA, May 1988.

[35] Ellen Spertus. Dataflow Computation on the J-Machine. AI technical report 1233, MIT Artificial Intelligence Laboratory, 1990.

[36] Ellen Spertus and William J. Dally. Experiments with Dataflow on a General-Purpose Parallel Computer. In *Proceedings of International Conference on Parallel Processing*, pages II231–II235, Aug 1991.

TRIPTYCH: An FPGA Architecture with Integrated Logic and Routing

Scott Hauck, Gaetano Borriello, Carl Ebeling
Department of Computer Science and Engineering
University of Washington
Seattle, WA 98195

Abstract

We describe Triptych, a new FPGA architecture, that blends logic and routing resources to achieve efficient implementation of a wide range of circuits in both area and speed. The physical structure of Triptych attempts to match the structure of factored logic functions, thus providing an efficient substrate in which to implement these circuits. This approach both requires and takes advantage of an integrated approach to the mapping, placement and routing process. We first describe the Triptych architecture in detail. This is followed by the development of a new method for architectural comparison of FPGAs that is free of irrelevant implementation effects. Then the Triptych, Xilinx, Algotronix, and Concurrent Logic architectures are compared using this method to obtain normalized area and performance figures for a wide range of circuits, including both datapath elements and control logic. Our results indicate that Triptych is more area-efficient (Xilinx mappings average 3.5 times larger than Triptych mappings) and has at least comparable delay characteristics.

1 Introduction

Field-programmable gate arrays (FPGAs) are chips that can be programmed to implement arbitrary circuit structures. This is achieved by using a combination of logic and interconnection resources, which are generally treated separately. That is, the FPGA is partitioned into a set of logic blocks interconnected by programmable wiring in the form of routing channels or a cross-bar matrix. The target circuit is mapped to an FPGA by dividing it into a set of logic functions and registers, each of which can be implemented by one logic block, and a netlist indicating how these elements are connected. The circuit is then realized by placing each logic function and register in a logic block, and routing the logic blocks together according to the netlist. This clear separation between the logic and interconnection resources is attractive because the mapping, placement and routing decisions are decoupled. The price for this separation is the large area and delay costs incurred for the flexible interconnection needed to support arbitrary routing requirements. This leads to architectures like Xilinx [4], where the routing resources consume more than 90% of the chip area . Even so, it is well known that the largest Xilinx FPGA (3090) seldom achieves more than 50% logic block utilization for random logic. A lack of interconnect resources also leads to decreased performance as critical paths are forced into more indirect routes.

Domain-specific FPGAs like the Algotronix CAL1024 [2] and the Concurrent Logic CFA6000 [5] increase the chip area devoted to logic by reducing the routing to nearest-neighbor communication (this categorization is not completely accurate since Xilinx FPGAs have some nearest-neighbor connections and the CFA 6000 has introduced local busses). The result is that these architectures are restricted to highly pipelined dataflow applications, for which they are more efficient than the general-purpose FPGAs. Implementing circuits with these FPGAs requires close attention to routing during placement, which is generally accomplished by manual placement and routing of regular structures, or using specialized tools optimized to specific circuit domains [9]. Some research has been done on integrating placement and routing, but this work is very preliminary [3].

It is clear that domain-specific FPGAs can benefit greatly from an integrated approach to mapping, placement and routing. It is also the case that general-purpose FPGAs can take advantage of the increased flexibility offered by this approach. While it is conceptually appealing to view routing as a separate process which can always be completed, this does not coincide with reality, and attempting to achieve this extracts a large cost in terms of flexible routing resources. We have taken this argument one step further to establish that the FPGA architecture itself can take advantage of an integrated approach to mapping. In this paper, we present a new FPGA architecture called Triptych, that blends logic and routing resources. That is, each routing and logic block (RLB) in the Triptych array can be used both to compute a logic function and to route signals. More importantly, the array is structured to match the inherent fanin/fanout tree structure of circuit graphs. This allows the physical layout of a mapped circuit to follow its logical structure, reducing the need for extensive routing resources. It also turns routing resources into a much more optimizable resource. Taking full advantage of this structure will require integrating mapping, placement and routing tools, but we show in Section 3 that this structure has significant potential across a wide range of circuit types when compared to existing FPGA architectures.

Triptych defines a structure that is amenable to both highly regular circuits with local communication and more general circuits including the random logic circuits associated with finite state machines. As such, Triptych falls somewhere between general-purpose and domain-specific FPGAs. For applications with local communication, RLBs are used primarily for logic, allowing implementations that are competitive with domain-specific FPGAs. For more general applications such as FSMs, a higher percentage of RLBs are used for signal routing. Thus, although logic density decreases with greater global routing demands, a tradeoff between routing and logic resources can be reached on a per-circuit basis instead of as an architecturally dictated percentage.

We begin this paper by describing the Triptych architecture in detail, giving some implementation examples to illustrate its use. We then compare the Triptych architecture to that of the Xilinx, Algotronix and Concurrent Logic FPGAs by measuring the implementation cost and performance of each for a variety of interesting circuits. We have not included Actel in this study because it uses an anti-fuse technology, which makes architectural comparison difficult [1, 6]. We conclude by summarizing the results of this research and discussing possible future directions.

2 The Triptych Architecture

The FPGA architecture we present in this paper differs from other FPGAs by matching the structure of the logic array to that of the target circuits, rather than providing an array of logic cells embedded in a general routing structure. By matching the physical structure to the logical structure, we reduce the amount of "random" routing that is otherwise required.

Figure 1 shows a high-level view of a typical multi-level combinational logic circuit. The flow is shown as unidirectional, from inputs to outputs. From the point of view of each input, the data flow forms a fanout tree (shown with solid arrows) to those outputs that the input affects. From the point of view of each output, the data flow forms a fanin tree (shown with dashed arrows) from those inputs it depends upon. It is this fanin/fanout tree form that Triptych emulates architecturally by arranging RLBs (routing and logic blocks) into columns, with each RLB having a short, hard-wired connection to its nearest neighbors in adjacent columns (see Figure 2).

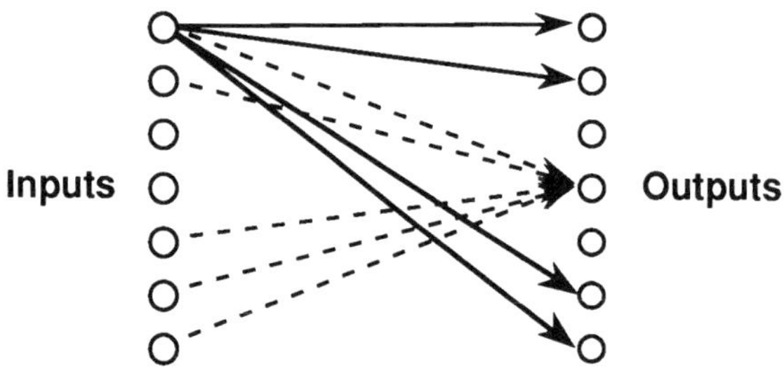

Figure 1. View of a multi-level combinational logic circuit as interleaved fanin/fanout trees.

The basic structure is augmented with segmented routing channels between the columns that facilitate larger fanout structures than is possible in the basic array. Finally, two copies of the array, flowing in opposite directions, are overlaid. Connections between the planes exist at the crossover points of the short diagonal wires. It is clear that this array does not allow arbitrary point-to-point routing like that associated with the Xilinx FPGA. However, we claim that this array matches the form of a large class of circuits, and that a mapping strategy that takes this structure into account can produce routable implementations.

Each RLB in the array has three inputs and three outputs, and may perform an arbitrary logic function of the three inputs, with the result optionally held by a master/slave D-latch. Routing in the Triptych array is in three forms: horizontally through the RLBs (by selecting an input to be routed to an output), diagonally through short wires to neighbors, and vertically through the segmented channels between columns of RLBs. Only one input and one output can be connected to the vertical wires; the other two must be on the local diagonal interconnect.

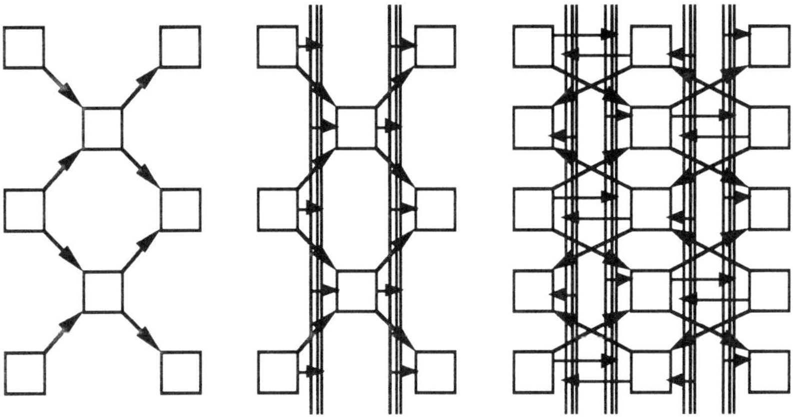

Figure 2. The overall structure of the Triptych FPGA shown in a progression of steps. The basic fanin/fanout structure on the left is augmented with segmented routing channels attached to a third RLB input and output. The structure on the right is obtained by merging two copies of the middle structure, with data flowing in opposite directions in the two copies. Not shown are the connections between the two copies, which permit internal feedback.

Circuits can be mapped onto this array by partitioning the logic into circuit DAGs containing nodes with at most three inputs. These DAGs are then mapped to the physical structure, with the inputs taken on one side and the outputs generated at the other. The nodes of the DAGs are placed such that input signals are available from the neighbor nodes or along a vertical connection. As suggested in [13], delay can be minimized by using mostly direct, hard-wired connections for the critical path. It is not the goal of Triptych implementations to achieve 100% logic utilization. Many RLBs are used to provide routing, either to fanout a signal or to pass it forward to the next level. Sometimes a mapping will even leave some cells unused to achieve a more routable placement of nodes. Examples of mappings are provided below.

2.1 RLB structure

A logical schematic of the Triptych RLB is show in Figure 3. As can be seen, the cell is designed to handle both function calculation and signal routing simultaneously (hence the name *routing and logic block*, RLB). It takes input from three sources and feeds them into a function block capable of computing any function of these three inputs, and the output can then be used in latched or unlatched form. The RLB's three outputs can choose from any of the three inputs and either the latched or unlatched version of the function block output. One last feature is the loopback from the master/slave D-latch, which enables the function to be dependent on its previous value. This is included for state machine implementation, although it may be used to output both the latched and unlatched versions of the function block. Again, only one of the inputs and one of the outputs can be connected to the vertical wires; the other two of each type are connected to the local diagonal wires.

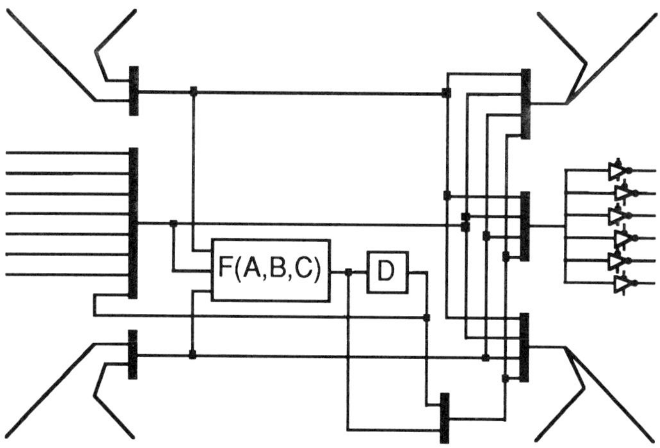

Figure 3. Triptych routing and logic block (RLB) design. The RLB consists of: 3 multiplexers for the inputs, a 3-input function block, a master/slave D-latch, a selector for the latched or unlatched result of the function, and 3 multiplexers for the outputs.

2.2 Typical RLB utilization

A Triptych RLB is capable of performing both function calculation and routing tasks simultaneously, which leads to several different uses of the RLB (see Figure 4). The three most obvious are: (a) a routing block with each input connected to one of the outputs; (b) a splitter with one of the inputs going to two or three of the outputs; and (c) as a function calculator with the three inputs going to the function block and the function going out the outputs. However, there are two important classes of hybrids that help produce more compact designs. The first comes from the observation that in blocks used to calculate a three-input function, the function block will most likely not go out all three outputs, and one or two of the input signals could be sent out the unused output connection(s), as in (d). Secondly, a function of two inputs can be implemented by making the function insensitive to the third input, thus allowing the unused input to be used to route an arbitrary signal, as in (e). An important observation is that the RLBs will never need to be used for one-input functions (i.e., an inverter), since any output signal will only be used either as an input to another arbitrary function block, where the inverter could be merged into the function computed, or to an output pin, where an optional inversion can be applied.

As was shown earlier, the Triptych FPGA has no global routing for moving signals horizontally. Instead, there is a heavy reliance on unused RLBs and unused portions of RLBs to perform these routing tasks.

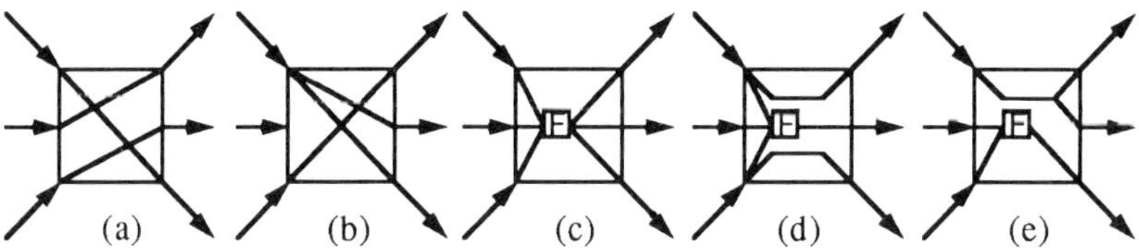

Figure 4. Five typical uses of Triptych routing and logic blocks (RLBs).

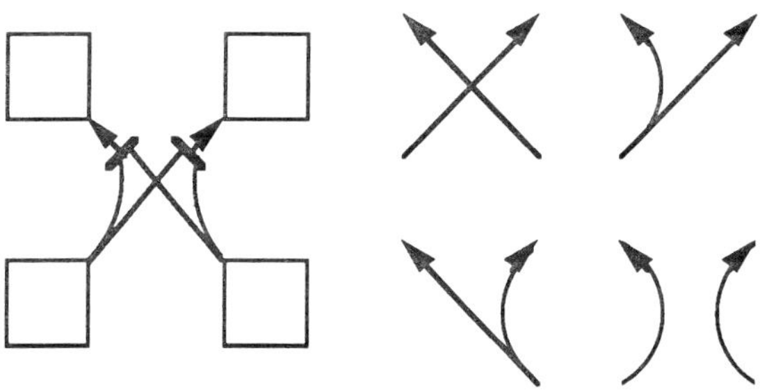

Figure 5. Schematic view of a pair of diagonals and the routing combinations they allow (implemented by a multiplexer at each diagonal input).

2.3 Interconnection

The Triptych RLBs are connected by three separate interconnection schemes. The first is for horizontal interconnect, and is accomplished through the RLBs as described above. The second is for local high-speed communication between neighboring RLBs and is achieved through "diagonals". The detailed structure of the diagonals is shown in Figure 5. They allow outputs to be sent to the four RLBs immediately above and below them; two in the next column, which flow in the same direction, and two in the same column, which flow in the opposite direction. Diagonals are important for two reasons. Diagonals permit the construction of multilevel functions of more than three inputs without the speed penalty of general-purpose interconnect. They also allow signal flow to change direction both so that circuits can be more tightly packed and feedback can be provided for the implementation of sequential logic.

The third type of interconnect is used for longer range connections and large fanout nodes. It is implemented as a set of segmented "channel wires" between adjacent columns (see Figure 6) that connect middle outputs of RLBs to the middle inputs of RLBs flowing in the same direction in the next column. Needless to say, this flexibility leads to slower signal propagation, and speed-critical designs will avoid using the vertical channels for critical paths. There are 7 tracks in a vertical channel, with 6 handling inter-cell RLB routing and a seventh to carry a pin input. The 6 inter-cell tracks are broken up into two tracks each of 8, 16, and 32 RLB high segments.

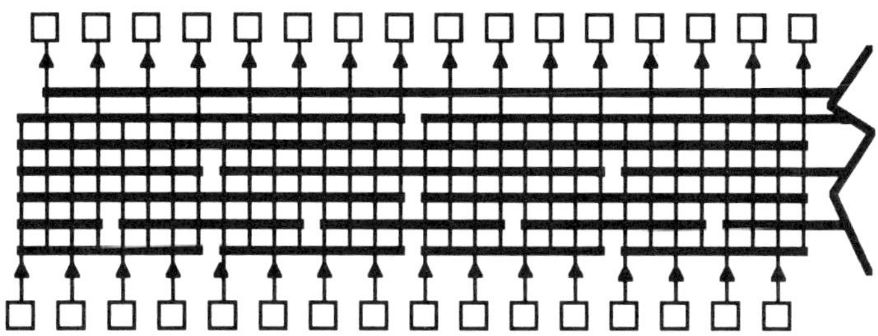

Figure 6. Top half of a segmented channel (on its side). The bottom half is a mirror image of the top.

One last important feature of the interconnect structure is how it handles the array borders. Since there are no RLBs beyond the right and left edges for the channel wires to route to, the channels on the edges tie the two directions of RLBs together. This way of handling the border cases leads to a different way of looking at the array, namely as a cylinder of RLBs. If the diagonals leading to the opposite direction of RLBs were cut except for those at the edges, the chip would appear to be a folded cylinder of RLBs. In fact, it is often helpful to think of the array as containing many smaller cylinders. For example, a six by six square of RLBs can be broken off from the rest of the array and considered to be a cylinder three RLBs high and twelve RLBs in circumference. This is not quite true, since the vertical channel for the left and right edges of the original six by six square will be unusable on the cylinder, but it can still be a useful abstraction for mapping. In fact, the current Triptych chip is an array of 64x8 RLBs, designed to yield a 32x16 cylinder.

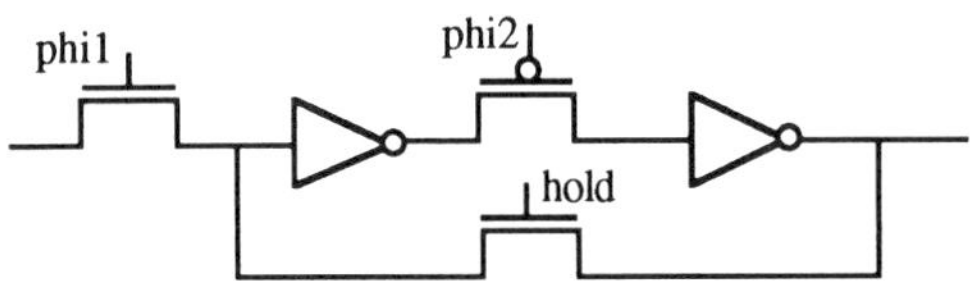

Figure 7. Programming bit cell schematic. The clock is only active during programming. A separate clock is used for the latches after the FPGA is configured.

2.4 Programming bit implementation and the scan path

Triptych is a RAM-based reprogrammable gate array with 26 memory bits per RLB, including those bits used for all three types of routing. The memory cells are implemented pseudo-statically with a "hold" signal asserted during normal operation and unasserted during programming. We found that this gave a much smaller layout than a fully static design (including the space needed for this extra hold line), especially when it was realized that the hold signal was necessary for disabling RLB output drivers during programming. However, since the pass gates controlled by phi2 and hold are non-complimentary, and thus the normally off transistors in the inverters will be slightly on, there will be extra static power dissipation during chip operation. These cells could easily be replaced by 5-transistor static RAM cells, and the overall RLB size would not change appreciably.

The memory cells are connected by a scan path running throughout the chip, allowing it to be programmed by cycling data through the bits. This scan path is also attached to the RLBs' master/slave D-latches. This not only allows the chip to start in any combination of latch states, but also allows the contents of the latches to be shifted out after the chip has run an arbitrary number of cycles, thus aiding in debugging and chip testing.

2.5 Vital statistics

The speed of a path in a Triptych mapping can be calculated from the numbers given below in Table 1. For example, a path using 4 RLBs, 2 for routing and 2 for function calculation, and 1 channel wire would take 13.3 to 14.5 nanoseconds ($4\times1.6 + 2\times2.2 + 2.5$ to $3.7 = 13.3$ to 14.5). Note that being able to use such a simple speed calculation method is due both to the simplicity of the interconnect and also to the design philosophy of "independence of paths". Simply put, "independence of paths" means that gate logic is used in place of switch logic in much of the FPGA, making signal propagation insensitive to the amount of fanout contained in the current mapping.

Resource Used	Delay
RLB	1.6ns
Function Block	additional 2.2ns
Channel Wire	2.5-3.7ns

Table 1. Speed of important features, estimated using HSPICE with parameters for the 1.2μm CMOS n-well process available from MOSIS.

Routing resources, their programming bits, buffers and tri-state drivers account for only 60% of the total area, as compared to at least 90% for the Xilinx FPGA. Approximately one-fourth (26%) of a Triptych RLB is devoted to the 26 programming bit memory cells.

3 Using Triptych

In this section we present two examples of circuits mapped to Triptych. The purpose of these examples is to demonstrate the constraints on routing and how multilevel logic circuits do indeed map to the physical structure provided by Triptych. In these examples, each RLB is shown as a cell with three input entries and three output entries. Each entry indicates an incoming or outgoing signal. Note that each block may create a new signal by computing a logic function over the inputs. Diagonals and reverse diagonals that are used in the implementation are highlighted, as are connections to the channel wires. For clarity, only those vertical wires carrying signals are shown.

3.1 8-bit rotate function

The power of using columns of RLBs for routing only is shown in this example, which rotates a set of 8 bits 4 positions. Each level can be used to send one signal from each RLB to a neighbor of the final position. Since each RLB carries two outputs, one intermediate RLB column and two vertical channels are required to route the signals to their final destination (see Figure 7). This generalizes to the case where three signals are routed per RLB, which requires two intermediate RLB columns and three channels.

Generalizing this use of the vertical channels suggests a naive place and route algorithm that alternates columns of RLBs used for routing with columns used to compute logic functions. Subject to a sufficient number of routing tracks, this leads to a viable routing of arbitrary logic functions. However, as the next example shows, this scheme is much less area-efficient than is generally achievable.

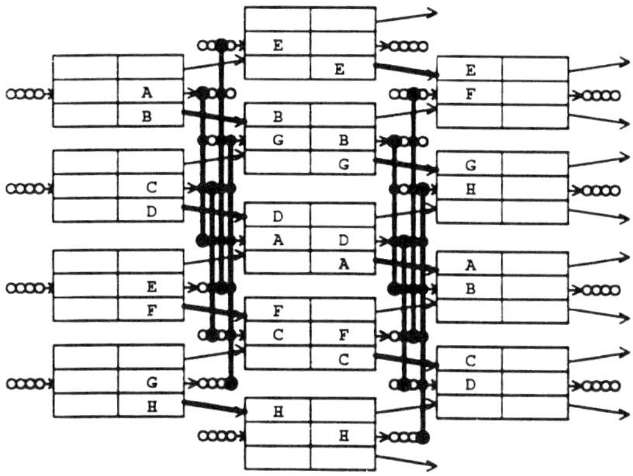

Figure 7. Triptych mapping of a 4-bit rotate of 8 bits.

3.2 State machine example

Figure 8 shows the factored logic equations and Figure 9 the corresponding Triptych implementation for the ubiquitous traffic light example. This example shows that circuit mappings can be very compact if the individual logic blocks are correctly placed. The inputs and outputs of this circuit are all connected at the left and right of the array, except for three signals that use the pin input track of the vertical channels (shown dangling off the bottom). In this example 16 RLBs are used to compute logic functions, 2 RLBs are used only for routing, and 6 RLBs are left unused (these 6 RLBs must be counted in order to achieve a rectangular mapping; they might be used in neighboring circuits). Also, this circuit is assumed to be placed along the left edge of the chip, so the vertical tracks at that edge are used to connect RLBs in the same column. This is about as compact a Triptych layout as can be achieved for a random logic function.

```
INORDER = s1 s2 d1 st SB0 SB1;
OUTORDER = NSB0 NSB1 r1 y1 g1 r2 y2 g2 sd;
NSB0 = !st * !r2;        NSB1 = !st * !g1 * !g2;
r1 = NSB0;               r2 = !st * (SB0 * !9 + !SB0 * 9);
y1 = r2 * 51;            y2 = 53 + 45;
g1 = r2 * !51;           g2 = !st * !r2 * !y2;
sd = 12 + 45;            9  = !SB1 + !d1;
12 = !SB1 * 18;          18 = !SB0 * s2 * !st;
46 = !st * SB0;          45 = s1 * !SB1 * 46;
52 = !d1 * SB1;          51 = s2 * !SB1 + !SB0 * SB1;
53 = 52 * 46;
```

Figure 8. Factored logic equations for the traffic light controller finite state machine.

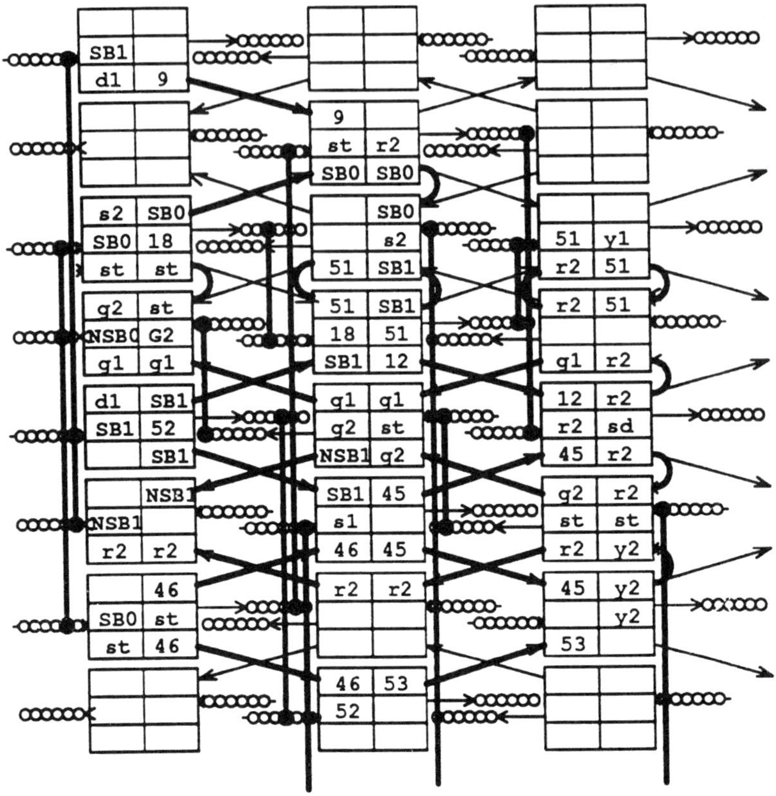

Figure 9. Triptych realization of the traffic light controller.

4 Area comparisons with other FPGAs

In this section, we present a comparison of the area efficiency of four different FPGA architectures, based on several representative examples. The FPGAs compared are CLI's CFA, Algotronix's CAL1024, the Xilinx 3000 series, and Triptych. The example circuits were chosen across a wide range of domains, rather than a specific one for which an architecture might have been optimized (e.g., bit-serial computations). The circuits were mapped by hand and only optimized for area. Automatic tools were not used as this would make the comparison as much about the CAD algorithms as the architectures themselves.

Even after eliminating these two factors, domain and CAD support, performing an architectural comparison between FPGAs is difficult because of complicating factors due to particular implementations. The most significant of these include the manufacturing process, the design rules employed, and the quantity and quality of designer effort. These factors make the straightforward approach of comparing actual chip area inadequate for purposes of architectural comparison. Simply using bits per cell also fails because it assumes that all FPGAs have equal per bit logic complexity, which is not true in current FPGAs (see discussion of table 2 below).

Our approach to architectural comparison is based on the premise that we can normalize cell area (the basic tiling structure in the FPGA) based on the size of the programming bits. There are two assumptions underlying this claim. First, a programming bit should occupy the same area if two different architectures are implemented on the same fabrication line, using the same design rules, and with the same amount of designer expertise and effort. It is not obvious that this is true, since the programming bits may be designed quite differently in different FPGAs. In the case of Triptych, the bits are 7-transistor pseudo-static shift-register cells interconnected in a scan-path, while Xilinx and CAL use 5-transistor static RAM cells (CFA's bit cell implementation is unspecified). However, even with these differences, we claim that the programming bits would take up the same area, due to the fact that the Triptych design can avoid ratioed transistors and use minimum-size devices (there is no feedback inverter to overpower through a pass-gate), and that it does not have the overhead of addressing and decode logic.

The second assumption is that the rest of the cell's logic and routing will scale proportionally to the programming bit cell size. This second assumption can be stated as the following relationship:

$$\frac{\text{total cell area in process } \alpha}{\text{total RAM bit area in process } \alpha} = \frac{\text{total cell area in process } \beta}{\text{total RAM bit area in process } \beta} \quad (1)$$

There are at least two ways in which this assumption may not hold true. First, speed impacts area, and one could design a very small cell using minimum-size devices that would be unreasonably slow. Since speed of programming is not a major issue, programming bit cells are almost always made as small as possible, so decreasing the logic size couldn't decrease the programming bit size, and the ratio given above would change. This means that any area comparison must also include a speed comparison of the circuit mappings in the sample. We present such a comparison below. Second, programming bit cells are the easiest and most advantageous cell components to optimize, due to the lack of speed requirements and their high degree of replication, and thus receive a much higher share of optimization efforts in less optimized designs. Also, as the programming bits shrink, the number of bit-equivalents the cell occupies increases. Therefore, less optimized implementations will compare less favorably to more highly optimized implementations (this is in fact the case for Triptych, thus handicapping it somewhat in the comparison).

Given these assumptions, we compare cell areas using the following normalization to a hypothetical standard process and design effort, Π:

$$\text{total cell area in process } \Pi \times \frac{1}{\text{single RAM bit area in process } \Pi} \qquad (2)$$

Note that this is simply the area of the cell scaled by the area of a single RAM bit, which our first assumption says will be the same size for any architecture. Equation 2 can be rewritten to yield:

$$= \;\# \text{ of RAM bits per cell} \times \frac{\text{total cell area in process } \Pi}{\text{total RAM bit area in process } \Pi} \qquad (3)$$

And finally, by applying equation 1, we obtain the following expression, which consists of three parameters easily obtained from specific FPGA implementations (where α is whatever process, design rules, and designer effort were used to implement that particular FPGA):

$$= \;\# \text{ of RAM bits per cell} \times \frac{\text{total cell area in process } \alpha}{\text{total RAM bit area in process } \alpha} \qquad (4)$$

Equations 2 to 4 allow us to scale all FPGA cell areas to our hypothetical process Π. Table 2 shows the results of applying equation 4 to the four FPGAs, with Triptych as the baseline for comparison.

	CFA	CAL	Xilinx	Triptych
RAM bits per cell	18	18	201	26
Percentage of cell area for bits	40%	40%	22%	26%
Normalization factor	0.45	0.45	9.1	1.0

Table 2. Area normalization figures calculated by applying equation 4, and the data used to obtain them. Triptych is used as the baseline. The numbers for CFA, CAL, and Xilinx were obtained from [7, 10, 14] respectively.

Normalization is also necessary for speed comparisons. In this case, we use the delay values provided by the manufacturers (Xilinx values are for the 3020-50 [12]), and scale them based on the feature sizes of the processes used. As shown in [15], FPGA logic speed scales linearly with process size, while routing delay scales quadratically. However, since the division of FPGA timings into routing-based and logic-based delays isn't always clear, we have scaled CFA linearly from 1.0μm to 1.2μm, and Xilinx quadratically from 1.25μm to 1.2μm, both of which are generous to the given FPGAs. For CAL, which has no long-distance routing from cell outputs, all delays were scaled linearly. Table 3 shows the normalization factors calculated for the FPGAs in question.

	CFA	CAL	Xilinx	Triptych
Technology feature size	1.0	1.5	1.25	1.2
Scaling method	Linear	Linear	Quadratic	—
Normalization factor	1.2	0.8	0.92	1.0

Table 3. Speed normalization figures and the data used to obtain them. A 1.2μm process is used as a baseline.

4.1 Summary of mapping results

Table 4 shows the results of area comparisons using this methodology for a wide range of circuits. They are collected into the following general categories: systolic, arithmetic, linear-growth bit-parallel, exponential-growth bit-parallel, and finite state machines (random logic). The systolic circuits are two versions of a string comparison circuit that computes the "edit distance" of two strings [11], the two versions differing on whether both strings move or if one is fixed in place, as well as a FIFO [8]. The arithmetic circuits chosen are a counter with a 4-stage carry look-ahead and an adder with a 2-stage carry look-ahead. The linear-growth bit-parallel circuits (circuits whose interconnection tends to scale linearly with problem size) include a simple equal/not-equal comparison and a shift register, while the exponential-growth bit-parallel circuits are a decoder and a multiplexor. Finally, for the random logic/FSMs, we chose the two smallest ISCAS benchmarks (s27 and s208), as well as the traffic light controller example.

Table 5 shows the results of delay comparisons for the circuits of Table 4. As discussed above, an FPGA architecture yielding much faster or slower circuits should have its area decreased or increased respectively to achieve a more accurate architectural comparison. Note also that these numbers do not form a true comparison of each FPGA's inherent delay requirements for the given circuits, since the circuits were optimized purely for area. Instead, these speed numbers are a part of the area comparison, making sure that no FPGA completely ignores speed in the quest for even smaller cells.

Even under the model given above, which for several reasons is biased towards the other FPGAs, Triptych does quite well in both area and speed. For all but the counter, whose CFA mapping is only 10% smaller, Triptych mapping are consistently smaller than CFA mappings, quite often by a factor of 2 or more, and Xilinx mappings are usually three to four times larger. As to CAL mappings, the best are only about 10% smaller than Triptych mappings, with the worst being more than twice as large. At the same time, delays in these circuits are almost always worse, up to a factor of more than 4 in the larger multiplexors. All of these factors indicate that Triptych is a viable architecture. This is especially true since the only FPGA surveyed that comes close to matching Triptych's area efficiency is CAL, which has no global routing facilities. This absence implies that larger, more complex circuits than those surveyed would be more difficult to map to CAL than to Triptych. They will be larger and slower, due to more cells being utilized for routing.

	size	Triptych area	CFA factor	CAL factor	Xilinx factor
Systolic					
Single-flow string compare (# of bits per character)	2	7	3.14	1.14	3.86
	4	10	2.90	1.10	3.60
	8	16	2.69	1.00	3.69
Double-flow string compare (# of bits per character)	2	20	2.30	1.00	2.75
	4	24	2.38	1.04	3.42
	8	36	2.17	1.00	3.81
FIFO (width in bits)	8	58	—	1.05	2.36
	16	90	—	1.00	2.32
	32	154	—	.96	2.31
	64	282	—	.93	2.29
Arithmetic					
Counter (width in bits)	8	24	.92	.92	2.67
	16	48	.90	.90	2.65
	32	96	.90	.90	2.66
Adder (width in bits)	8	40	1.63	.90	3.65
	16	80	1.63	.90	3.64
	32	160	1.62	.90	3.64
Linear-growth bit-parallel					
Comparator (width in bits)	8	8	2.25	.88	4.50
	16	16	2.25	.88	4.56
	32	32	2.38	.91	4.56
	64	64	2.36	.91	4.56
Shift Register (number of bits)	8	8	2.75	2.25	4.50
	16	16	2.69	2.25	4.56
	32	32	2.69	2.25	4.56
	64	64	2.70	2.25	4.55
Exponential-growth bit-parallel					
Decoder (inputs:outputs)	2:4	4	—	1.00	4.50
	3:8	10	—	1.00	3.60
	4:16	22	—	1.14	3.32
	5:32	46	—	1.35	3.57
Multiplexor (inputs:outputs)	4:1	3	2.67	1.67	4.67
	8:1	8	2.13	1.63	4.00
	16:1	18	2.06	1.67	3.78
	32:1	41	2.00	1.78	3.44
Random Logic/FSM					
s27		11	—	—	4.18
s208		59	—	—	2.93
traffic light controller		24	—	—	3.04
Geometric Means			2.06	1,17	3.53

Table 4. Area costs of circuits hand-mapped to four FPGAs using normalized area values. For each circuit, the Triptych area and its relative size in the three other FPGAs is listed (e.g., the Triptych s27 mapping is 11 normalized area units, while the Xilinx mapping is 4.18 times as big, or 46 area units).

	size	Triptych speed	CFA factor	CAL factor	Xilinx factor
Systolic					
Single-flow string compare	2	14.6	2.66	1.77	.91
(# of bits per character)	4	20.9	2.15	1.69	.64
	8	33.5	1.34	1.63	.59
Double-flow string compare	2	26.3	1.92	1.09	.76
(# of bits per character)	4	31.7	1.60	1.02	.84
	8	64.2	.88	.70	.62
FIFO	—	34.0	—	3.01	1.52
(controller speed only)					
Arithmetic					
Counter	8	21.2	1.62	1.52	.96
(width in bits)	16	32.0	1.63	1.56	1.27
	32	53.6	1.63	1.59	1.51
Adder	8	42.3	1.63	1.92	.88
(width in bits)	16	73.9	1.43	1.77	.92
	32	137.1	1.31	1.67	.94
Linear-growth bit-parallel					
Comparator	8	30.4	1.56	1.35	.87
(width in bits)	16	60.8	1.22	1.31	.87
	32	121.6	1.26	1.29	.87
	64	243.2	1.18	1.27	.87
Exponential-growth bit-parallel					
Decoder	2:4	7.9	—	1.29	.96
(inputs:outputs)	3:8	7.9	—	2.25	1.34
	4:16	19.0	—	1.47	.73
	5:32	24.0	—	1.88	.76
Multiplexor	4:1	7.6	3.47	3.24	1.72
(inputs:outputs)	8:1	13.9	2.68	3.05	1.50
	16:1	17.7	3.02	3.94	1.72
	32:1	26.5	2.88	4.32	1.52
Geometric Means			1.74	1.73	.99

Table 5. Delay of circuits hand-mapped to 4 FPGAs, in nanoseconds, using normalized delay values. The Triptych column contains actual speed numbers, while the other columns are the factor by which the given FPGA's mapping is slower (or faster) than Triptych's.

Three more points need to be made about these comparisons. First, the Xilinx mappings are the only ones for which we did not perform the assignment of routing resources. This was not done due to the complexity of hand-mapping Xilinx routing, and it was simply assumed that the circuit could be routed without leaving unreachable CLBs. This is not always the case in the 3000 series, where large designs can often only achieve 50% cell utilization due to the scarcity of resources. Thus, the area numbers for Xilinx should be considered as a lower bound, and the mappings may actually require significantly more space. Since we did not do routing, we have no direct method for measuring the speed of Xilinx circuits. However, by using a routing delay model for Xilinx [12], we were able to compute estimated routing delay, and these numbers were presented in Table 5. Second, since all the circuits were mapped by hand, we were only able to use small examples. No comparisons of large, complex random logic circuits is presented. Third, since FPGAs normally come in families, with the amount of cells in a chip variable, the issue of I/O blocks has been ignored and the circuits mapped to an unbroken plane of cells with no chip edges (except for a few circuits who actually map easier along a chip boundary, in which case an arbitrarily long boundary was assumed).

Clearly, more comparisons need to be made on larger circuits. We are currently developing automatic mapping tools to make this task tractable, although it will be more difficult to factor out CAD technology effects across the different architectures. We also need to compare the performance of circuits specifically optimized for speed instead of area. However, the results of Table 4 and Table 5 indicate the Triptych architecture is promising as it consistently yields smaller and faster circuits than the others.

5 Conclusions

The new FPGA architecture presented in this paper was motivated by three needs: including data-path and control elements in the same array; minimizing the space devoted to routing resources; and permitting the realization of delay-critical circuits. We believe that Triptych achieves these goals given the experience gained so far with many example circuits, a few of which have been presented above. The examples have proven to be more densely packed and to have delay characteristics at least comparable to the other FPGAs.

The most interesting and challenging future direction for research is automatic mapping. Triptych requires that the functional and interconnect elements not be treated separately. Combining the considerations for covering, placement, and routing should allow us to develop mapping tools that more precisely predict the performance of circuits and more accurately trade off density for speed. In fact, such an approach could benefit any FPGA.

In summary, we have a viable new FPGA architecture for circuits that demonstrates a higher area-efficiency, with comparable delay, for a wide range of circuits that include both data-path elements and control logic. There is much work to be done, especially in the area of automatic mapping, and promising directions are just beginning to be pursued.

Acknowledgements

Thanks to Christopher Hébert and the other students of CSE568, Winter quarter 1991 for contributing to the early design phases of the Triptych layout and mapping techniques. This research was funded in part by the Defense Advanced Research Projects Agency under Contract N00014-J-91-4041. Gaetano Borriello and Carl Ebeling were supported in part by NSF Presidential Young Investigator Awards, with matching funds provided by IBM Corporation and Sun Microsystems.

References

[1] Actel Corporation, "ACT Family Field Programmable Gate Array Data Book", 1991.

[2] Algotronix Limited, "CAL1024 Preliminary Datasheet", 1991.

[3] J. F. Beetem, "Simultaneous Placement and Routing of the Labyrinth Reconfigurable Logic Array", International Workshop on Field-Programmable Logic and Applications, Oxford, 1991.

[4] W. Carter et al., "A User Programmable Reconfigurable Gate Array", Proceedings of the IEEE Custom Integrated Circuits Conference, May 1986.

[5] Concurrent Logic, Inc., "CFA6006 Field Programmable Gate Array", March 1991.

[6] K. A. El-Ayat, A. El-Gamal, R. Guo, J. Chang, R. K. H. Mak, F. Chiu, and E. Z. Hamdy, "A CMOS Electrically Configurable Gate Array", IEEE Journal of Solid-State Circuits, Vol. 24, No. 3, June 1989, pp. 752-761.

[7] F. Furtek, Personal communication, October 1991.

[8] L. J. Guibas and F. M. Liang, "Systolic Stacks, Queues, and Counters", Second MIT Conference on Advanced Research in VLSI, 1982.

[9] T. Kean, "Configurable Logic: A Dynamically Programmable Cellular Logic Architecture and its VLSI Implementation", PhD Thesis, Dept. of Computer Science, University of Edinburgh, 1989.

[10] T. Kean, Personal communication, October 1991.

[11] R. J. Lipton and D. Lopresti, "A Systolic Array for Rapid String Comparison", Chapel Hill Conference on VLSI, 1985.

[12] M. Schlag, P. K. Chan, and J. Kong, "Empirical Evaluation of Multilevel Logic Minimization Tools for a Field Programmable Gate Array Technology", International Workshop on Field-Programmable Logic and Applications, Oxford, 1991.

[13] S. Singh, J. Rose, D. Lewis, K. Chung, and P. Chow, "Optimization of Field-Programmable Gate Array Logic Block Architecture for Speed", Proceedings of the IEEE Custom Integrated Circuits Conference, May 1990.

[14] S. Trimberger, Personal communication, October 1991.

[15] J. Vuillamy, Z. G. Vranesic, and J. Rose, "Performance Evaluation and Enhancement of FPGAs", International Workshop on Field-Programmable Logic and Applications, Oxford, September 1991.

Reducing Multi-Processor Memory Contention with Orthogonal Redundant Memory

Robert Berger
MIT Lincoln Laboratory
Lexington, MA 02173-9108, USA
bb@vlsi.ll.mit.edu

Abstract

A multi-processor computer with shared main memory is usually limited in speed of execution by memory latency. Part of the latency time is attributable to contention that arises when two or more processors attempt to access a single-port memory device. The Orthogonal Redundant Memory described in this paper can reduce the latency time caused by contention at the cost of duplicate storage and more complex hardware for control. The strategy is to store each piece of data twice, once in a primary memory and again in an orthogonally organized secondary memory. The data in a given module of the primary memory is distributed evenly among all the modules of the secondary memory. When there are two requests for data in the same module of the primary memory, the copies will very likely be in different modules in the second memory, in which case both pieces of data will be accessible simultaneously.

The main portion of this paper is a description of the *allocator*, a hybrid digital/analog CMOS device that will perform the assignment of memory modules to processors. It consists of an artificial neural net, with three neurons for each processor, and an array of digital comparator circuits whose outputs configure the neural net. The allocator has been simulated for the case of 16 processors. Results of the simulation are presented to show decreases in latency caused by contention for a range of read-to-write ratios and a range of total traffic.

1 Introduction

A multi-processor computer with shared main memory is usually limited in speed of execution by the memory latency, the time during which a processor must wait while loading data from or storing data to the main memory. Because the cycle time of a main memory module is usually much greater than the processor clock period, and because data must pass through some kind of communication network, an access to main memory usually takes several tens of processor clock periods. The accesses to main memory are slowed still more by contention, the attempt by more than one processor to access a main memory module that can serve only one request at a time. While the first requesting processor is being served, the others must wait. Even when there are cache memories at each processor to reduce the number of accesses to main memory, processors typically spend 70 percent of the time idled while waiting for access to data locations in main memory (Gupta et al [1]) .

The following section describes the *Orthogonal Redundant Memory*, a scheme for reducing the latency caused by contention. Following that are two sections describing the *allocator*, a hybrid digital/analog CMOS design, whose function is to assign to each requesting processor the correct memory module. The final section describes a transistor-level simulation of the allocator and its predicted decreases in average memory latency for typical values of total traffic and ratios of read requests to write requests.

2 Orthogonal Redundant Memory

Assume that the shared memory is made up of a number of single-port memory modules. Furthermore assume that the whole memory system can run synchronously, i.e. with all the processors' requests being sampled by the allocator at the same time. The latter assumption is not essential to the utility of the architecture, but it simplifies the design and gives slightly improved performance. The basic idea is to organize the memory modules into two sets, the primary memory and the secondary memory. Each piece of data is stored twice, once in the primary and once in the secondary memory. The addressing scheme for the two memories should be chosen so that the duplicates of all the data in one module in the primary memory would be distributed uniformly among all the modules of the secondary memory. For example, if each memory comprised 16 modules, the module in the primary memory might be selected by the highest four bits of the address, while the module in the secondary memory would be selected by a different four bits of the address.

Such a memory system is represented schematically in Figure 1 for the case of eight memory modules, with the primary memory modules shown as horizontal rectangles and the secondary memory modules as vertical rectangles; hence the name Orthogonal Redundant Memory. Two orthogonal rectangles intersect in a square region, which should be thought of as containing the memory addresses common to both modules. A request to access one of those memory addresses is represented by a circle in the square region. For simplicity in explaining the operation, assume all the requests are to read from memory. A read request is easier to allocate with Orthogonal Redundant Memory than with the primary memory alone, because data may be read from *either* of two modules; a write request is harder, because the data must be written to *both* modules. Fortunately read misses in the cache memory tend to outnumber write misses, often heavily, and there are

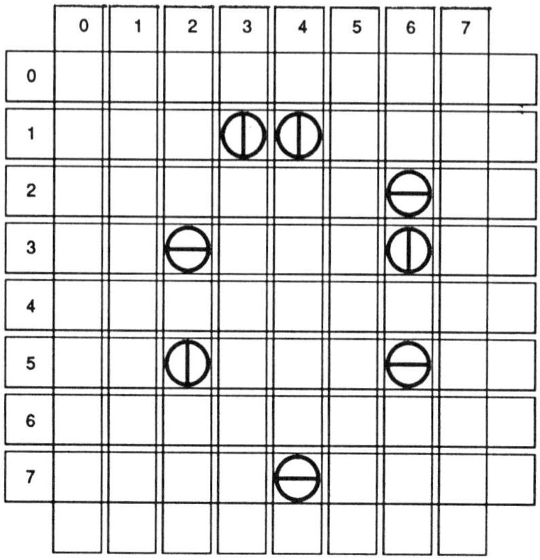

Figure 1: **Orthogonal Redundant Memory.**

relaxed schemes for cache consistency which allow write latency to be hidden [1]. Handling of write requests will be discussed when the allocator is described. In Figure 1 a horizontal (vertical) bar through the circle indicates that the data is to be read from the horizontal (vertical) module. The pattern of requests in Figure 1 can be completely allocated by the Orthogonal Redundant Memory, since no horizontal memory module contains more than one horizontal bar and no vertical memory module more than one vertical bar. It could not have been completely allocated by either the horizontal or vertical modules alone. Of course some patterns of requests are impossible to allocate completely, such as one with three requests in the same square intersection.

When there are only reads, Orthogonal Redundant Memory can dramatically increase the ability to allocate *completely* a pattern of requests for memory access. When there is only a primary memory, with N modules, the probability of there being some contention (and therefore incomplete allocation) for random, independent requests by N processors is $1 - N!/N^N$. This probability is plotted in the upper part of Figure 2. The lower part of that figure shows the results of a simulation in which N processors each produce independent, random requests to read from a memory address in an Orthogonal Redundant Memory system of N primary and N secondary memory modules. Each pattern of requests was operated on by an algorithm which determined whether Orthogonal Redundant Memory could possibly allocate it completely. For N between and 4 and 128 a complete allocation is highly unlikely with non-redundant memory and highly likely with Orthogonal Redundant Memory. A less dramatic measure of the effect of

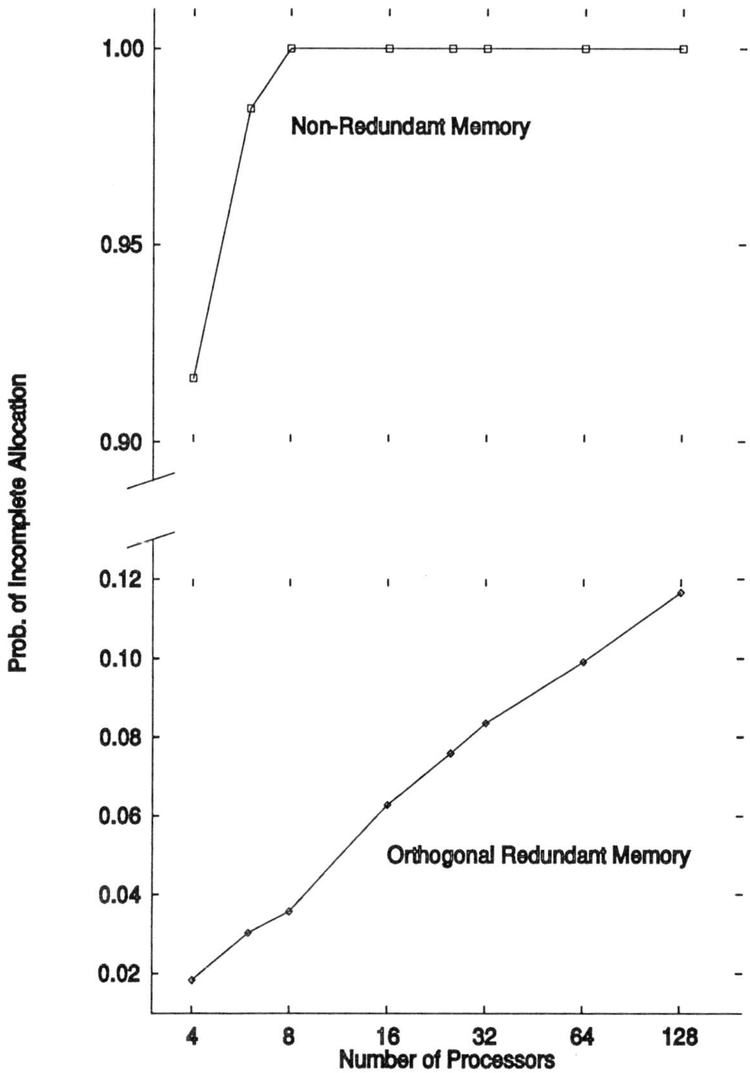

Figure 2: Reduction in Incomplete Allocations by Orthogonal Redundant Memory in the Read-Only Case.

Orthogonal Redundant Memory, but one more important to the throughput of a multi-processor computer, is its reduction of the average memory latency for patterns containing both reads and writes. This will be discussed in the section describing the simulation of the allocator.

3 The Basic Allocator

Orthogonal Redundant Memory requires a practical way to allocate the memory modules to the requesting processors. The algorithm that was used in generating Figure 2 could be made to generate those allocations that are possible, but it is sequential and does not translate simply into parallel hardware. Instead the allocator was designed using a modification of a scheme originally proposed by Yih and Mazumder [3] for selecting spare cells to replace faulty ones in an array of CMOS cells. They described an artificial neural network with two neurons per faulty cell. If one neuron became active, the spare was chosen from a row of spares at one edge of the array; if the other became active, from a column of spares at another edge. That sparing problem is almost isomorphic to the allocator problem, where the row and column of spares correspond to the primary and secondary memory modules, and the faulty cells correspond to the circles in Figure 1. However the neural network they used cannot represent request patterns with more than one circle in a square intersection, a common occurrence in the simulations that generated Figure 2.

Figure 3 shows an upper-left portion of the design for the allocator. In the following discussion the terms "row" and "column" refer to the rows and columns of memory in Figure 1, not to the rows and columns of the array in Figure 3. The allocator consists of *row rival detectors*, RD_{ij}, and *column rival detectors*, CD_{ij}, which determine whether or not requests from any pair of processors are rivals for the same row or column memory module, and an artificial neural net, which resolves the rivalries. At the start of an allocator cycle each processor selects the row and column memory module it wishes to access. Suppose, for example, that processor 2 wishes to access data that is in row memory 4, with its duplicate in column memory 7. Then it would impress the binary value 100 on the row-select bus RS_2 at the top edge and the binary value 111 on the column-select bus CS_2 at the left edge.

The row select buses extend downward from the top, "reflect" off the diagonal, and continue horizontally to the right. In this way each row select bus has an intersection with each other row select bus in exactly one place above the diagonal. At each such intersection is one of the row rival detectors; e.g. RD_{12} has inputs from RS_1 and RS_2. Similarly each column select bus goes right and reflects down, crossing each other column select bus in exactly one place below the diagonal, at which crossing is a column rival detector. The rival detector is simply a digital comparator whose output is active if and only if its two input buses carry identical selection codes. An allocator

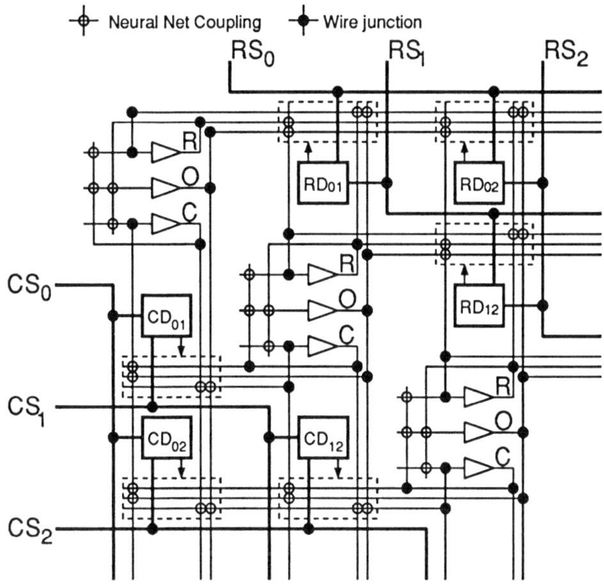

Figure 3: **Part of the Allocator Array.**

serving N processors, M row memories, and M column memories, would need $N(N-1)/2$ row rival detectors, each about $\log_2 M$ bits wide, and an equal number of column rival detectors.

On the diagonal there is a group of three neurons for each processor: R(row), C(column), and O(orphan). The row neuron for processor 2, to continue the above example, is intended to be active at the end of the decision time, if and only if processor 2 is to be granted access to row memory module 4; similarly for the column neuron of processor 2 and column memory module 7. The role of the orphan neuron will be discussed later. The input and output lines of the row neuron for processor i are extended along the same path as row select bus RS_i and therefore cross the inputs and outputs of each other row neuron in exactly one place above the diagonal; similarly for column neurons below the diagonal. The output lines of orphan neurons extend the full length and width of the array, but the inputs to orphan neurons come only from their local row and column neurons. In some of the places where the output line of a neuron crosses the input line of a neuron, there are coupling circuits, indicated by open circles in Figure 3. Where these couplings are grouped by dashed boxes, the couplings within a box are gated by the adjacent rival detector. Where the rival detector is inactive, there is no coupling. In other words, a row neuron, for example, receives coupling to its input from other row neurons only if they belong to processors that are its row rivals.

The coupling strengths were not chosen by first devising an "energy" function over the neural outputs and then determining coupling strengths that would cause the neural net to minimize that energy (see Hopfield [2]). In fact they could not have been, because that method relies on a symmetry of coupling: neuron x coupling to neuron y with the same strength that neuron y couples to neuron x. Later it will be seen that the orphan neurons violate that symmetry. The actual complement of neurons and the conditions governing the strengths of their couplings were obtained directly from the conditions of the allocator problem and then the actual coupling strengths were determined on the basis of simulations of the neural net.

First consider the simple case of only read requests, and ignore for now the orphan neurons. Because one should allocate memory modules only to those processors actually requesting access, only the row and column neurons for processors making requests receive on their inputs excitatory biases (see discussion of Figure 4), all of the same strength, C. Because no more than one processor may be granted access to a given row memory module, the row rival detectors activate inhibitory couplings of strength A from each row neuron to each of its row rivals; similarly for column neurons and their rivals. That is to say, when a row neuron is fully on, it couples an inhibition of strength A onto the input of each of its row rivals. If the neuron is only partially on, the inhibition is smaller, and if it is fully off the inhibition is zero. By choosing $A > C$ one can assure that a fully-on row neuron will keep all of its row rivals fully off; similarly for columns. Because one wants to keep a processor from being granted access to read both its row and column memory modules, there is also an inhibitory coupling of strength B from each processor's row neuron to that processor's column neuron, and the same coupling from the column neuron to the row neuron. By choosing $B > C$ one can assure that a fully-on row neuron will keep its own column neuron fully off, and vice versa. Choosing both A and B greater than the coupling strength, F, of a neuron's feedback transistor (see discussion of Figure 4) assures that the competition between rivals or between row and column neurons are each won decisively by one neuron, rather than ending with more than one competing neuron partially on.

The circuit diagram for a neuron is shown in Figure 4. The "body" of the neuron (inside the dashed box) is simply a pair of cascaded CMOS inverters with a small pulldown transistor providing negative feedback. When the corresponding processor is requesting access to memory, a neuron's input line is excited (pulled upward in voltage) by its bias transistor. The input may also be inhibited (pulled downward in voltage) by other neurons, through coupling transistors. In Figure 4 neuron x is inhibiting neuron y, while itself being inhibited by neuron z. If the net excitation is negative, the output will be at ground, the feedback transistor will be off, and the external circuit will draw only a minute leakage current from the neuron's input terminal. When the net excitation goes slightly positive, the neuron output voltage will jump to just above the threshold voltage of the feedback transistor, which will then sink a small current from the external circuit, just

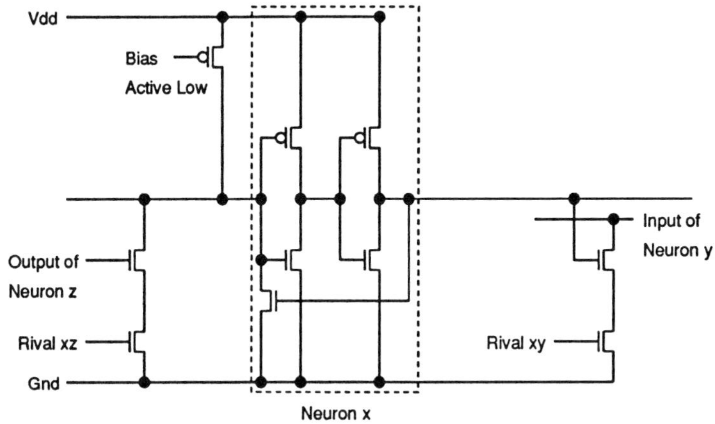

Figure 4: **Circuit Diagram for a Neuron.**

enough to hold the neuron input voltage at the low edge of the inverter's transition range. Further increases in net excitation will drive the neuron output voltage higher, allowing the feedback transistor to sink more current, while the neuron input voltage rises through the inverter's narrow transition range. Strong net excitation will drive the neuron output voltage to the positive supply voltage, after which the feedback transistor will no longer be able to sink enough current to hold the neuron input voltage within the inverter's transition range.

Each coupling contains a pulldown transistor, such as the one gated by "Output of Neuron z" in Figure 4, whose channel width-to-length ratio is proportional to its chosen coupling strength. Except on the diagonal the coupling circuit is conditioned by the corresponding rival detector, which gates a series transistor with a much larger width-to-length ratio (see "Rival xz" in Figure 4). Where the rival detector output is LOW, the coupling will be zero.

4 Additional Aspects of the Allocator

Most patterns of only read requests can be fully allocated, if that is possible, by the couplings described above. However simulations of the allocator (see next section) reached stable, but sub-optimal allocations on certain patterns. For example, Figure 5 shows a complete allocation for a loop of length 6. Figure 6, however, shows a stable but incomplete allocation of the same pattern, where row neuron 3 is kept off by row neuron 4 and column neuron 3 is kept off by column neuron 2. Neuron pair 2 cannot *flip*, i.e. change from "column" to "row", because it is restrained by 1; similarly 4 is restrained by 5. Neuron pair 1 is not restrained from flipping, however. If something could

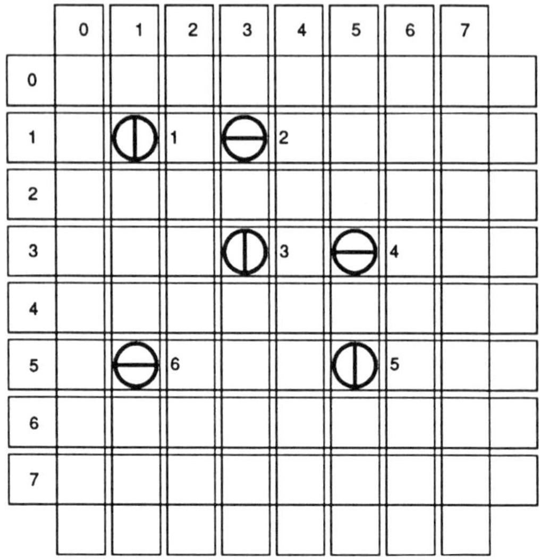

Figure 5: **Full Allocation of a Loop of Length 6.**

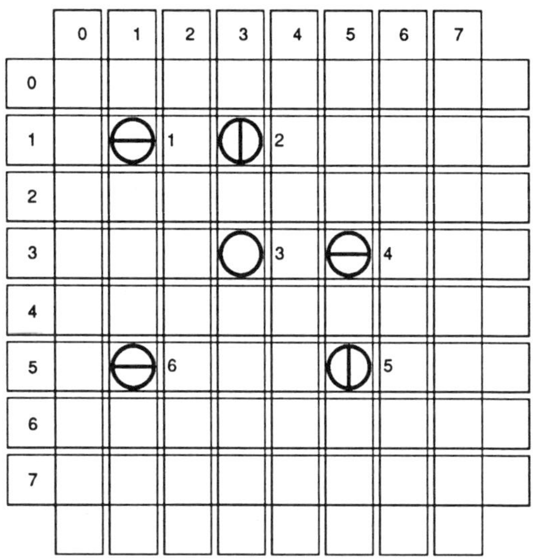

Figure 6: **Sub-Optimal Allocation of a Loop of Length 6.**

make it flip, then 2 could flip to "row", and then 3 could go to "column", completing the allocation. The orphan neuron was added to the row and column neuron to encourage such flipping. The orphan operates by turning on when its associated row neuron and column neuron are both off. The orphan then inhibits the rivals of both its row and its column neuron. The idea is to perturb the neural net in the vicinity of an orphan in the hopes of coaxing it into an optimum allocation. In Figure 6, for instance, orphan neuron 3 will become active. Should orphan 3 succeed in suppressing column 2, for instance, then column 3 could become active. But then both row 2 and column 2 would be off, so orphan 2 would turn on. Should orphan 2 then suppress row 1, then row 2 could turn on. There is nothing to prevent column 1 from turning on, and when it does so Figure 6 will look like Figure 5. The input to an orphan neuron has an excitatory bias of strength P and inhibitory couplings of strength Q from its row and column neurons. By choosing $P < Q$ one can assure that the orphan will be fully off if either its row neuron or its column neuron is fully on. The orphan neuron inhibits all the rivals of its row and column neurons with strength H. Judicious guessing and subsequent searching found certain regions in the six-dimensional space of biases and coupling strengths where the simulation (see next section) showed the orphan neurons indeed fixing allocation errors.

The allocator as described so far does not support a mixture of read and write requests. In an Orthogonal Redundant Memory each piece of data must be written both to a row memory and to a column memory. The couplings described so far prohibit both a row and a column memory from being accessed simultaneously by the same processor, i.e. prohibit both a row neuron and a column neuron for the same processor from being turned on at the same time. This restriction can be removed by extra digital logic at each diagonal position that gates the B coupling off if the processor's request is a write. Another kind of request that can occur is the *half-write*. This is a request to write explicitly to just the row memory or just the column memory. It would occur when one of the two halves of a write operation had been granted access on a previous memory cycle, but the other had not. The half-write can be requested by gating on the C bias only for the neuron desired (say the row neuron). In that case the column neuron could still be detected as a rival by other column neurons, but since it can never turn on, that would have no effect.

One might wonder whether the allocator function could be performed by a digital logic circuit instead of by the neural net. This is more a question of semantics than of circuitry. The "neuron" in Figure 4 could have been referred to as an "inverter". The input circuit to the inverter could have been described as a wired NOR function of the outputs of other inverters, for which the bias serves as a pull-up. The inequalities imposed on the choices of coupling strengths assured that any stable state of the circuit would show each inverter output at either the HIGH voltage or the LOW voltage, not at some intermediate value. One might say that designing the

Table 1: **The Model for Coupling Transistors.**

subthreshold	$V_{gs} < V_{th}$	$I_{ds} = 0$
saturation	$V_{th} < V_{gs} < V_{ds} + V_{th}$	$I_{ds} \propto (V_{gs} - V_{th})^2$
resistive	$V_{gs} > V_{ds} + V_{th}$	$I_{ds} \propto (V_{gs} - V_{th})V_{ds}$

allocator around a neural network model just made explicit the importance of the proper choices of transistor size that was implicit in the same circuit designed around a digital logic model.

5 The Simulation of the Allocator

The operation of the allocator was simulated on a workstation, so that the interaction of the neurons could be visualized and the statistics concerning average memory latency collected. An Orthogonal Redundant Memory with 16 processors and 16 primary and 16 secondary memory modules was simulated. The simulation model chosen was the simplest that would still reflect the nonlinearity in the combining of couplings to a neuron's input. The feedback and coupling transistors in Figure 4 were modeled in one of three states, as specified in Table 1: subthreshold, saturation, and resistive (see Mead and Conway [4]). The inverters were modeled as ideal inverters with very high gain and a transition midway between the positive and negative supply voltages. The gating transistors, such as the "rival" transistor, were modeled as ideal switches.

At the start of a simulation all neurons were off, the pattern of row and column addresses were presented to the (digital) rival detectors, and the selected neurons received their excitatory biases. The subsequent evolution of the circuit state was obtained by a number of iterations. During each iteration the net excitation at the input of each neuron was computed as if all the neuron outputs were held fixed. Then the neuron outputs were computed with the just-computed inputs held fixed. Finally the outputs were relaxed a fraction of the way from their present state toward the computed output state. The fraction was chosen to be 0.2, just small enough to make the iteration process stable; therefore five iterations corresponded to one charging time constant for a neuron driving the long wire coupling it to other neurons. The actual charging time will depend on compromises to be made when the circuit is laid out, but 1-2 ns is a reasonable value. For request patterns that don't need the orphan neurons, the neural net typically settles in about 60 iterations, which could be expected to take about 12-24 ns. For the worst patterns that can be fully allocated, settling can take several hundred iterations. Of course the neural net will never settle if the pattern cannot be fully allocated. Therefore the orphan neurons had to be suppressed (excitatory bias turned off) after a fixed time to allow the network to settle into an incomplete allocation.

After the neurons settled, the allocation pattern was evaluated for its correctness and completeness. The first set of simulations was done with read requests only to compare the performance of the allocator with the theoretical limit in Figure 2. Patterns of 16 read requests with randomly generated addresses were used. Without the orphan neurons the allocator was able to completely allocate 88% of these patterns, compared to the 94% that were possible to allocate completely. With orphan neurons and a well-chosen set of coupling strengths, the allocator was able to completely allocate all 94%. One effective set of coupling strengths, normalized to the feedback coupling, F, were: $A = B = 2.5, C = 2.0, P = 1.1, Q = 1.2, H = 0.6$. For this choice there was a limited spreading of the orphan state to the rivals of the orphaned request. Then this group of orphans propagated among the requests, flipping one at a time as it moved, until it reached a request like 1 in Figure 6, whose flipping does not create another orphan. Then the group of orphan neurons decayed to the off state, and the assignment was complete. The orphan group did not move in a random walk about the loop; it started in a particular direction and continued in the same direction until the misallocation was fixed. For other choices of coupling, the motion can be random or the orphan group can oscillate forever between two positions. For still another range of choices, the orphan state can spread in both directions around the loop until the misallocation is corrected. Then the corrections propagate back, turning the orphans off as allocations are flipped. When the orphan coupling is too weak, orphans can turn on but do not propagate far enough to correct the misallocation.

Another set of simulations was designed to predict the decrease in memory latency that might be achieved with Orthogonal Redundant Memory. Again random requests were generated for 16 processors to form a pattern, but this time the requests could be one of three types: read, write, or no request, each generated with a given probability. If a request was not satisfied for a given pattern, the next pattern repeated the unsatisfied request instead of using a new random choice for that processor. In other words, the processor whose request was not satisfied was assumed to be waiting until the request was satisfied on a later memory cycle. In the case of a write, the request could be half satisfied, in which case the other half-write was carried over to the next pattern. Statistics were kept of how many requests were kept waiting zero cycles, one cycle, two cycles, etc. The wait for a write was the wait for both halves to be allocated.

The simulations were done for two different ratios of reads to writes, 1:1 and 7:2, and for different levels of combined read and write traffic. For example, a 7:2 read-to-write ratio and a traffic level of 0.9 meant that a processor requested a read, a write, or nothing, with probability 0.7, 0.2, or 0.1, respectively. The results are shown in Figure 7. The ordinate is average memory latency added by contention, measured in memory cycles. Curves are shown for the average latency of combined read and write traffic. Included for comparison are simulations of added latency if only the primary

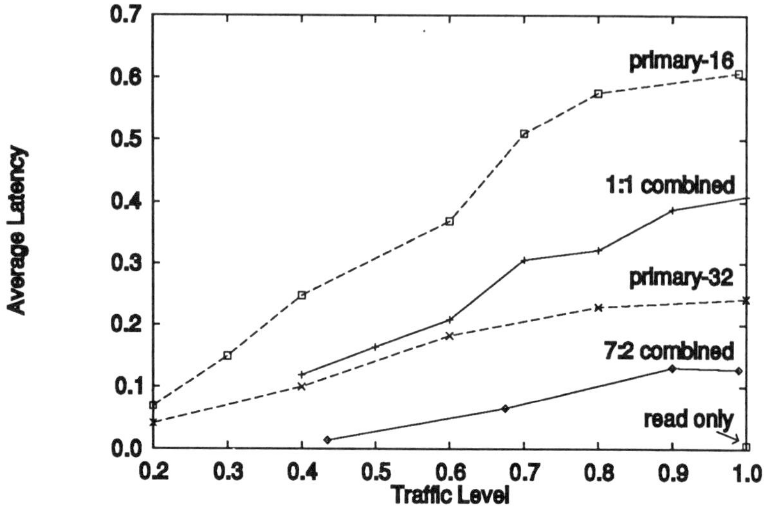

Figure 7: **Average Memory Latency Added by Contention.**

memory were available, for 16 and 32 memory modules. In those cases there is no distinction between reads and writes. The 32-module case is probably a fairer comparison, since it uses the same number of memory modules as the Orthogonal Redundant Memory, but organized differently. Orthogonal Redundant Memory with a 7:2 read-to-write ratio has about one half the latency of the 32-module primary memory. Of course in read-only or read-mostly applications, virtually all the latency cause by contention can be eliminated with Orthogonal Redundant Memory. For instance, a traffic level of 1.0 in the read-only case has a simulated average latency of 0.006 memory cycles.

6 Summary

Orthogonal Redundant Memory has been introduced as a scheme for decreasing memory contention in a multiprocessor system with shared memory. It uses duplicate storage of data. A design was shown for a CMOS integrated circuit that could allocate memory modules to the requesting processors to take advantage of Orthogonal Redundant Memory. The allocator is composed of an artificial neural net and an array of digital comparators. Simulations of the allocator were described and their results indicate that Orthogonal Redundant Memory can eliminate much of the memory latency caused by contention, especially when read requests outnumber writes.

Acknowledgements

This work was sponsored by the Department of the Air Force and the Defense Advanced Research Projects Agency.

References

[1] Anoop Gupta, John Hennessy, Kourosh Gharachorloo, Todd Mowry, and Wolf-Dietrich Weber. Comparative Evaluation of Latency Reducing and Tolerating Techniques. *ACM SigArch News, 18th Comp. Architecture Symp.*, 254-263, Toronto, May 1991.

[2] J. J. Hopfield. Neural networks and physical systems with emergent collective computational abilities. *Proc. Natl. Acad. Sci. USA*, 79:2554-2558, April 1982.

[3] P. Mazumder and J.S. Yih. Processor Array Self-Reconfiguration by Neural Networks. *Proc. IEEE Intl. Conf. on Wafer Scale Integration*, Jan. 22-24, 1992, San Francisco.

[4] Carver Mead and Lynn Conway. *Introduction to VLSI Systems*,1-5, Addison-Wesley, Reading, Mass., 1980.

Fast Permuting on Disk Arrays
(Extended Abstract)

Thomas H. Cormen
MIT Laboratory for Computer Science
545 Technology Square
Cambridge, MA 02139
thc@theory.lcs.mit.edu

Abstract

We present fast algorithms to accomplish common classes of permutations on parallel disk systems. Vitter and Shriver introduced a model in which N records are stored on D disks and each parallel I/O operation can transfer a block of B records between disk and a memory capable of holding M records. They showed a tight bound of $\Theta\left(\min\left(\frac{N}{D}, \frac{N}{BD}\frac{\lg(N/B)}{\lg(M/B)}\right)\right)$ parallel I/Os to perform a general permutation, but they demonstrated that some permutations can be performed faster: an $R \times S$ matrix can be transposed using only $\Theta\left(\frac{N}{BD}\left(1 + \frac{\lg\min(B,R,S,N/B)}{\lg(M/B)}\right)\right)$ parallel I/Os.

This paper generalizes the Vitter-Shriver matrix-transposition result, showing that other classes of permutations can be performed faster than the general permutation bound in many cases. We show how to perform bit-permute/complement (BPC) permutations—a class including many common permutations—using at most $\frac{2N}{BD}\left(2\left\lceil\frac{\rho(A)}{\lg(M/B)}\right\rceil + 1\right)$ parallel I/Os, where $\rho(A)$, the cross-rank, is a measure of the bit movement in the characteristic matrix A of the BPC permutation, relative to the machine parameters B and M. We also show how to perform bit-matrix-multiply/complement (BMMC) permutations (which are based on multiplication over $GF(2)$ by nonsingular matrices) using at most $\frac{2N}{BD}\left(2\left\lceil\frac{\lg M - r}{\lg(M/B)}\right\rceil + H(N,M,B)\right)$ parallel I/Os, where r is the rank of the leading $\lg M \times \lg M$ submatrix of the characteristic matrix of the BMMC permutation and

$$H(N,M,B) = \begin{cases} 4\left\lceil\dfrac{\lg B}{\lg(M/B)}\right\rceil + 11 & \text{if } M \le \sqrt{N}\ , \\ 4\left\lceil\dfrac{\lg(N/B)}{\lg(M/B)}\right\rceil + 2 & \text{if } \sqrt{N} < M < \sqrt{NB}\ , \\ 6 & \text{if } \sqrt{NB} \le M\ . \end{cases}$$

The algorithms for these permutations are built from restricted classes of permutations that we define, each requiring only $2N/BD$ parallel I/Os. We also show how to perform arbitrary block permutations off-line using $2N/BD$ parallel I/Os.

We give asymptotic lower bounds for BPC and BMMC permutations, proving that our upper bound for BPC permutations is asymptotically optimal. All the permutation algorithms presented in this paper are deterministic and, except for arbitrary block permutations, are easily performed on-line.

1 Introduction

In both serial and parallel computing, there is often more data than will fit in the random-access memory of the computer. A common solution to this problem is to store the data on a disk system and move it into memory for processing a portion at a time. If the disk system contains just one disk, moving data between the memory and the disk may present a formidable bottleneck. Disk arrays present an efficient and cost-effective solution to the disk I/O bottleneck problem.

Disk arrays are not just a theoretical solution. The RAID (Redudant Arrays of Inexpensive Disks) project at Berkeley (see [4, 11] for example) has demonstrated that disk arrays can improve performance and reliability for very little cost. Moreover, disk arrays are appearing as commercial products in parallel systems. Tera Computer [3] will offer a disk array in its shared-memory machine, and the Thinking Machines Corporation Datavault has been on the market for several years.

Performance in conventional serial and parallel machines depends on algorithms as well as system design, and the same is true of disk arrays. This paper presents algorithms for performing specific classes of permutations quickly on disk arrays. We assume that each of N records, where N is an exact power of two, has a $(\lg N)$-bit address that gives its position in the disk array. A permutation maps the "source address" of each record to its "target address."

One class that we consider in this paper is the bit-permute/complement, or BPC, permutations, in which we form a record's target address $y = (y_0, y_1, \ldots, y_{\lg N-1})$ from its source address $x = (x_0, x_1, \ldots, x_{\lg N-1})$ by applying a fixed permutation π to the address bits and then complementing a fixed subset of bits of the result. The complementing is equivalent to exclusive-oring by a $(\lg N)$-bit "complement vector" $c = (c_0, c_1, \ldots, c_{\lg N-1})$. A source address x maps to a target address y by the equation

$$y_j = x_{\pi(j)} \oplus c_j \tag{1}$$

for $j = 0, 1, \ldots, \lg N - 1$, where $\oplus$ denotes the exclusive-or (XOR) operation.

Many familiar permutations fit into the class of BPC permutations. For example, matrix transposition is a type of BPC permutation. Suppose that the N records are the entries of an $R \times S$ matrix stored in row-major order. The $(\lg N)$-bit address of each record is comprised of a $(\lg R)$-bit row number followed by a $(\lg S)$-bit column number. To transpose this matrix, we map the (i, j) entry to the (j, i) position. We can do so by performing the permutation in which the upper $\lg R$ bits and the lower $\lg S$ bits of the source address are swapped to form the target address. Here, the fixed permutation of address bits is a cyclic rotation by either $\lg R$ positions in one direction or $\lg S$ positions in the other, and no bits are complemented. (Cyclic rotation by one bit position is equivalent to the perfect shuffle and inverse perfect shuffle permutations.) In the context of equation (1), we have $\pi(j) = (j - \lg R) \bmod \lg N = (j + \lg S) \bmod \lg N$ and $c_j = 0$ for

$j = 0, 1, \ldots, \lg N - 1$. Other examples of BPC permutations include bit-reversal permutations, hypercube permutations, and vector-reversal permutations. In bit-reversal permutations, which are often used in performing FFTs, we reverse the order of the bits of each record's source address to form its target address. In hypercube permutations, we complement a single address bit. In vector-reversal permutations, we complement all address bits, thus reversing the order of records.

Another class of permutations that we consider are the bit-matrix-multiply/complement, or BMMC, permutations. Here, we are given a $\lg N \times \lg N$ matrix $A = (a_{ij})$ with 0-1 entries that is nonsingular over $GF(2)$, and a complement vector c. (Matrix multiplication over $GF(2)$ is like standard matrix multiplication over the integers but with all arithmetic performed modulo 2.) A source address $x = (x_0, x_1, \ldots, x_{\lg N-1})$ maps to a target address $y = (y_0, y_1, \ldots, y_{\lg N-1})$ by matrix multiplication over $GF(2)$ and complementation: $y = Ax \oplus c$. That is,

$$y_i = \left(\bigoplus_{0 \le j \le \lg N - 1} a_{ij} x_j \right) \oplus c_i \ ,$$

where the product $a_{ij} x_j$ is simply the logical AND of the two bits. The class of BMMC permutations includes the class of BPC permutations. It also includes permutations such as the Gray code and inverse Gray code permutations.

BMMC permutations are also useful for reducing data-access conflicts in parallel systems. Using a scheme developed by Norton and Melton [10], a sequence of record accesses can sometimes be performed faster than by simply accessing the records in their given locations. We first permute the records according to a BMMC permutation with matrix A, then access the records in their new locations, and finally restore the original record ordering by performing the BMMC permutation with matrix A^{-1}. We refer the interested reader to [10] for details on computing the matrix A.

Having seen examples of the classes of permutations we shall perform, we now present the disk-array model that we shall use. It was originally proposed by Vitter and Shriver [12, 13], who also gave asymptotically optimal algorithms for sorting, matrix multiplication, matrix transposition, permuting, and pebbling FFT graphs and permutation networks.

In the Vitter-Shriver model, N *records* are stored on D *disks* $\mathcal{D}_0, \mathcal{D}_1, \ldots, \mathcal{D}_{D-1}$, with N/D records stored on each disk. The records on each disk are organized in *blocks* of B records each. When a disk is read from or written to, an entire block of records is transferred. Disk I/O transfers records between the disks and a *memory* capable of holding M records. In a *parallel read operation*, we read a block from any subset of the D disks into the memory. In a *parallel write operation*, we write blocks from the memory to any subset of the disks. (Each parallel I/O operation used in this paper will write to all D disks.) The complexity measure of interest is the number of *parallel I/Os*, i.e., parallel read or write operations, used to accomplish a given task.

There are some restrictions placed on the parameters N, D, M, and B. We assume that each is an exact power of two. In order for the memory to accomodate the records transferred in a parallel I/O operation to all D disks, we require that $BD \leq M$. Also, we assume that $M < N$, since otherwise we can just perform all operations in memory.

The parallel I/O operations can be either independent or striped. In a *striped* I/O operation, all blocks read or written must reside at the same location on each disk; we call such a set of blocks a *track*. In an *independent* I/O operation, disk blocks read or written may reside at any location as long as only one block per disk is accessed. Although striped I/O facilitates reliability, it has the constraining effect of transforming the block size B and number of disks D to the values $B' = BD$ and $D' = 1$. All algorithms in this paper use independent I/O.

Since each parallel I/O operation accesses at most BD records, any algorithm that must access all N records requires $\Omega(N/BD)$ parallel I/Os. The equivalent of linear time in the Vitter-Shriver model is therefore $O(N/BD)$ parallel I/Os. Vitter and Shriver give a randomized sorting algorithm that uses $\Theta\left(\frac{N}{BD}\frac{\lg(N/B)}{\lg(M/B)}\right)$ parallel I/Os, and Nodine and Vitter [8, 9] present a determininstic sorting algorithm with the same asymptotic bound. Vitter and Shriver also give an algorithm for permuting that uses $\Theta\left(\min\left(\frac{N}{D}, \frac{N}{BD}\frac{\lg(N/B)}{\lg(M/B)}\right)\right)$ parallel I/Os. Here, the second term comes from the sorting bound, and the first term comes into play when the block size B is small. Although this bound applies to permuting in general, Vitter and Shriver demonstrate that a specific permutation can sometimes be performed faster, sketching how to transpose an $R \times S$ matrix using $\Theta\left(\frac{N}{BD}\left(1 + \frac{\lg \min(B,R,S,N/B)}{\lg(M/B)}\right)\right)$ parallel I/Os. All the above bounds are asymptotically tight.

Vitter and Shriver actually show only the upper bounds. The lower bounds are derived from the slightly more powerful model of Aggarwal and Vitter [2], which has just one disk and D independent read/write heads. Aggarwal and Vitter consider the same problems as Vitter and Shriver and, perhaps surprisingly, arrive at the same asymptotic bounds. Since their model is more powerful, lower bounds for their model apply to the Vitter-Shriver model as well.

In this paper, we show how to perform BPC and BMMC permutations faster than sorting in many cases. In the full paper, we also prove an asymptotic lower bound on the number of parallel I/Os required to perform BMMC permutations and use it to show that our BPC algorithm is asymptotically optimal. The algorithms in this paper have several advantages:

- They are asymptotically efficient and, moreover, the constants in the upper bounds are small.
- They are deterministic.
- They are easily programmed and can be performed "on-line" in the sense that they take little time and space. The data structures are

vectors of length $\lg N$ or $\lg N \times \lg N$ matrices, and serial algorithms for the harder computations take time polynomial in $\lg N$.

The remainder of this paper is organized as follows. Section 2 defines the classes of permutations used in this paper. These include the BPC and BMMC permutations we have already seen, as well as block BMMC, block BPC, MRC, and arbitrary block permutations. Section 3 presents what is perhaps the most generally useful result of this paper: an asymptotically optimal algorithm for performing BPC permutations. Section 4 describes a technique that is used in Sections 5 and 6 to perform MRC and block BMMC permutations. Using results from the previous sections, we see how to perform BMMC permutations quickly in Section 7. Finally, Section 8 contains some concluding remarks. The full paper also presents an off-line algorithm for arbitrary block permutations, proves that our BPC algorithm is asymptotically optimal, and proves a lower bound for BMMC permutations.

2 Classes of Permutations

In this section, we present notational conventions, precisely define the classes of permutations we shall work with, and describe some of their characteristics.

For convenience, we use the following notation extensively:

$$b = \lg B, \quad d = \lg D, \quad m = \lg M, \quad t = \lg(N/BD), \quad n = \lg N .$$

We shall assume that b, d, m, t, and n are nonnegative integers, which implies that B, D, M, and N are exact powers of two. Note that $n = b + d + t$ and that, since $BD \leq M < N$, we have $b + d \leq m < n$.

The Vitter-Shriver model lays out data on a disk array as shown in Figure 1. Record indices vary most rapidly within a block, then among disks, and finally among tracks. Thus, we can parse the address of a record in a disk array as shown in Figure 2: the least significant b bits give the offset of the record within its block, the next most significant d bits give the number of the disk that the record resides on, and the most significant t bits give the number of the track containing the record. We indicate the address of a record as an n bit vector x with the least significant bit first: $x = (x_0, x_1, \ldots, x_{n-1})$. The offset is given by bits $x_0, x_1, \ldots, x_{b-1}$, the disk number by bits $x_b, x_{b+1}, \ldots, x_{b+d-1}$, and the track number by bits $x_{b+d}, x_{b+d+1}, \ldots, x_{n-1}$. Figure 2 also shows the relationship of m to b, d, and t, and the bit positions $x_b, x_{b+1}, \ldots, x_{m-1}$, which will be important later on.

A permutation of records is a one-to-one mapping of input addresses from the set $\{0, 1, \ldots, N-1\}$ onto itself. We specify a permutation by considering a record's *source address* $x = (x_0, x_1, \ldots, x_{n-1})$ and its *target address* $y = (y_0, y_1, \ldots, y_{n-1})$. Table 1 shows the classes of permutations that we shall be looking at in this paper and summarizes the upper-bound results herein.

	$\mathcal{D}_0$		$\mathcal{D}_1$		$\mathcal{D}_2$		$\mathcal{D}_3$		$\mathcal{D}_4$		$\mathcal{D}_5$		$\mathcal{D}_6$		$\mathcal{D}_7$	
track 0	0	1	2	3	4	5	6	7	8	9	10	11	12	13	14	15
track 1	16	17	18	19	20	21	22	23	24	25	26	27	28	29	30	31
track 2	32	33	34	35	36	37	38	39	40	41	42	43	44	45	46	47
track 3	48	49	50	51	52	53	54	55	56	57	58	59	60	61	62	63

Figure 1: The layout of $N = 64$ records in a disk array with $B = 2$ and $D = 8$. Each box is one block. The number of tracks needed is $N/BD = 4$. Numbers indicate record indices.

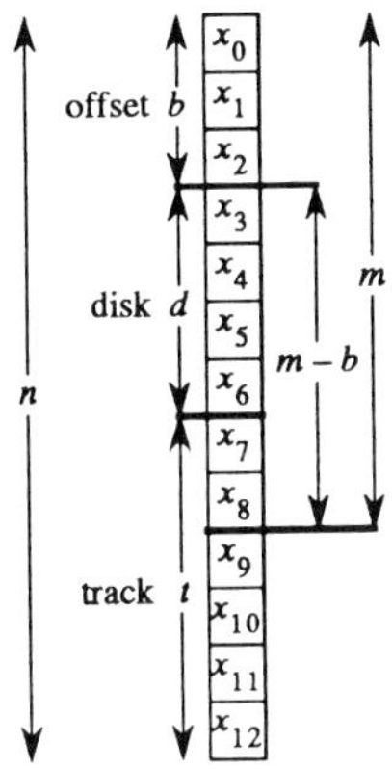

Figure 2: Parsing the address $x = (x_0, x_1, \ldots, x_{n-1})$ of a record in a disk array. Here, $n = 13$, $b = 3$, $d = 4$, $t = 6$, and $m = 9$. The least significant b bits contain the offset of a record within its block, the next d bits contain the disk number, and the most significant $t = n - (b + d)$ bits contain the track number. The least significant m bits and bits $x_b, x_{b+1}, \ldots, x_{m-1}$ are also indicated.

The most general class is the *bit-matrix-multiply/complement*, or *BMMC*, permutations. In a BMMC permutation, we have an $n \times n$ *characteristic matrix* $A = (a_{ij})$ whose entries are drawn from $\{0, 1\}$ and is nonsingular (i.e., invertible) over $GF(2)$, and we have a *complement vector* $c = (c_0, c_1, \ldots, c_{n-1})$ of length n. Treating a source address x as an n-vector (equivalent to an $n \times 1$ matrix, as in Figure 2), we perform matrix-vector multiplication of $GF(2)$ and then complement some subset of the resulting bits to form the corresponding target address y. That is, $y = Ax \oplus c$, or

$$\begin{bmatrix} y_0 \\ y_1 \\ y_2 \\ \vdots \\ y_{n-1} \end{bmatrix} = \begin{bmatrix} a_{00} & a_{01} & a_{02} & \cdots & a_{0,n-1} \\ a_{10} & a_{11} & a_{12} & \cdots & a_{1,n-1} \\ a_{20} & a_{21} & a_{22} & \cdots & a_{2,n-1} \\ & \vdots & & \ddots & \vdots \\ a_{n-1,0} & a_{n-1,1} & a_{n-1,2} & \cdots & a_{n-1,n-1} \end{bmatrix} \begin{bmatrix} x_0 \\ x_1 \\ x_2 \\ \vdots \\ x_{n-1} \end{bmatrix} \oplus \begin{bmatrix} c_0 \\ c_1 \\ c_2 \\ \vdots \\ c_{n-1} \end{bmatrix} .$$

Permutation	Characteristic matrix	Number of passes
BMMC (bit-matrix-multiply/ complement)	nonsingular matrix A	$2\left\lceil \frac{\lg M - r}{\lg(M/B)} \right\rceil + H(N,M,B)$
BPC (bit-permute/ complement)	permutation matrix A	$2\left\lceil \frac{\rho(A)}{\lg(M/B)} \right\rceil + 1$
Block BMMC	$\begin{array}{c} \begin{array}{cc} b & n-b \end{array} \\ \left[\begin{array}{c\|c} \text{nonsingular} & 0 \\ \hline 0 & \text{nonsingular} \end{array}\right] \begin{array}{l} b \\ n-b \end{array} \end{array}$	1
Block BPC	$\begin{array}{c} \begin{array}{cc} b & n-b \end{array} \\ \left[\begin{array}{c\|c} \text{nonsingular} & 0 \\ \hline 0 & \text{permutation} \end{array}\right] \begin{array}{l} b \\ n-b \end{array} \end{array}$	1
MRC (memory-rearrangement/ complement)	$\begin{array}{c} \begin{array}{cc} m & n-m \end{array} \\ \left[\begin{array}{c\|c} \text{nonsingular} & \text{arbitrary} \\ \hline 0 & \text{nonsingular} \end{array}\right] \begin{array}{l} m \\ n-m \end{array} \end{array}$	1
Arbitrary block		1 (off-line scheduling)

Table 1: Classes of permutations, their characteristic matrices, and upper bounds shown in this paper on the number of passes needed to perform them. A pass uses exactly $2N/BD$ parallel I/Os. For block BMMC, block BPC, and MRC permutations, submatrix dimensions are shown on matrix borders. For BMMC permutations, the term r is equal to $\text{rank}(A_{0..\lg M-1, 0..\lg M-1})$, and the function $H(N,M,B)$ is given by equation (5). For BPC permutations, the function $\rho(A)$ is defined in equation (3). Each of these classes, with the exception of arbitrary block permutations, can be performed on-line.

By restricting the characteristic matrix A of a BMMC permutation to be a permutation matrix—having exactly one 1 in each row and each column—we obtain the class of *bit-permute/complement*, or *BPC*, permutations. (Johnsson and Ho [7] call BPC permutations *dimension permutations*, and Aggarwal, Chandra, and Snir [1] call BPC permutations without complementing *rational permutations*.) We can also view BPC permutations as being defined by a permutation $\pi : \{0,1,\ldots,n-1\} \overset{1\text{-}1}{\rightarrow} \{0,1,\ldots,n-1\}$ on address bits and a complement vector c. The characteristic matrix $A = (a_{ij})$ and the address-bit permutation π used in equation (1) are related in the usual way for permutation matrices:

$$a_{ij} = \begin{cases} 1 & \text{if } j = \pi(i) , \\ 0 & \text{otherwise} \end{cases} \tag{2}$$

for $i, j = 0, 1, \ldots, n-1$. We shall characterize BPC permutations by both matrices and address-bit permutations according to which is more convenient at the time.

The next three permutation classes to be defined are useful as subroutines in performing BMMC and BPC permutations. Each is a restricted BMMC permutation and is described by a characteristic matrix and a complement vector. First is the class of *block BMMC* permutations, which are BMMC permutations with the restrictions shown in Table 1. We require that both the leading $b \times b$ submatrix and the trailing $(n-b) \times (n-b)$ submatrix be nonsingular and that the rest of the characteristic matrix be 0.

The next class is the *block BPC* permutations, which are block BMMC permutations with the additional restriction shown in Table 1: the trailing $(n-b) \times (n-b)$ submatrix must be a permutation matrix.

The third and final class of restricted BMMC permutations are *memory-rearrangement/complement*, or *MRC*, permutations. Table 1 shows the form of an MRC characteristic matrix. Both the leading $m \times m$ and trailing $(n-m) \times (n-m)$ submatrices are nonsingular, the upper right $m \times (n-m)$ submatrix can contain any 0-1 values at all, and the lower left $(n-m) \times m$ submatrix is all 0. As we shall see more formally in Section 5, an MRC permutation can be performed by reading in a *memoryload* (i.e., M records) of M/BD tracks, permuting the records read within memory, and writing them out to M/BD tracks.

The full paper also deals with *arbitrary block permutations*, in which all records within a source block remain together in a target block, and the blocks may be permuted arbitrarily.

We shall use some more notational conventions in this paper. Matrix row and column numbers are indexed from 0 starting from the upper left. We index rows and columns by sets to indicate submatrices, using ".." notation to indicate sets of contiguous numbers. We denote an identity matrix by I and a matrix whose entries are all 0s by 0; the dimensions of such matrices will be clear from their contexts. We will frequently switch between bit vectors and their interpretations as base-2 integers; we interpret the vector $x = (x_0, x_1, \ldots, x_{n-1})$ as the integer $\sum_{i=0}^{n-1} x_i 2^i$. Vectors are treated as 1-column matrices in context.

We will characterize the I/O complexity of BPC permutations by their *cross-ranks*, which we can define on the characteristic matrix or the equivalent address-bit permutation. Intuitively, for any bit position k, the *k-cross-rank*, denoted ρ_k, is the number of bits that cross a dividing line between bit positions $k-1$ and k one way. More formally, for a bit permutation π or the equivalent characteristic matrix A, and for $k = 0, 1, \ldots, n-1$, we define

$$
\begin{aligned}
\rho_k(\pi) &= |\{j : 0 \le j \le k-1 \text{ and } k \le \pi(j) \le n-1\}| \\
&= |\{j : k \le j \le n-1 \text{ and } 0 \le \pi(j) \le k-1\}| \ , \\
\rho_k(A) &= \operatorname{rank}(A_{0..k-1,k..n-1}) \\
&= \operatorname{rank}(A_{k..n-1,0..k-1}) \ ,
\end{aligned}
$$

and we define the *cross-rank* ρ by

$$
\rho(\pi) = \max(\rho_m(\pi), \rho_b(\pi)) \ ,
$$

$$\rho(A) = \max(\rho_m(A), \rho_b(A)) . \tag{3}$$

The cross-rank, therefore, is the maximum of the m- and b-cross-ranks and is thus dependent on the memory size and the block size.

3 BPC Permutations

In this section, we examine how to perform BPC permutations, which, as we have seen, include many commonly used permutations. We are given an $n \times n$ characteristic matrix A that is a permutation matrix and a complement vector c of length n. Our algorithm performs at most $\frac{2N}{BD}\left(2\left\lceil\frac{\rho(A)}{\lg(M/B)}\right\rceil + 1\right)$ parallel I/Os.

It is more convenient to think in terms of address-bit permutations than in terms of permutation matrices. Given a characteristic matrix A for a BPC permutation, we shall work with the equivalent address-bit permutation π that satisfies equation (2). For now, we ignore the complement part of the BPC permutation.

We shall perform BPC permutations by alternately performing MRC and block BPC permutations. As we shall see in Sections 5 and 6, we can perform any MRC or any block BPC permutation using exactly $2N/BD$ parallel I/Os.

The characteristic matrices of the MRC and block BPC permutations we use have particular forms. For the MRC permutations, the leading $m \times m$ submatrix is a permutation matrix and the trailing $(n-m) \times (n-m)$ submatrix is an identity matrix. These permutations are equivalent to bit permutations π_{MRC} in which $\pi_{\text{MRC}}(j) = j$ for $j = m, m+1, \ldots, n-1$. For the block BPC permutations, the leading $b \times b$ submatrix is an identity matrix and the trailing $(n-b) \times (n-b)$ submatrix is a permutation matrix. These permutations are equivalent to bit permutations π_{BlockBPC} in which $\pi_{\text{BlockBPC}}(j) = j$ for $j = 0, 1, \ldots, b-1$.

The idea in performing BPC permutations is to sort the bits according to the permutation π. We alternately determine and perform MRC permutations π_{MRC} and block BPC permutations π_{BlockBPC}. We maintain a running "remaining permutation" π_{rem} and a running "permutation performed so far" π_{perf}, updating them each time we perform an MRC or block BPC permutation. Initially, π_{perf} is the identity permutation and $\pi_{\text{rem}} = \pi$; the algorithm terminates when π_{rem} is the identity permutation, at which time $\pi_{\text{perf}} = \pi$. The following pseudocode uses $\circ$ to indicate function composition—$(f \circ g)(j) = f(g(j))$—which is associative. The input to the procedure PERFORM-BPC is the permutation π to be performed, and the procedure's side effect is to perform π. Parameters are passed by reference, so that changes to the parameters π_{perf} and π_{rem} in the subroutines PERFORM-MRC and PERFORM-BLOCK-BPC are seen by PERFORM-BPC. Inspection of the code reveals that the permutation π_{perf} is not really needed; we include it to facilitate the proof of correctness below.

PERFORM-BPC(π)

$\pi_{\text{perf}} \leftarrow$ identity permutation
$\pi_{\text{rem}} \leftarrow \pi$
PERFORM-MRC($\pi_{\text{perf}}, \pi_{\text{rem}}$)
while π_{rem} is not the identity permutation
    do PERFORM-BLOCK-BPC($\pi_{\text{perf}}, \pi_{\text{rem}}$)
        PERFORM-MRC($\pi_{\text{perf}}, \pi_{\text{rem}}$)

PERFORM-MRC($\pi_{\text{perf}}, \pi_{\text{rem}}$)

for $j \leftarrow 0$ **to** $m - 1$
    do $\pi_{\text{MRC}}(j) \leftarrow i$, where $\pi_{\text{rem}}(j)$ is the ith smallest member
        (starting from $i = 0$) of $\{\pi_{\text{rem}}(0), \pi_{\text{rem}}(1), \ldots, \pi_{\text{rem}}(m-1)\}$
for $j \leftarrow m$ **to** $n - 1$
    do $\pi_{\text{MRC}}(j) \leftarrow j$
perform the permutation π_{MRC}
$\pi_{\text{perf}} \leftarrow \pi_{\text{MRC}} \circ \pi_{\text{perf}}$
$\pi_{\text{rem}} \leftarrow \pi_{\text{rem}} \circ \pi_{\text{MRC}}^{-1}$

PERFORM-BLOCK-BPC($\pi_{\text{perf}}, \pi_{\text{rem}}$)

for $j \leftarrow 0$ **to** $b - 1$
    do $\pi_{\text{BlockBPC}}(j) \leftarrow j$
for $j \leftarrow b$ **to** $n - 1$
    do $\pi_{\text{BlockBPC}}(j) \leftarrow b + i$, where $\pi_{\text{rem}}(j)$ is the ith smallest member
        (starting from $i = 0$) of $\{\pi_{\text{rem}}(b), \pi_{\text{rem}}(b+1), \ldots, \pi_{\text{rem}}(n-1)\}$
perform the permutation π_{BlockBPC}
$\pi_{\text{perf}} \leftarrow \pi_{\text{BlockBPC}} \circ \pi_{\text{perf}}$
$\pi_{\text{rem}} \leftarrow \pi_{\text{rem}} \circ \pi_{\text{BlockBPC}}^{-1}$

We now sketch a proof of correctness, argue that the algorithm achieves the claimed upper bound on the number of parallel I/Os, and show how to extend the algorithm for the case in which the complement vector is nonzero.

Lemma 1 *When* PERFORM-BPC *terminates, it has performed the permutation* π.

Proof sketch: It is straightforward to show that between subroutine calls, PERFORM-BPC preserves the invariant that $\pi_{\text{rem}} \circ \pi_{\text{perf}} = \pi$. When π_{rem} is the identity permutation, therefore, $\pi_{\text{perf}} = \pi$. ■

Lemma 2 PERFORM-BPC *achieves the termination condition that* π_{rem} *is the identity permutation after at most* $\lceil \rho(\pi)/(m-b) \rceil$ *iterations of the* **while** *loop.*

Proof sketch: We start by noting that PERMUTE-BPC works by sorting the π_{rem} permutation, calling PERFORM-MRC to sort the first m items and

PERFORM-BLOCK-BPC to sort the last $n-b$ items. There is an $(m-b)$-sized area of overlap. Thus, we can view PERMUTE-BPC as sorting the numbers $\{0, 1, \ldots, n-1\}$ in an array of size n in which we alternately sort the first m numbers and the last $n-b$ numbers until the entire array is sorted.

Let κ_m be the set of numbers that start in the first m positions but belong in the last $n-m$ positions, so that $\rho_m(\pi) = |\kappa_m|$. If we consider the combined effect of calling PERFORM-MRC followed by calling PERFORM-BLOCK-BPC, we can show that after $\lceil |\kappa_m| / (m-b) \rceil$ pairs of calls, all the numbers of κ_m have been moved into the last $n-m$ positions. Similarly, we can show that if κ_b is the set of numbers that start in the last $n-b$ positions but belong in the first b positions (so that $\rho_b(\pi) = |\kappa_b|$), after $\lceil |\kappa_b| / (m-b) \rceil$ pairs of calls of PERFORM-BLOCK-BPC and PERFORM-MRC, all the numbers in κ_b have been moved into the first b positions.

Therefore, one call of PERFORM-MRC followed by the calls of PERFORM-BLOCK-BPC and PERFORM-MRC made during $\lceil \max(|\kappa_m|, |\kappa_b|)/(m-b) \rceil$ iterations of the **while** loop moves all the numbers in κ_m and κ_b into their correct ranges. The last sorts performed place all the numbers into their actual correct positions. Since $\rho(\pi) = \max(\rho_m(\pi), \rho_b(\pi)) = \max(|\kappa_m|, |\kappa_b|)$, we have proven the lemma. ■

Theorem 3 *A BPC permutation with characteristic matrix A and complement vector c can be performed using at most $\frac{2N}{BD}\left(2\left\lceil\frac{\rho(A)}{\lg(M/B)}\right\rceil + 1\right)$ parallel I/Os.*

Proof: We note that $m - b = \lg(M/B)$ and that each call of PERFORM-MRC and PERFORM-BLOCK-BPC performs $2N/BD$ parallel I/Os. Thus, Lemmas 1 and 2 (letting π and A satisfy equation (2)) prove the theorem for cases in which the complement vector c is all 0. To handle cases in which $c \neq 0$, we simply modify PERFORM-BPC so that the last call of PERFORM-MRC also complements according to c. ■

4 Permutable Sets of Blocks

In this section, we sketch the technique of decomposing a permutation into permutable sets of blocks. We shall use this technique in later sections to perform MRC and block BMMC permutations.

We assume that the records to be permuted initially reside in areas on their disks that we call the *source portion* of the disk array. Each disk holds N/D records. We also assume that each disk in the disk array has enough spare storage to hold another copy of the data it holds, that is, that in addition to the N/D records initially on each disk, there is room for another N/D records as well; we call this spare storage area the *target portion*. The track number in a record address is relative to the beginning of the disk portion in which the record resides. A *pass* consists of repeatedly reading *source blocks* from the source portion of each disk into memory, permuting the records of these blocks in memory, and then writing the records out in *target blocks* to the target portion of each disk. The roles of the source and target portions can then be reversed for the next pass.

Each record, and thus each block, is read exactly once and written exactly once during a pass. One pass, therefore, performs exactly $2N/BD$ parallel I/Os.

To perform a pass, we need to decompose the permutation into disjoint sets of blocks that we can read in k blocks per disk at a time, permute their records within memory, and then write out k blocks per disk. For a permutation of source addresses to target addresses and a positive integer k, we define a *k-permutable* set of blocks as a set of kD source blocks such that

1. each disk contains exactly k of these source blocks,

2. the kBD records in the source blocks are mapped to exactly kD full target blocks, and

3. each disk has exactly k of these target blocks mapped to it.

The following theorem, whose proof appears in the full paper, shows how k-permutable sets of blocks can be used to perform certain permutations in one pass.

Theorem 4 *Suppose that we wish to permute N records according to a permutation $\pi : \{0, 1, \ldots, N-1\} \stackrel{1\text{-}1}{\to} \{0, 1, \ldots, N-1\}$. Suppose further that for some positive integer $k \leq M/BD$, the N/B blocks containing the records to be permuted can be decomposed into a series of k-permutable sets of blocks such that each source address appears in exactly one source block (and thus each target address appears in exactly one target block). Then the permutation π can be performed in one pass, using a total of $2N/BD$ parallel I/Os.* ■

We call such a sequence of N/kBD reads and writes of k-permutable sets of blocks a *schedule.* In Sections 5 and 6, we shall see how to devise schedules for MRC and block BMMC permutations.

5 MRC Permutations

In this section, we show how to perform any MRC permutation in just one pass, thus using exactly $2N/BD$ parallel I/Os. We shall do so by decomposing the permutation into M/BD-permutable sets of blocks to form a schedule.

We construct the schedule as follows. For $i = 0, 1, \ldots, N/M - 1$ and $j = 0, 1, \ldots, M-1$, define the source address $x^{(i,j)}$ so that

$$\begin{aligned} x^{(i,j)}_{0..m-1} &= j \, , \\ x^{(i,j)}_{m..n-1} &= i \, . \end{aligned}$$

That is, the least significant m bits of $x^{(i,j)}$ are the binary representation of j, and the most signficant $n-m$ bits are the binary representation of i. Then the schedule is comprised of the following N/M sets:

$$\begin{gathered}\{x^{(0,0)}, x^{(0,1)}, \ldots, x^{(0,M-1)}\} , \\ \{x^{(1,0)}, x^{(1,1)}, \ldots, x^{(1,M-1)}\} , \\ \vdots \\ \{x^{(N/M-1,0)}, x^{(N/M-1,1)}, \ldots, x^{(N/M-1,M-1)}\} .\end{gathered}$$

The full paper proves that these sets form a schedule for any MRC permutation.

6 Block BMMC and Block BPC Permutations

In this section, we show how to perform any block BMMC permutation in just one pass, using exactly $2N/BD$ parallel I/Os. We also show how some of the mathematics of the block-BMMC algorithm can be avoided for the more restricted class of block BPC permutations.

The algorithm for block BMMC permutations is easier to understand if we assume for now that $B = 1$, and thus $b = 0$ and $n = d + t$. That is, we assume that blocks contain just one record each and that we perform a BMMC permutation on them. We are given an $n \times n$ characteristic matrix A and a complement vector c of length n.

The algorithm has two parts. First, we find a 1-permutable set of blocks. Second, we construct a schedule for the entire permutation given any 1-permutable set of blocks. After presenting these two parts, we shall see how to remove the restriction that $B = 1$.

We find a 1-permutable set of blocks in four steps:

1. Find a set S of d "basis" columns such that the submatrix $A_{0..d-1,S}$ is nonsingular. Such a set exists by the following argument. Because A is nonsingular, all of its rows are linearly independent. In particular, rows $0, 1, \ldots, d-1$ are. Thus, the submatrix $A_{0..d-1,0..n-1}$ has full row rank. Hence, there exists a subset of column indices $S \subseteq \{0, 1, \ldots, n-1\}$ such that $|S| = d$ and the $d \times d$ submatrix $A_{0..d-1,S}$ is nonsingular. Let us define $Q = A_{0..d-1,S}$, so that Q^{-1} exists. The basis S can be determined by performing Gaussian elimination to create an LUP decomposition of A^{T}; the full paper contains the details.

2. Given a set of basis columns S, we define

$$\begin{aligned} T &= \{0, 1, \ldots, d-1\} - S , \\ U &= S \cap \{0, 1, \ldots, d-1\} , \\ V &= \{0, 1, \ldots, n-1\} - (S \cup T) . \end{aligned}$$

   The sets T and U form a partition of $\{0, 1, \ldots, d-1\}$, and the sets S, T, and V form a partition of $\{0, 1, \ldots, n-1\}$. We order T and U into increasing order so that

$$T = \{t_0, t_1, \ldots, t_{|T|-1}\}\ , \quad t_j > t_{j-1} \quad \text{for } j = 1, 2, \ldots, |T| - 1$$
$$U = \{u_0, u_1, \ldots, u_{|U|-1}\}\ , \quad u_j > u_{j-1} \quad \text{for } j = 1, 2, \ldots, |U| - 1\ .$$

3. For the sake of clarity, we shall be more explicit in the remainder of this section about conversions between integers and their binary representations. Let us denote the d-bit representation of an integer $i \in \{0, 1, \ldots, D-1\}$ by $\mathrm{bin}(i)$; this representation can be thought of as a vector of length d. Let us also denote the jth bit of this representation, for $j = 0, 1, \ldots, d-1$, by $\mathrm{bin}_j(i)$. (Thus, $\mathrm{bin}_j(i) = \lfloor i/2^j \rfloor \bmod 2$.) We extend this notation from individual bits to sets of bits in the natural way.

   We define the permutation R on $\{0, 1, \ldots, D-1\}$ as follows. Let $i \in \{0, 1, \ldots, D-1\}$, and let the binary representation of i be $\mathrm{bin}(i) = (i_0, i_1, \ldots, i_{d-1})$. Then $R(i) = k$, where the binary representation of k is $\mathrm{bin}(k) = (k_0, k_1, \ldots, k_{d-1})$ and

$$\begin{aligned} k_j &= i_{u_j} \quad \text{for } j = 0, 1, \ldots, |U| - 1\ , \\ k_{|U|+j} &= i_{t_j} \quad \text{for } j = 0, 1, \ldots, |T| - 1\ . \end{aligned}$$

4. Define the set $\{x^{(0)}, x^{(1)}, \ldots, x^{(D-1)}\}$ according to bit indices:

$$\begin{aligned} x_S^{(i)} &= Q^{-1} A_{0..d-1,T}\, \mathrm{bin}_T(i) \oplus \mathrm{bin}(R(i)) \oplus c_{0..d-1}\ , \qquad (4) \\ x_T^{(i)} &= \mathrm{bin}_T(i)\ , \\ x_V^{(i)} &= 0 \end{aligned}$$

   for $i = 0, 1, \ldots, D-1$.

The full paper proves that both $\{x^{(0)}_{0..d-1}, x^{(1)}_{0..d-1}, \ldots, x^{(D-1)}_{0..d-1}\}$ and $\{y^{(0)}_{0..d-1}, y^{(1)}_{0..d-1}, \ldots, y^{(D-1)}_{0..d-1}\}$ are permutations of $\{0, 1, \ldots, D-1\}$, thereby showing that the set $\{x^{(0)}, x^{(1)}, \ldots, x^{(D-1)}\}$ is a 1-permutable set of blocks.

Having seen how to find a 1-permutable set of blocks, we now show how to use this set to decompose the BMMC permutation into 1-permutable sets of blocks that together form a schedule. We still assume that $B = 1$ and thus $b = 0$.

The algorithm is simple. We define a set of $N/D - 1$ n-vectors $\{p^{(0)}, p^{(1)}, \ldots, p^{(N/D-1)}\}$ such that $p^{(j)}_{0..d-1} = 0$ and $p^{(j)}_{d..n-1}$ is the binary representation of j for $j = 0, 1, \ldots, N/D - 1$. Given a 1-permutable set of blocks $\{x^{(0)}, x^{(1)}, \ldots, x^{(D-1)}\}$, we form the following N/D sets:

$$\begin{gathered} \{x^{(0)} \oplus p^{(0)}, x^{(1)} \oplus p^{(0)}, \ldots, x^{(D-1)} \oplus p^{(0)}\}\ , \\ \{x^{(0)} \oplus p^{(1)}, x^{(1)} \oplus p^{(1)}, \ldots, x^{(D-1)} \oplus p^{(1)}\}\ , \\ \{x^{(0)} \oplus p^{(2)}, x^{(1)} \oplus p^{(2)}, \ldots, x^{(D-1)} \oplus p^{(2)}\}\ , \\ \vdots \\ \{x^{(0)} \oplus p^{(N/D-1)}, x^{(1)} \oplus p^{(N/D-1)}, \ldots, x^{(D-1)} \oplus p^{(N/D-1)}\}\ . \end{gathered}$$

The full paper proves that these sets form a schedule for the BMMC permutation. Assuming that $b = 0$, therefore, once we have found *any* 1-permutable set of blocks, we can construct an entire schedule by simply XORing the most signficant t bits by $0, 1, \ldots, N/D - 1$ in turn.

Removing the restriction that $B = 1$ is easy. By the structure of block BMMC matrices, the the b least significant bits of a target address depend only on the b least significant bits of the corresponding source address. (Recall that the lower left $(n-b) \times b$ submatrix of the characteristic matrix is 0.) Moreover, the leading $b \times b$ submatrix of the characteristic matrix is nonsingular. Therefore, records that start in the same block end up in the same block. Likewise, the most significant $n - b$ bits of a target address depend only on the most significant $n - b$ bits of the corresponding source address. Blocks, therefore, are permuted according only to their block addresses.

We have thus shown how to perform block BMMC permutations in one pass, using exactly $2N/BD$ parallel I/Os.

For block BPC permutations, which are included in the class of block BMMC permutations, some of the mathematics in finding an initial 1-permutable set of blocks becomes considerably simpler. Assuming that we are given a permutation matrix A and that $b = 0$, finding the set of basis columns S is easy. We don't need to perform Gaussian elimination. We only need to find the columns that have a 1 within the first d rows:

$$S = \{j : a_{ij} = 1 \text{ for some } 0 \leq i \leq d - 1\} .$$

Since $S \cap T = \emptyset$ and all d of the 1s in the first d rows of A are contained in $A_{0..d-1,S}$, we have that $A_{0..d-1,T} = 0$. The computation of $x_S^{(i)}$, for $i = 0, 1, \ldots, D - 1$, becomes simpler, involving no matrix operations. Starting from equation (4), we have

$$\begin{aligned} x_S^{(i)} &= Q^{-1} A_{0..d-1,T} \operatorname{bin}_T(i) \oplus \operatorname{bin}(R(i)) \oplus c_{0..d-1} \\ &= \operatorname{bin}(R(i)) \oplus c_{0..d-1} . \end{aligned}$$

The definitions of $x_T^{(i)} = \operatorname{bin}_T(i)$ and $x_V^{(i)} = 0$ remain the same.

7 BMMC Permutations

In this section, we show how to perform BMMC permutations using MRC, block BPC, and BPC permutations. As usual, we assume that the BMMC permutation is given by an $n \times n$ characteristic matrix A and a complement vector c of length n. The number of parallel I/Os is at most

$$\frac{2N}{BD}\left(2\left\lceil\frac{\lg M - \operatorname{rank}(A_{0..\lg M-1,0..\lg M-1})}{\lg(M/B)}\right\rceil + H(N, M, B)\right) ,$$

where

$$H(N,M,B) = \begin{cases} 4\left\lceil \dfrac{\lg B}{\lg(M/B)} \right\rceil + 11 & \text{if } M \le \sqrt{N}\ , \\ 4\left\lceil \dfrac{\lg(N/B)}{\lg(M/B)} \right\rceil + 2 & \text{if } \sqrt{N} < M < \sqrt{NB}\ , \\ 6 & \text{if } \sqrt{NB} \le M\ . \end{cases} \tag{5}$$

Our strategy is to factor the matrix A into a product of matrices, each of which is the characteristic matrix of an MRC, block BPC, or BPC permutation. For now, we ignore the complement vector c. We read the factors right to left to determine the order in which to perform these permutations. For example, if we factor $A = V\,W$, then we perform A by performing the permutation with characteristic matrix W and then performing the permutation with characteristic matrix V.

We start by permuting the columns of A so that the leading $m \times m$ submatrix is nonsingular. That is, we factor $A = \widehat{A}\,\Pi$, where Π is a permutation matrix and the submatrix $\widehat{A}_{0..m-1,0..m-1}$ is nonsingular. The full paper shows how to determine the matrix Π and that the corresponding BPC permutation can be performed with at most $\frac{2N}{BD}\left(2\left\lceil\frac{\lg M - \operatorname{rank}(A_{0..\lg M-1,0..\lg M-1})}{\lg(M/B)}\right\rceil + 1\right)$ parallel I/Os.

Our next task is to factor the matrix $\widehat{A}$, which is nonsingular since A and Π are. We divide $\widehat{A}$ between the lower m rows and columns and the upper $n-m$ rows and columns:

$$\widehat{A} = \begin{array}{c} \begin{array}{cc} m & n-m \end{array} \\ \left[\begin{array}{c|c} \alpha & \beta \\ \hline \gamma & \delta \end{array}\right] \end{array} \begin{array}{l} m \\ n-m \end{array} ,$$

so that the leading $m \times m$ submatrix α is nonsingular. We factor

$$\widehat{A} = V\,W\,Y = \left[\begin{array}{c|c} I & 0 \\ \hline \gamma\,\alpha^{-1} & I \end{array}\right] \left[\begin{array}{c|c} I & 0 \\ \hline 0 & \gamma\,\alpha^{-1}\,\beta \oplus \delta \end{array}\right] \left[\begin{array}{c|c} \alpha & \beta \\ \hline 0 & I \end{array}\right] .$$

(Submatrix dimensions are the same as for $\widehat{A}$.) The matrices W and Y each characterize MRC permutations, so they take only one pass each.

We still have to factor the matrix V, which would be an identity matrix if the lower left submatrix, equal to $\gamma\,\alpha^{-1}$, were 0. The location of this submatrix is exactly the location that is required to be all 0s in both MRC and block BMMC permutations. If $\gamma = 0$, then $\gamma\,\alpha^{-1} = 0$, and we don't need to factor any further. In general, however, we factor V by applying BPC permutations to move the entries of this submatrix to locations that allow us to perform MRC or block BPC permutations. The way in which we factor V depends on whether $m \le n-m$, which is equivalent to $m \le n/2$ or $M \le \sqrt{N}$.

First, we handle the case in which $m > n/2$. We divide the $(n-m) \times m$ submatrix product $\gamma\,\alpha^{-1}$ as

$$\gamma\,\alpha^{-1} = \begin{array}{c} \begin{array}{cc} 2m-n & n-m \end{array} \\ \left[\begin{array}{c|c} \sigma & \tau \end{array}\right] \end{array} n-m$$

so that we divide V as

$$V = \begin{array}{c} \begin{array}{ccc} 2m-n & n-m & n-m \end{array} \\ \left[\begin{array}{c|c|c} I & 0 & 0 \\ \hline 0 & I & 0 \\ \hline \sigma & \tau & I \end{array}\right] \end{array} \begin{array}{l} \\ 2m-n \\ n-m \\ n-m \end{array} .$$

We factor V into

$$V = \Pi' \, Z \, \Pi' = \left[\begin{array}{c|c|c} I & 0 & 0 \\ \hline 0 & 0 & I \\ \hline 0 & I & 0 \end{array}\right] \left[\begin{array}{c|c|c} I & 0 & 0 \\ \hline \sigma & I & \tau \\ \hline 0 & 0 & I \end{array}\right] \left[\begin{array}{c|c|c} I & 0 & 0 \\ \hline 0 & 0 & I \\ \hline 0 & I & 0 \end{array}\right] .$$

(Submatrix dimensions are the same as for V.) The matrix Z characterizes an MRC permutation, so it can be performed in one pass.

There are two cases in the analysis for the permutation matrix Π'. If $b \le 2m - n$ or, equivalently, $\sqrt{NB} \le M$, then Π' is the characteristic matrix of a block BPC permutation, so Π' can be performed in only one pass. Otherwise, $\rho(\Pi') = n - m$ (see the full paper for details), and we use at most $2\left\lceil \frac{\lg(N/B)}{\lg(M/B)} \right\rceil - 1$ passes to perform Π'.

Now we factor V for the case in which $m \le n/2$ or, equivalently, $M \le \sqrt{N}$. We divide $\gamma\,\alpha^{-1}$ as

$$\gamma\,\alpha^{-1} = \begin{array}{c} m \\ \left[\begin{array}{c} \phi \\ \hline \psi \end{array}\right] \end{array} \begin{array}{l} \\ m \\ n-2m \end{array}$$

so that we divide V as

$$V = \begin{array}{c} \begin{array}{ccc} m & m & n-2m \end{array} \\ \left[\begin{array}{c|c|c} I & 0 & 0 \\ \hline \phi & I & 0 \\ \hline \psi & 0 & I \end{array}\right] \end{array} \begin{array}{l} \\ m \\ m \\ n-2m \end{array} .$$

We factor V into

$$V = \Pi'' \, Q \, R \, \Pi'' = \left[\begin{array}{c|c|c} 0 & I & 0 \\ \hline I & 0 & 0 \\ \hline 0 & 0 & I \end{array}\right] \left[\begin{array}{c|c|c} I & \phi & 0 \\ \hline 0 & I & 0 \\ \hline 0 & 0 & I \end{array}\right] \left[\begin{array}{c|c|c} I & 0 & 0 \\ \hline 0 & I & 0 \\ \hline 0 & \psi & I \end{array}\right] \left[\begin{array}{c|c|c} 0 & I & 0 \\ \hline I & 0 & 0 \\ \hline 0 & 0 & I \end{array}\right] .$$

(Submatrix dimensions are the same as for V.) The matrix Q characterizes an MRC permutation, and R characterizes a block BMMC permutation. Therefore, each of Q and R can be performed in one pass. For the permutation matrix Π'', we have $\rho_b(\Pi'') = b \le m = \rho_m(\Pi'')$, which implies that the number of passes to perform Π'' is at most $2\left\lceil \frac{\lg B}{\lg(M/B)} \right\rceil + 3$.

Putting all the factors together, we get the following theorem.

Theorem 5 *A BMMC permutation with characteristic matrix A and complement vector c can be performed using at most*

$$\frac{2N}{BD}\left(2\left\lceil \frac{\lg M - \operatorname{rank}(A_{0..\lg M-1,0..\lg M-1})}{\lg(M/B)} \right\rceil + H(N,M,B)\right)$$

parallel I/Os, where the function $H(N, M, B)$ is given by equation (5).

Proof sketch: If the complement vector c is all 0s, counting up the passes in the above argument gives the stated bound. To handle cases in which the complement vector c is nonzero, we include c in the last BPC permutation performed, either Π' or Π''. If $V = I$, in which case we perform neither Π' nor Π'', we complement when performing the block BMMC permutation W. ∎

We conclude this section with an observation about the function $H(N, M, B)$. It is $\Theta\left(\frac{\lg B}{\lg(M/B)} + 1\right)$ if $M \leq \sqrt{N}$ and $\Theta\left(\frac{\lg(N/B)}{\lg(M/B)} + 1\right)$ if $\sqrt{N} < M < \sqrt{NB}$. Although the latter bound is asymptotically no better than we get by sorting, we can actually write the asymptotic bound as $\Theta\left(\frac{\lg \min(B,N/B)}{\lg(M/B)} + 1\right)$ for $M < \sqrt{NB}$. (The full paper contains the argument.) Although our method for performing BMMC permutations is not asymptotically faster than sorting when $\sqrt{N} < M < \sqrt{NB}$, the value of $\lg(N/B)$ is relatively small, being asymptotically equal to $\lg B$ in this range.

8 Conclusions

We have shown that many permutations can be performed on disk arrays faster than sorting. In particular, we generalized the matrix-tranpose algorithm of Vitter and Shriver to BPC permutations and BMMC permutations. These are broad classes that include many commonly used permutations: matrix transpose (or cyclic rotation of address bits), bit-reversal, hypercube, vector-reversal, Gray and inverse Gray codes, and permutations that help reduce data-access conflicts. Along the way, we came up with restricted permutations—MRC, block BPC, and block BMMC—that can be performed in just one pass over the records. The full paper also shows how to perform arbitrary block permutations with off-line scheduling and also proves lower bounds for BPC and BMMC permutations, showing that our scheme for BPC permutations is asymptotically optimal.

Acknowledgements

This research was supported in part by the Defense Advanced Research Projects Agency under Grant N00014-91-J-1698.

Thanks to Michelangelo Grigni, Michael Klugerman, James Park, Eric Schwabe, Burton Smith, Cliff Stein, and especially Charles Leiserson for many helpful discussions. Thanks also to Steve Heller, Mark Nodine, and Lisa Tucker for comments on an earlier draft of the manuscript.

References

[1] Alok Aggarwal, Ashok K. Chandra, and Marc Snir. Hierarchical memory with block transfer. In *Proceedings of the 28th Annual Symposium on Foundations of Computer Science*, pages 204–216, October 1987.

[2] Alok Aggarwal and Jeffrey Scott Vitter. The input/output complexity of sorting and related problems. *Communications of the ACM*, 31(9):1116–1127, September 1988.

[3] Robert Alverson, David Callahan, Daniel Cummings, and Brian Koblenz. The Tera computer system. In *Proceedings of the 1990 International Conference on Supercomputing*, pages 1–6, June 1990.

[4] Peter Chen, Garth Gibson, Randy H. Katz, David A. Patterson, and Martin Schulze. Two papers on RAIDs. Technical Report UCD/CSD 88/479, Computer Science Division (EECS), University of California, Berkeley, December 1988.

[5] Robert W. Floyd. Permuting information in idealized two-level storage. In Raymond E. Miller and James W. Thatcher, editors, *Complexity of Computer Computations*, pages 105–109. Plenum Press, 1972.

[6] Harold N. Gabow and Oded Kariv. Algorithms for edge coloring bipartite graphs and multigraphs. *SIAM Journal on Computing*, 11(1):117–129, February 1982.

[7] S. Lennart Johnsson and Ching-Tien Ho. Generalized shuffle permutations on boolean cubes. Technical Report TR-04-91, Harvard University Center for Research in Computing Technology, February 1991.

[8] Mark H. Nodine and Jeffrey Scott Vitter. Greed sort: An optimal external sorting algorithm for multiple disks. Technical Report CS-90-04, Department of Computer Science, Brown University, February 1990.

[9] Mark H. Nodine and Jeffrey Scott Vitter. Large-scale sorting in parallel memories. In *Proceedings of the 3rd Annual ACM Symposium on Parallel Algorithms and Architectures*, pages 29–39, July 1991.

[10] Alan Norton and Evelyn Melton. A class of boolean linear transformations for conflict-free power-of-two stride access. In *Proceedings of the 1987 International Conference on Parallel Processing*, pages 247–254, August 1987.

[11] David A. Patterson, Garth Gibson, and Randy H. Katz. A case for redudant arrays of inexpensive disks (RAID). In *ACM International Conference on Management of Data (SIGMOD)*, pages 109–116, June 1988.

[12] Jeffrey Scott Vitter and Elizabeth A. M. Shriver. Algorithms for parallel memory I: Two-level memories. Technical Report CS-90-21, Department of Computer Science, Brown University, September 1990.

[13] Jeffrey Scott Vitter and Elizabeth A. M. Shriver. Optimal disk I/O with parallel block transfer. In *Proceedings of the Twenty Second Annual ACM Symposium on Theory of Computing*, pages 159–169, May 1990.

Computer Aided Design

Compiling Joy to Silicon

Sam Weber[1]
Cornell University
samuel@cs.cornell.edu

Bard Bloom[2]
Cornell University
bard@cs.cornell.edu

Geoffrey Brown[3]
Cornell University
gbrown@ee.cornell.edu

January 2, 1992

Abstract

We present the highlights of an algorithm and correctness proof for compiling programs written in Joy, a process algebra, into delay-insensitive circuits. This is the among the first formally verified algorithms for compiling a process algebra into silicon. The implementation of our algorithm produces a netlist suitable for input to the standard place-and-route tools.

1 Introduction

Joy is a pared-down but complete concurrent programming language, with basic conditional, looping, and parallel control constructs, Boolean variables, and synchronization along communication channels. Most programs could be written in Joy, albeit clumsily, and we expect our methods to extend to a variety of other programming constructs: procedure and function calls, data-bearing channels, and so forth. We describe a scheme for compiling Joy programs to delay-insensitive circuits, present the highlights of the proof that the scheme is correct, and briefly discuss the implementation.

Our circuits are quite simple. A circuit consists of a set of primitive *components* connected by *wires* attached at *input* and *output* ports. A wire connects a single input port to a single output port. The unconnected ports of the constituent components of a circuit are the inputs and outputs of the circuit and connect to the outside environment. The components of a circuit communicate by initiating events on their output ports or by absorbing events on their input ports. One possible physical interpretation of an event is a voltage transition. For physical reasons, circuits cannot refuse input – even if that input may cause them to fail. Furthermore, input and output events on a wire must strictly alternate to prevent possible electrical interference. A wire connecting an output port to an input port is *active* if an event has been initiated on the output port but not yet absorbed on the input port.

We give a compilation scheme from Joy programs to circuits. Our set of primitive components is quite small, and most of them are straightforward to implement. Indeed, most components turn out to be patterns of wires. For related work on implementing delay-insensitive circuits see [1, 4, 16, 20, 5, 9]. The resulting circuits are amenable to a simple low-level mathematical model of wires and impulses, which we use to verify the compilation scheme. The full correctness proof is too large to be presented; we describe its mathematical foundations (the theory of bisimulation), and sketch the main lemmas describing how programs and circuits execute.

As the translation scheme is syntax-directed, we found it straightforward to implement with `lex` and `yacc`. Our Joy compiler produces netlists of basic components. The output of the compiler is suitable for input to any of the many place-and-route tools available for silicon compilation, given a VLSI cell-library of the basic components. The development of such a cell-library is fairly standard; see [1, 4, 6, 13]. As the implementation is routine compiler technology, we concentrate on the description of the compilation scheme and correctness proof.

1.1 Related Work

Previous efforts at verifying compilers for VLSI have been restricted to weaker languages than Joy. An early work by Milne verified the correctness of a compiler for NOR expressions. [10] A more recent work described a compiler for translating path expressions into self-timed circuits. [18] The target architecture in this article was a synchronizer consisting of an arbiter circuit and a sequencer for each of the component path expressions. The resultant circuits are not free from timing assumptions; however, the construction for path recognizers (a component part of each sequencer) is very close in spirit to the algorithm used by our compiler to generate delay-insensitive cicuits.

A related approach was taken by Straunstrup and Greenstreet. [17] They present a method for describing the behavior of a circuit in terms of a set of transition rules and define a set of "implementation conditions" which a circuit must meet in order to be delay-insensitive. It is the designer's obligation to prove, for each circuit, that these conditions are met. Our compiler guarantees that all circuits which it generates are delay-insensitive.

Martin's work, like that of Strauntrup and Greenstreet, provides a methodology for verifying delay-insensitive circuits with respect to an idealized model of hardware behavior [9]. Although Martin has developed a compiler from a process algebra to delay-insensitive circuits, he has not published a proof that his compilation algorithm is correct [5].

2 Joy

In this section we present Joy. In the following BNF-style syntax description, c is drawn from a set of channel identifiers *CID*, $\mathcal{C}$ is a set of channel identifiers, and v is drawn from a set of variable identifiers *VID*.

$$\begin{array}{llll} & c & \in & CID \\ & v & \in & VID \\ \\ (expr) & E & ::= & v \mid \neg E \mid E_1 \wedge E_2 \mid E_1 \vee E_2 \mid tt \mid ff \\ (process) & P & ::= & \mathtt{skip} \mid v := expr \mid c! \mid P_1 ; P_2 \\ & & \mid & \mathtt{if}\, G\, \mathtt{fi} \mid \mathtt{do}\, G\, \mathtt{od} \mid P_1 \parallel_{\mathcal{C}} P_2 \\ (guard) & G & ::= & E \rightarrow P \mid c? \rightarrow P \mid E \,\&\, c? \rightarrow P \mid G_1 [\!] G_2 \end{array}$$

Informally, the process terms stand for the following processes:

`skip`: is a process which simply terminates.

$v := expr$: is a process which assigns $expr$ to v.

$c!$ is a process which is ready to output along channel c. Send commands are blocking; that is, they will wait for the signal to be received before allowing the sender to continue. (Receives are done as part of guarded commands, and are thus non-blocking.)

$P_1; P_2$: is the sequential composition of P_1 and P_2.

`if` G `fi`: attempts to execute the guarded command set G until it succeeds. A guarded command set consists of one or more guarded commands separated by $[\!]$ and is executed by executing one enabled command from the set. If no command is enabled, execution fails. A guarded command can be of the form $E \rightarrow P$, $c? \rightarrow P$, or $E\&c? \rightarrow P$. In the first case the command is enabled if E is true. In the second case the command is enabled if some process is prepared to output on channel c, and, in the third case the command is enabled if E is true and some process is prepared to output on c. The command is executed by inputting from channel c (for the second and third cases) and then executing P.

`do` G `od`: repeatedly executes the guarded command set G until execution fails.

$P_1 \parallel_{\mathcal{C}} P_2$: is the concurrent composition of P_1 and P_2, communicating along the channels in $\mathcal{C}$.

Legal Joy programs must satisfy some further syntactic restrictions. In particular, the set of input channels used by two concurrently executing processes must be disjoint; similarly for output channels. Furthermore, concurrently active processes may not share any variables.

3 Compiling Joy Programs

The compilation method results in connection of primitive circuit components, which we specify by finite state machines. For circuit realizations of similar components see [1, 8, 16].

It is possible to realize all of our basic components with four building blocks [3]. The implementations of our components can then be verified with respect to models of these building blocks using an automated tool such as that developed by Dill [7]. The next level of the proof, verifying the four building blocks, is likely to be extremely difficult as any implementation of these is dependent upon analog transistor effects (e.g. transistor threshold voltages and gate capacitance).

In order to ensure that these components are used in a manner which ensures the requirements on input and output events are met, we use a simple communication convention between components. Wires are always used in pairs or triples. This convention is illustrated in the following figure.

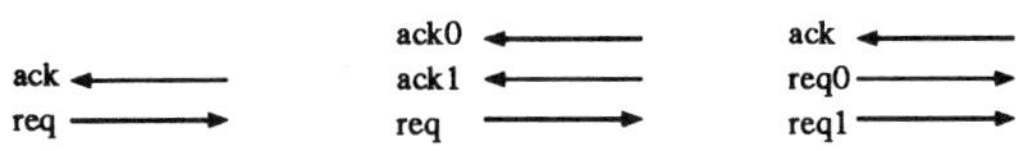

In the two-wire interface, every transition on the request wire `req` must be explicitly followed by a transition on the acknowledgement wire `ack`. There are two types of three wire interfaces. The read interface allows a binary value to be returned with an acknowledgment by allowing a transition on the request wire to be followed by a transition on either of the acknowledgement wires `ack0, ack1`. The write interface allows a value to be sent by allowing a transition on either of the request wires `req0, req1` followed by a transition of the `ack` wire.

The idea of rigidly following this communication convention was developed by the Philips group [20] who call the resulting components "handshake circuits." In the remainder of this section, we develop a set of basic handshake components and use them to implement Joy programs.

3.1 Compiling Expressions

We illustrate the use of the handshake convention by developing a set of components to perform Boolean expression evaluation. A Boolean expression is implemented as a component with a three wire read interface.

We present each primitive component as a wiring diagram and a finite-state automaton. Input ports are pictured as arrows pointing into the component. Output ports are pictured as arrows pointing out of the component. The specification of a component consists of a finite-state machine with its initial state labeled `Init`. All legal state transitions are indicated by arcs in the specification, labeled by the input or output event associated with the state transition. Output events are distinguished by overlined names, e.g. $\overline{ack}$.

The simplest expressions are the Boolean constants `tt` and `ff` which return true and false respectively. `tt` is illustrated and specified in the following figure; `ff` is similar.

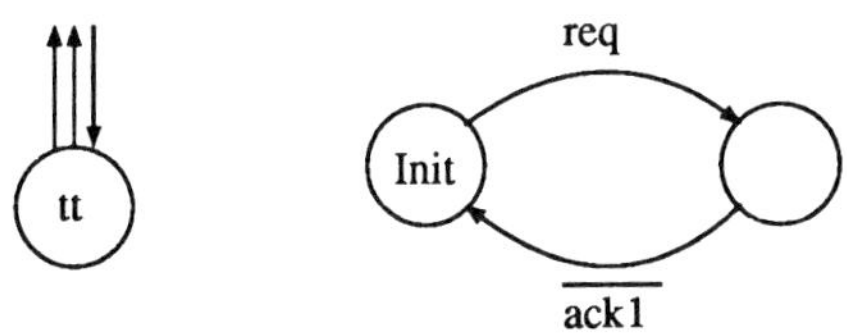

The simplest Boolean operator is negation (˜)

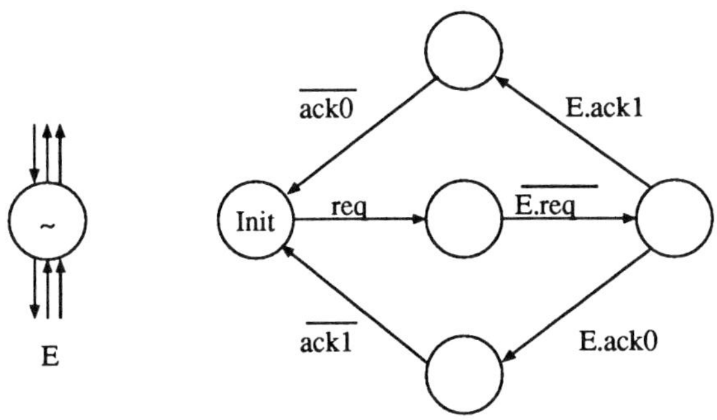

The binary operator $\wedge$ is a short-circuiting and; it first evaluates the left expression and then, if necessary, the right expression. $\vee$ is similar. (For simplicity, we use sequential rather than parallel Boolean operations. Concurrent evaluation of Booleans would in general require concurrent read accesses to variables. Doing this should be routine; however, the details remain to be worked out.)

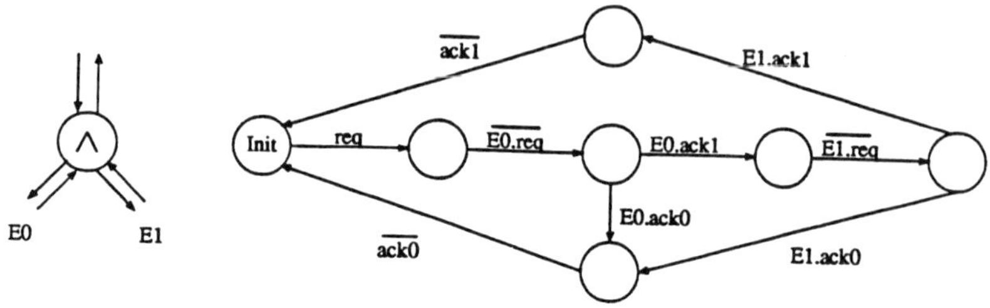

The final expression type is the Boolean variable. Boolean variables have two three-wire interfaces: a read interface for reading the variable, and a write interface for writing.

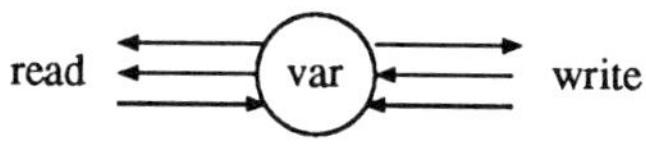

In practice more than one connection to each of these interfaces may be necessary (e.g. the process $v :=$ **expr1**$; v :=$ **expr2**); however, our rules for circuits prohibit multiple connections to a port. We solve this problem by introducing a set of components for sharing a single interface – the call $CALL$, call-read $RCALL$, and call-write $WCALL$ components.

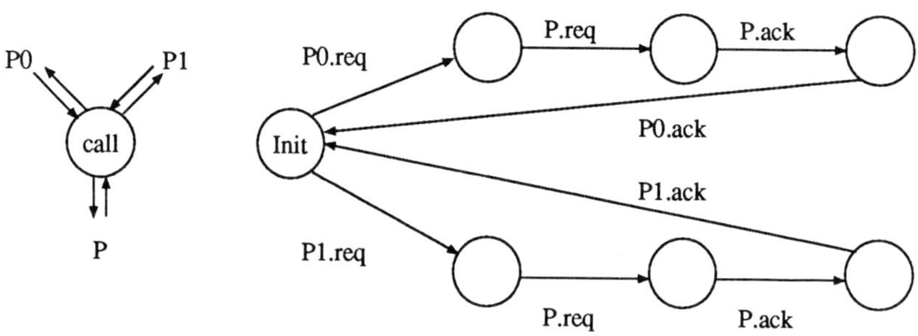

The call component routes requests from interfaces $P0$ and $P1$ to P returning subsequent acknowledgements from P to $P0$ or $P1$ as appropriate.

Clearly, requests must be mutually exclusive; we shall prove this follows from our linguistic constraints. The call component can easily be extended into a multiway call component by cascading call components in a tree or a linear array.

The call-read and call-right components are similar, but allow receiving and sending a bit of information on return and call respectively. Here is the read call component; the write call component is similar.

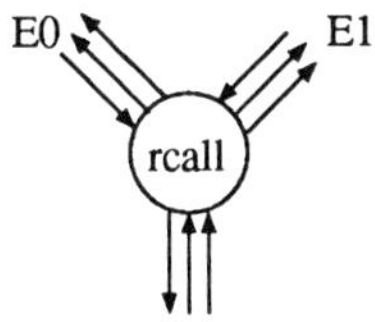

3.2 Primitive Processes

Each process P has a two-wire interface, allowing the P's user to invoke P and P to signal that it has finished.

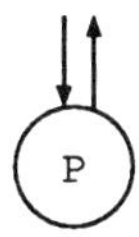

The simplest process is `skip`, which acknowledges each request immediately.

The assignment process provides a simple interface between the expression to be assigned and the variable to be assigned to.

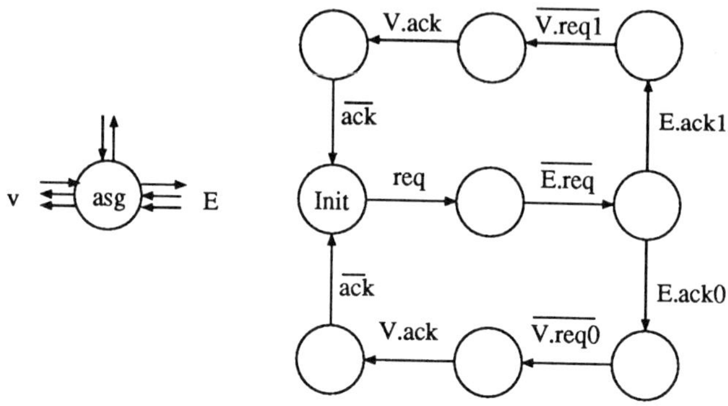

3.3 Compound Processes

Compound processes are formed by sequential composition, parallel composition, iterative (do) and conditional (if) constructs.

Sequential composition is implemented with the seq (;) component, which first requests its left child, and when that has finished its right child. The labels in states will be used in Section 5.4.1.

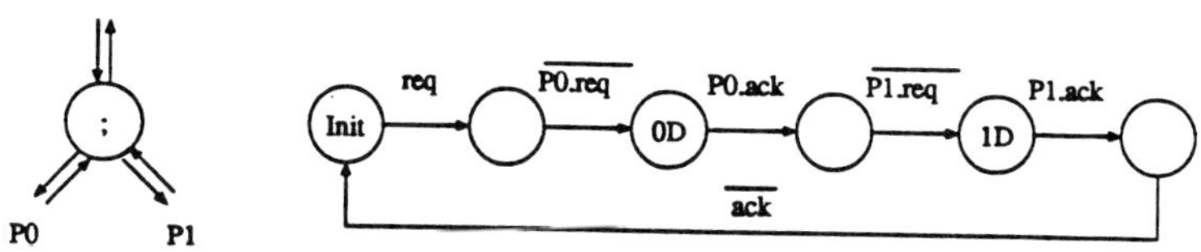

A similar component is the par (||) component used for implementing parallel composition, requesting both children in parallel.

The implementation of the iterative and conditional constructs is somewhat complex with expressions, and more complex with communication. We treat a communication-free guarded command $E \rightarrow P$ as an expression with a side-effect. P is executed if and only if E returns true. We use a guard (G) component to provide the connection between E and P.

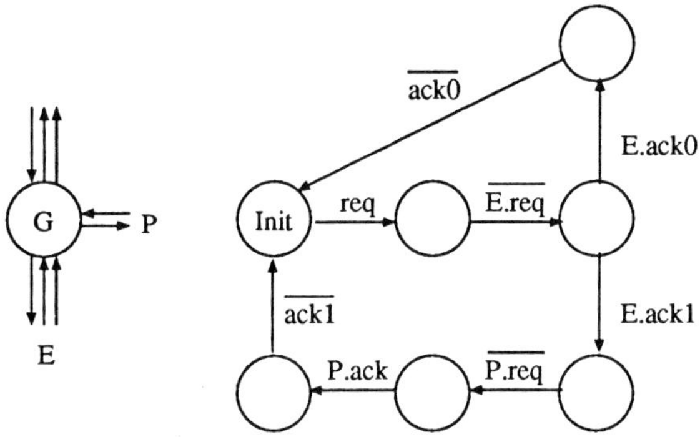

A guarded command set is implemented by joining the individual guarded commands with ∨ components. Thus

$$G_0 [] G$$

is implemented as

$$G_0 \vee G$$

Recall that ∨ only evaluates its right side if its left side is false. Thus, guard evaluation involves attempting to execute each guarded command until one is successfully executed or all of them fail. The value returned from executing a guarded command set is false if no guards were enabled, or true if some guarded command was executed.

We implement the iterative construct with a `DO` component which repeatedly executes a guarded command set until all the guards are false. Implementing the conditional construct `if` G `fi` is similar; however, the `if` component repeatedly executes the guarded command set until one guarded command is successfully executed, which gives us the blocking behavior called for in the semantics.

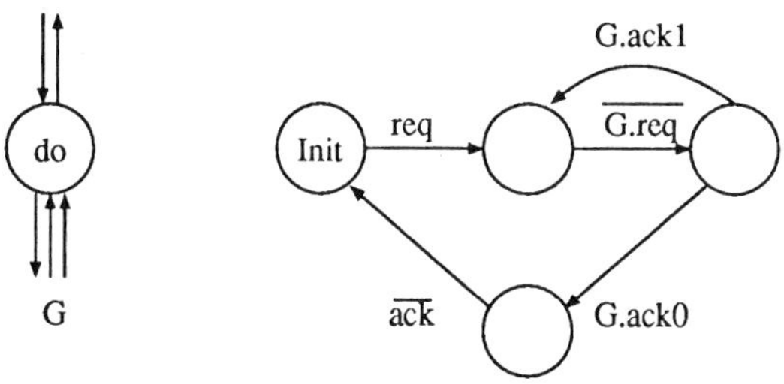

4 Implementing Communication Channels

We implement communication channels with a two wire interface for the sender and a three-wire interface for the receiver. (The arc label $\overline{S.ack}, \overline{R.ack1}$ denotes performing the two output events in either order.)

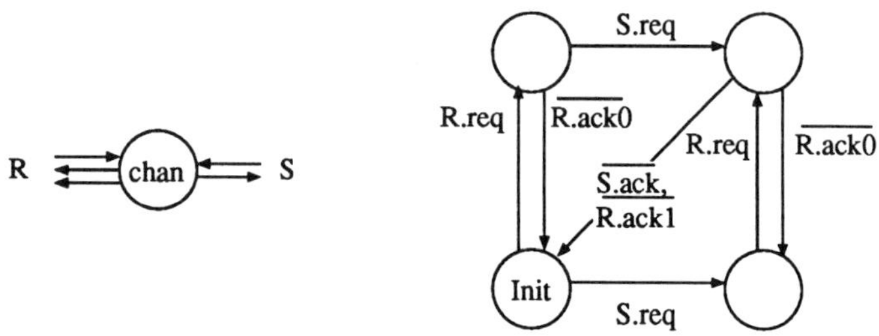

The sender requests communication on its request wire and waits for completion of the communication on its acknowledge wire. The receiver may test the sender's readiness to communicate by initiating an event on its request wire. The receiver receives a "true" acknowledgement if communication is completed; otherwise it receives a "false" acknowledgement.

One peculiarity of our specification is that the sender's request may be ignored. This is necessary for the specification to be delay-insensitive – a delay-insensitive circuit which has its inputs buffered (hence delaying the "arrival" of an input event) must be indistinguishable from the unbuffered circuit. The fact that our channel component is delay-insensitive has an effect on our operational semantics for Joy. In particular, in evaluating a receive guard, the guard may fail even if there is a sender active on the same channel. We require that an implementation always have some minimum nonzero probability of attention to the request. This guarantees that a communication checked repeatedly will be accepted with probability 1, which we believe to be an adequate liveness condition.

Notice that the receivers and expressions have the same (read) interface. The use of this interface in a guard should be clear from the previous section. However, since evaluation of the receive command has a potential side-effect (viz., completion of the communication), in guards of the form $E\&c?$, we evaluate E first. This happens to be the same behavior as the $\wedge$ component.

Similarly, the sender's interface is of the form of a process. The two interfaces may be shared among multiple processes (which are not concurrently active) using the call and call-read components. [4]

5 Proof of Correctness

We verify our compilation scheme by a formal proof of correctness. We give a fairly abstract operational semantics of Joy, and a fairly concrete operational semantics of the circuits that the compilation scheme produces. We then show that the two are equivalent. More precisely, we show that there is a precise correspondence between the possible actions of a program and the possible gate-firings of the circuit; as discussed below, we establish a *bisimulation* (one of the strongest forms of process equivalence, and hence a strong guarantee of correctness) between the two.

5.1 Semantics of Joy

We sketch a structured operational semantics for Joy, along the lines of [15]. This lets us execute Joy programs without compiling them; our main theorem will be that they execute the same before and after compilation.

We must keep track of the control state of the process (which it is convenient to represent as a Joy program P), and the state of the variables (stored as a function η from variable names to their (Boolean) values). So, our states are pairs, written $P + \eta$. We need to include a special state done in programs, representing a program which has finished.

We also have *actions*: receives c? and sends c! on channel c, and a special action "-" for internal computation.[5] Processes will produce "-" actions when the results of their action are not visible outside the circuit; for example, a **skip** statement simply produces a "-" action.

The main operational semantic relation is $P + \eta \xrightarrow{\alpha} P' + \eta'$, which is intended to mean that a process in state $P + \eta$ can perform the action α and thereafter behave like $P' + \eta'$. For example, the rather silly program $(x := \mathrm{tt}\ ;\ x := \mathrm{ff})$ started in state $\{x = \mathrm{ff}\}$ would evolve as follows:

$$(x := \mathrm{tt}; x := \mathrm{ff}) + \{x = \mathrm{ff}\} \xrightarrow{-} (\mathsf{done}; x := \mathrm{ff}) + \{x = \mathrm{tt}\} \xrightarrow{-} \mathsf{done} + \{x = \mathrm{ff}\} \quad (1)$$

In the operational semantics, the program is gradually consumed as it executes. This is a standard technical convenience for the operational semantics – it allows us to avoid explicit program counters, or more accurately to use the program term itself as the program counters. It is not intended to suggest that pieces of the circuit will flake off as it executes!

Processes without subprocesses are given behavior by axioms. For example, **skip** quietly evolves to a finished process in any state: **skip** $\xrightarrow{-}$ done. (When the variable state component does not change, we omit it from the semantic relation.) Assignments behave in the obvious way: $(x := E) + \eta \xrightarrow{-}$ done $+ \eta'$, where η' is the same as η except that $\eta'(x)$ is the value of E in state η.

Most composite processes are given behavior based on the behavior of their components. For example, a sequential composition executes the first process:

$$\frac{P_1 + \eta \xrightarrow{\alpha} P_1' + \eta'}{(P_1 ; P_2) + \eta \xrightarrow{\alpha} (P_1' ; P_2) + \eta'} \tag{2}$$

and when the first process finishes, the second one starts:

$$\mathsf{done} ; P_2 \xrightarrow{-} P_2 \tag{3}$$

Parallel composition is more complex. Either component may execute, or the two may coordinate on a send/receive pair. It is a theorem that visible communication does not happen at the same time as a change in the variable state, so the second rule does not need to worry about merging two differently-changed states.

$$\frac{P_1 + \eta \xrightarrow{\alpha} P_1' + \eta', \alpha \notin C}{\begin{array}{c}(P_1 \parallel_C P_2) + \eta \xrightarrow{\alpha} (P_1' \parallel_C P_2) + \eta' \\ (P_2 \parallel_C P_1) + \eta \xrightarrow{\alpha} (P_2 \parallel_C P_1') + \eta'\end{array}} \qquad \frac{P_1 \xrightarrow{c?} P_1', \quad P_2 \xrightarrow{c!} P_2', \quad c \in C}{P_1 \parallel_C P_2 \xrightarrow{-} P_1' \parallel_C P_2'}$$

A parallel composition finishes when both branches do: $(\mathsf{done} \parallel_C \mathsf{done}) \xrightarrow{-} \mathsf{done}$.

If and do statements require the evaluation of guarded commands. The semantics include a predicate GP **fails**. Basic guarded commands $(E\&\alpha \rightarrow P)$ evaluate E. If it is false, the guarded command fails. If it is true, then (nondeterministically) either the communication happens and P is executed or the guarded command fails. This rather peculiar behavior is required by delay-insensitivity as described on page 10.

Executing a list of guarded commands $G_1 [\!] G_2 [\!] \cdots [\!] G_n$ tries G_1, then G_2, and so on; if one succeeds, then the list succeeds; if all fail, then the list fails.

5.2 Semantics of Circuits

We model delay-insensitive circuits as directed graphs. Processing components are nodes; wires are edges. Each node has a set of *port names*, generally including distinguished *request* and *acknowledge* ports; wires have a name at each end. The state of the circuit has two components: a set of wires W which currently have signals on them, and a function κ giving the states of state-bearing components.

We define the operational semantics of circuits in a quite obvious way. To determine the behavior of $W + \kappa$, we look for some patterns of active wires on the inputs to a processing component. If one of the patterns is found, then the processing component can fire, making other wires active. For example, we have a sequencing component with six ports, $\{r, r1, a1, r2, a2, a\}$, corresponding to the ";" connective in Joy programs. If its request wire (leading to r) is active, then it can fire and send a signal on its $r1$ port: this is the pair of transitions from Init to 0D in 8. It then waits for an impulse on $a1$, which is the acknowledgement that the first subterm has finished. It then sends an impulse on $r2$, awaits a signal on $a2$, and finally sends its acknowledgement on a. These behaviors correspond to the transition graphs given in Section 3.

Communication within the circuit is straightforward. Communication with the outside environment is somewhat more confusing; we must keep track of what the circuit could do if the environment would cooperate, assuming that the environment obeys the basic conventions of communication.

The formalization of the circuit as a graph is straightforward but tedious.

5.3 Bisimulation

Concurrent entities are often described in terms of *labeled transition systems* (lts's): tuples $L = \left\langle \Sigma, \dot{\rightarrow}, \sigma_0 \right\rangle$ where Σ is a set of states, $\dot{\rightarrow}$ is a family of transition relations $\overset{\alpha}{\rightarrow} \subseteq \Sigma \times \Sigma$ for each α, and $\sigma_0 \in \Sigma$ is the initial state. In our case, Joy programs and circuits (plus states) determine lts's, with Σ given by the set of $(P + \eta)$'s or $(W + \kappa)$'s.

Two labeled transition systems L and L' are *strongly bisimilar* iff there is a relation $\sim$ between their states which relates the initial states, such that, if $\sigma \sim \tau$, then any action which one of the two can take, the other can also take and the resulting states are still related. That is, if $\sigma \overset{\alpha}{\rightarrow} \sigma'$, then there is a τ' such that $\tau \overset{\alpha}{\rightarrow} \tau'$ and $\sigma' \sim \tau'$; and vice versa. This implies (but is much stronger than) that σ and τ have the same set of traces. Strong bisimulation is too detailed a notion for most purposes, as it insists that the processes have the same silent moves as well as visible ones. There is a variant called *weak bisimulation*, which allows processes to take internal steps (- moves); see [12].

Bisimulation, strong or weak, is an extremely strong form of correctness. It has been shown [14, 11, 2] that bisimilar processes are indistinguishable in any reasonable setting, and in particular in Joy. Indeed, bisimulation is arguably too strong to be a good notion of specification of correctness, as in most settings there are non-bisimilar processes which are still indistinguishable. However, we are able to achieve bisimulation, and hence all weaker forms of correctness.

5.4 Highlights of the equivalence proof

We are now able to state the equivalence between Joy programs and their compilations. Let P_0 be any Joy program, and η_0 an initial state of its variables. Let C be the circuit into which P_0 is compiled, W_0 be its initial set of active wires (an impulse on the external request line and nothing else active), and κ_0 the initial state of variables corresponding to η_0. Then the lts's with initial state $P_0 + \eta_0$ and $W_0 + \kappa_0$ are bisimilar.

The major technical difficulty in the proof of correctness is that Joy programs and circuits execute in quite different ways. The operational semantics of Joy were designed to be understandable, to demonstrate that

Joy is a reasonable programming language. We have tried to minimize the amount of unnecessary information kept by Joy programs. Consequently, as a Joy program executes, it loses information about the original program.

Conversely, circuits are physical entities, and our model of them is intended to be reasonably realistic. In particular, the wires and gates of a circuit continue to exist throughout and after the circuit's execution. This disparity between the two languages makes proving a correspondence directly rather challenging.

We illustrate our method with the sequencing statement $P_1; P_2$. By Rule 2, P_1 will be executed until it completes; that is, reaches done[6]. Then the Joy rule $(\mathtt{done}; P_2) \xrightarrow{-} P_2$ will result in P_2 being executed. At this point the Joy program has forgotten the fact that it is executing a composite statement. However, in contrast, the circuit corresponding to this statement cannot forget that it is executing a composite; the ";" component and its connecting wires still exist after P_1 is finished.

5.4.1 Joy Prime

To connect the two disparate models, we introduce an intermediate language Joy Prime which has aspects of both. Joy Prime programs resemble Joy programs augmented with additional labels and markers which keep track of the information that circuits retain and Joy programs lose. We illustrate this construction for sequencing.

Each instance of each operation in Joy Prime is given a label, consisting of a unique identifier l and the operator symbol; for $P_1 \,;\, P_2$, it will have the form "l;". This label will remain with the statement as long as it is active. We will have to identify statements which have not been executed; we use a marker v. The Joy Prime statement corresponding to $P_1; P_2$ will be of the form $(P_1; P_2)^{vl;}$ This corresponds to being in state Init of the diagram on 8.

To start executing this, we use the Joy Prime rule $P^{vl*} \xrightarrow{-} P^{l*}$. where $*$ is any operator; in our case, this gives us the transition $(P_1 \,;\, P_2)^{vl;} \xrightarrow{-} (P_1 \,;\, P_2)^{l;}$. This corresponds to a component processing a signal on its request line; in this case, to the transitions from Init to 0D of the diagram on 8.

The Joy Prime rule for executing a sequence is identical to the Joy rule (2), except that the label does not change:

$$\frac{P_1 + \eta \xrightarrow{\alpha} P_1' + \eta'}{(P_1 \,;\, P_2)^{l;} + \eta \xrightarrow{\alpha} (P_1' \,;\, P_2)^{l;} + \eta'}$$

When P_1 finishes executing, we use the rule corresponding to the Joy rule (3) and to the 0D transitions between states 0D and 1D on page 8: when the first child finishes, the ";" component starts the second.

$$(\mathtt{done} \,;\, P_2)^{l;} \xrightarrow{-} (P_2)^{l;} \tag{4}$$

Notice that the label of P_2 says that P_2 is part of a sequencing statement, even though the sequencing operator is gone. P_2 will execute ordinarily, but the "$l;$" label will remain:

$$\frac{P_2 \xrightarrow{\alpha} P_2'}{(P_2)^{l*} \xrightarrow{\alpha} (P_2')^{l*}}$$

When P_2 finishes, it will be in the state $\left(\texttt{done}^{l'*}\right)^{l;}$ where "$l'*$" is the label of the main operator of P_2. The Joy Prime rule

$$\left(\texttt{done}^{l'*}\right)^{l*} \xrightarrow{-} \texttt{done}^{l*} \tag{5}$$

applies. This corresponds to the 1D to Init transitions on page 8: after the second child acknowledges that it is finished, the ";" component acknowledges as well.

5.4.2 Main Theorem

Once we have developed Joy Prime, it is possible to prove the main result: that the compilation scheme is correct.

Proposition 5.1 *Let P be a Joy program, P' the corresponding Joy Prime program, and C the corresponding circuit. Then:*

1. *P' behaves like P, except that it takes more silent moves. Formally P and P' are weakly bisimilar.*

2. *P' behaves just like C. Formally, P' and C are strongly bisimilar.*

As bisimulation is an equivalence relation, P and C are thus (weakly) bisimilar.

The first part of Prop. 5.1 is straightforward, as the rules for P' are essentially the same as those of P with a good deal of extra baggage added. The second part of Prop. 5.1 is proved by a detailed analysis of the behavior of programs and circuits, finishing with the precise correspondence demanded for strong bisimulation.

Briefly, we define a function $act(P')$ which lists the *active components* of P', *viz.* those in which computation could happen next. We prove two lemmas, which give a precise characterization of the behavior of terms. The *Activity Lemma* states roughly that, if $P \xrightarrow{\alpha} Q$, then there is an active component of P which is responsible for the action. R is responsible for the action if $R \xrightarrow{\alpha} S$ for some S, and Q is P with the instance of R replaced by S.[7] The *Converse Activity Lemma* states roughly that, if an active component R of P' is capable of an action, then R can take that action in P.[8]

We then define a function $w(R)$, which lists the wires that correspond to R being active. For active components, this is usually a single wire; for example, if R has yet to take its first step, $w(R)$ is R's circuit's request line. For general programs P', $w(P')$ is simply $\bigcup\{w(R) | R \in act(P')\}$.

We are now ready to apply the bisimulation methodology. We define $P' + \eta \sim W + \kappa$ if $W = w(P')$ and η and κ give the same values to all state-bearing components. We are obliged to show that, if $P' + \eta \xrightarrow{\alpha} P_1' + \eta_1$, then $W + \kappa \xrightarrow{\alpha} W_1 + \kappa_1$ where $P_1' + \eta_1 \sim W_1 + \kappa_1$; and that if $W + \kappa \xrightarrow{\alpha} W_1 + \kappa_1$, then $P' + \eta \xrightarrow{\alpha} P_1' + \eta_1$ where $W' = w(P_1')$ and η_1 and κ_1 give the same values to all state-bearing components. The labels and markers of Joy Prime were painstakingly chosen to give this exact correspondence between circuit and Joy Prime execution.

We need one Joy Prime rule for each input to each kind of component: some fifty rules.[9] Most of the subtlety of the proof of Theorem 5.1 is in the design of Joy Prime (which corresponds roughly to an induction hypothesis), and in careful choice of supporting lemmas such as the activity and converse activity lemmas. After much development of this mathematical machinery, the body of the proof consists of some eighty straightforward cases. This establishes Theorem 5.1.

The machinery we developed for the main correctness proof is useful in proving some other essential properties of the circuit. For example, we required that requests strictly alternate with acknowledgements for each component; proving this is straightforward with these tools. Also, the machinery is nicely modular; it should be possible to extend Joy (*e.g.*, by adding new types of variable, value-bearing communication channels, functions, and subroutines), and the correctness proof for extended Joy should be simply adding cases to the proof of Theorem 5.1.

6 Implementation

We have written a syntax-directed compiler using `yacc` and `lex`, compiling Joy to netlists of basic components as its output. The netlist format, suitable for input to standard route-and-place tools, is a list of component descriptions: tuples consisting of the type of the component and the names of the handshake interfaces to which it is connected, e.g. `(and a b c)`. Components listing the same handshake interface names are connected. The only difficult part of the compiler is keeping track of handshake interfaces for call components, which must be shared. Each component which is connected to a shared interface uses a unique name for that interface. These names are collected in a set and, in a postprocessing step, a balanced tree of call components is generated.

The components can be implemented by traditional VLSI design techniques, with some care. Although the circuits generated by our compiler are delay-insensitive, some of the basic components cannot be designed without consideration of internal delays [19]. Finally, while we have specified components communicating with "non-return-to-zero" signalling, the same compiler can be used with components which communicate with "return-to-zero" signalling.

7 Conclusion

We have presented Joy, a simple yet complete programming language for circuits. We have described the implementation of Joy programs in silicon, and stated and sketched the proof that the compilation scheme is correct.

Many extensions are desirable before this will be widely useful. Joy is a fairly sparse language. There are a number of obvious extensions: variables of types other than Boolean, procedures and functions, value-bearing communication channels, modules, and so forth. We have compilation techniques for many of these extensions, and expect our correctness proof to extend straightforwardly to cover them.

8 Notes

1. Supported by NSF grants CCR-9003441 and CCR-9058180.
2. Supported by NSF grant CCR-9003441.
3. Supported by a NSF grant CCR-9058180
4. The channel interface component is equivalent to the "atomic test and set" component introduced by Keller[8].
5. "-" is τ in [12], and does not appear explicitly in most models of CSP.
6. It may deadlock, in which case P_2 will never be executed.
7. Replacement of subterms is a delicate matter; one of the uses of the labels is to identify subterms so that we can replace them correctly.
8. Actually, both lemmas have second cases in which two components take steps: a send and receive along the same channel.
9. This may seem excessive for as simple a language as Joy. However, it is probably impossible to avoid. A proof that a circuit is correct which does not at some point validate the behavior of each wire and component is too abstract at best, and unlikely to be fully satisfactory.

References

[1] H. Bisseling, H. Eemers, M. Kamps, and A. Peeters. Designing delay-insensitive circuits. Technical report, 1990.

[2] B. Bloom. *Ready Simulation, Bisimulation, and the Semantics of CCS-Like Languages.* PhD thesis, Massachusetts Institute of Technology, Aug. 1989.

[3] G. Brown. Towards truly delay insensitive circuit realizations of process algebras. In *Proceedings of the Oxford Workshop on Designing Correct Circuits.* Springer-Verlag, September 1990.

[4] E. Brunvand and R. F. Sproull. Translating concurrent communicating programs into delay-insensitive circuits. Technical Report CMU-CS-89-126, Carnegie-Mellon University, 1989.

[5] S. M. Burns and A. J. Martin. Syntax-directed translation of concurrent programs into self-timed circuits. In *Advanced Research in VLSI: Proceedings of the 5th MIT Conference*, pages 35–50, 1988.

[6] W. A. Clark and C. E. Molnar. Macromodular computer systems. In B. Waxman and R. Stacey, editors, *Biomedical Research Vol. IV*, pages 45–85. Academic, New York.

[7] D. L. Dill. *Trace Theory for Automatic Hierarchical Verification of Speed-Independent Circuits.* MIT Press, 1989.

[8] R. M. Keller. Towards a theory of universal speed-independent modules. *IEEE Transactions on Computers*, C-23(1):21–33, January 1974.

[9] A. J. Martin. Compiling communicating processes into delay-insensitive VLSI circuits. *Distributed Computing*, 1:226–234, 1986.

[10] G. J. Milne. *The Correctness of a Simple Silicon Compiler.*, pages 1–12. North-Holland, 1983.

[11] R. Milner. A modal characterisation of observable machine-behaviour. In E. Astesiano and C. Böhm, editors, *CAAP '81: Trees in Algebra and Programming, 6th Colloquium*, volume 112 of *Lect. Notes in Computer Sci.*, pages 25–34. Springer-Verlag, 1981.

[12] R. Milner. *Communication and Concurrency.* Prentice Hall International Series in Computer Science. Prentice Hall, New York, 1989.

[13] C. E. Molnar, T.-P. Fang, and F. U. Rosenberger. Synthesis of delay-insensitive modules. In H. Fuchs, editor, *Chapel Hill Conference on VLSI*, pages 67–86, 1985.

[14] D. Park. Concurrency and automata on infinite sequences. In P. Deussen, editor, *Theoretical Computer Science*, Lect. Notes in Computer Sci., page 261. Springer-Verlag, 1981.

[15] G. Plotkin. A structural approach to operational semantics. Technical Report DAIMI FN-19, Aarhus University, Computer Science Department, Denmark, 1981.

[16] R. Sproull and I. E. Sutherland. *Asynchronous Systems*. Sutherland, Sproull & Associates, 1986.

[17] J. Staunstrup and M. Greenstreet. Atomicity, programs, and hardware. Technical Report ID-TR 198-46, Technical University of Denmark, 1990.

[18] M. J. F. T. S. Anantharaman, E. M. Clarke and B. Mishra. Compiling path expressions into vlsi circuits. *Distributed Computing*, 1:150–166, 1986.

[19] C. H. van Berkel. Beware the isochronic fork. Technical Report UR 003/91, Philips Research, 1991.

[20] C. H. K. van Berkel, C. Niessen, M. Rem, and R. W. Saeijs. VLSI programming and silicon compilation; a novel approach from philips research. In *ICCD*, 1988.

Optimal Joining of Compacted Cells

Andrew Lim[†] and **Siu-Wing Cheng**
Department of Computer Science
University of Minnesota
Minneapolis, Mn 55455

Sartaj Sahni
[†]Department of Computer and Information Sciences
University of Florida
Gainesville, Fl 32611

Abstract

We develop three algorithms to join together two compacted cells by using a combination of stretching and river routing. Each of these obtains the minimum area joining. One algorithm obtains a minimum area joining that also minimizes the length of the longest wire. Another obtains a minimum area joining that has the least possible total wire length. The simplest of our algorithms guarantees only a minimum area joining. All algorithms have a low order polynomial complexity. Experimental results indicate that our algorithms obtain joinings that are significantly superior to those obtained using the heuristic of [CD89].

1 Introduction

A popular design style involves tiling a two dimensional area with compacted basic cells such that all needed inter cell connections are between pairs of adjacent cells [Boy88, Wes81]. Furthermore, adjacent cells have the same number of pins on their common boundary and the i'th pins of the two cells are to be connected for all i. In the example of Figure 1 the adjacent cells A and B each have four pins on their common boundary (adjoining sides). Pin a_i of A is to be connected to pin b_i of B, $1 \leq i \leq 4$.

This connecting of pin pairs of the two adjacent cells is referred to as cell joining. Adjacent cells may be joined by stretching the cells so that the pin pairs line up. If cell A is stretched by y at any point x between a_1 and a_2, then the portion of A from x up is moved up by y units (Figure 1(b)). To join the pins of A and B by cell stretching it may be necessary to stretch A and B at several points. Figure 1(c) shows the effect of stretching A and B so as to join all pin pairs. Cell stretching clearly increases the area of the layout.

River routing [Pin82, Pin83, Mir87, LP83] is an alternative to cell stretching. Only one routing layer is available. Figure 1(d) shows how cells A and B can be joined using river routing. River routing also increases the layout area.

Cheng and Despain [CD89] have proposed using a combination of cell stretching and river routing to join together pairs of adjacent cells so that the area of the resulting layout is minimum. They, however, develop only a heuristic for this.

In this paper, we show that pairs of cells can be joined optimally (i.e., with the smallest possible area) using algorithms whose complexity is a low order polynomial

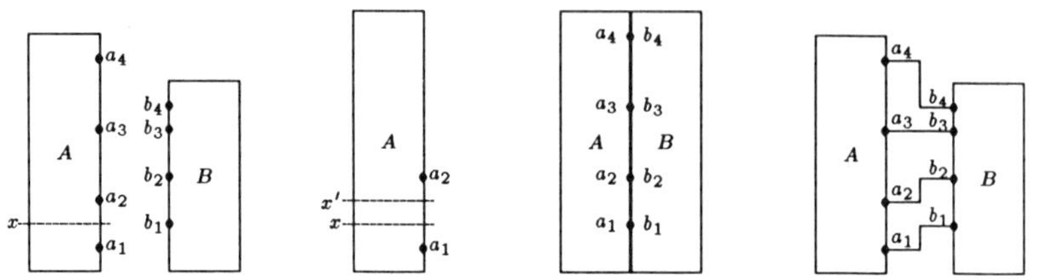

(a) Two adjacent cells (b) stretching at x (c) joining by stretching (d) joining by river routing

Figure 1: Joining Adjacent Cells

function of the number of pins in each of the two cells being joined. We actually develop three algorithms for this. These have differing capabilities and complexities. In the next section, we introduce our notation and terminology and also state the capabilities and complexities of our algorithms precisely. The remainder of the paper develops the algorithms and provides the results of our experimental studies.

2 Terminology

In this section, we formulate the joining problem for two cells L and R. Cell L is to the left of cell R and each cell has n terminals. The case when one cell is above another is symmetric to this. The terminals of L are on its right vertical boundary while those of R are on its left vertical boundary (Figure 2(a)). Let T_i^A denote terminal i of cell A, $A \in \{L, R\}$. Let $p_i^A, A \in \{L, R\}$ be the position of T_i^A, $1 \leq i \leq n$. Let $p_0^A = 0$ and $p_{n+1}^A = h_A$ where h_A is the height of cell A. Since we assume a virtual grid system, all the $n+2$ p_i^A's are integers. Let $\Delta_i^A = p_{i+1}^A - p_i^A, 0 \leq i \leq n$. Stretching the cell A is equivalent to repositioning the terminals of A on the virtual grid line corresponding to the right edge of A in the case $A = L$ and left edge of A in case $A = R$. Let q_i^A be the position of terminal i of A following stretching, $1 \leq i \leq n$. Let $q_0^A = 0$ and $q_{n+1}^A = q_n^A + \Delta_n^A$. Then, the height of the stretched cell A is q_{n+1}^A (Figure 2(b)). The q_i^A's define a *legal stretching* of cell A iff $q_{i+1}^A - q_i^A \geq \Delta_i^A$, $0 \leq i \leq n$.

For any set $Q = \{q_i^A | 1 \leq i \leq n, A \in \{L, R\}\}$ of terminal positions, we can determine whether all terminal pairs (q_i^L, q_i^R), $1 \leq i \leq n$ can be routed using s tracks using the following result from [LP83].

Lemma 1 *[LP83] The terminal pairs $(q_i^L, q_i^R), 1 \leq i \leq n$ can be river routed using s, $s \geq 0$ tracks iff*

a) $q_i^L \geq q_{i-s}^R + s, s < i \leq n$

b) $q_i^R \geq q_{i-s}^L + s, s < i \leq n$ □

Let s_{min} be the minimum number of tracks needed to route all terminals pairs (q_i^L, q_i^R), $1 \leq i \leq n$. The total area occupied by the stretched cells L and R as well as the inter cell routing is $(w_L + w_R + c(s_{min} + 1)) \times max\{q_{n+1}^L, q_{n+1}^R\}$ where

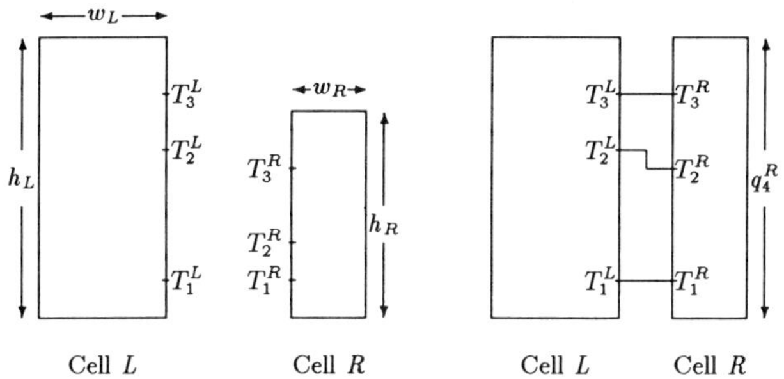

(a) Two compacted Cells (b) After stretching and routing

Figure 2: Cells L and R

w_L and w_R are, respectively, the widths of cells L and R, and c is a technology dependent constant that represents the minimum allowable track separation.

We define three versions of the joining problem.

1. **Minimum Area Joining (MAJ)**: Find a legal stretching Q such that the total area occupied by the stretched cells as well as the inter cell routing is minimum.

2. **Minimum Area Joining With Minimum Maximum Wirelength (MAJMMW)**: From the set of possible solutions to **MAJ** find one for which the maximum inter cell wire length is minimum. Note that for any solution Q, the maximum wire length is $max\{|q_i^L - q_i^R| + c(s_{min}+1)| \; 1 \le i \le n\}$.

3. **Minimum Area Joining With Minimum Total Wirelength (MAJMTW)**: From the set of possible solutions to **MAJ** find one for which the sum of the inter cell wire lengths is minimum. Note that for any solution Q, the total inter cell wire length is

$$\sum_{i=1}^{n} |q_i^L - q_i^R| + cn(s_{min}+1)$$

Algorithms to solve the **MAJ**, **MAJMMW** and **MAJMTW** problems in $O(n^2)$, $O(n^2)$, and $O(n^2 \log n)$ time, respectively, are developed in Sections 3, 4, and 5, respectively. These algorithms are based on an $O(n)$ time algorithm to obtain a minimum height legal stretching when the number of tracks available for river routing is fixed. Leiserson and Pinter [LP83] develop such an algorithm for the more general case when terminals are grouped into blocks. The special case considered here can be solved in $O(n)$ time by a simpler algorithm. This is developed in Section 3 . In Section 6 we compare the performance of our algorithms to that of the heuristic proposed in [CD89] for joining two cells.

```
procedure MinHeight(s);
   /* Determine a minimum height legal stretching Q
      that can be routed using s tracks */
begin
   q_0^L := 0; q_0^R := 0;
   for i:= 1 to s do begin
      q_i^L := q_{i-1}^L + Δ_{i-1}^L;
      q_i^R := q_{i-1}^R + Δ_{i-1}^R;
   end;
   for i:= s + 1 to n do begin
      q_i^L := q_{i-1}^L + Δ_{i-1}^L;
      q_i^R := q_{i-1}^R + Δ_{i-1}^R;
      if q_i^R < q_{i-s}^L + s then
         q_i^R := q_{i-s}^L + s
      else if q_i^L < q_{i-s}^R + s then
         q_i^L := q_{i-s}^R + s;
   end;
end;
```

Figure 3: Procedure to find minimum height legal stretch

3 $O(n^2)$ Algorithm For MAJ

Our algorithm for the **MAJ** problem is based on an $O(n)$ time algorithm that obtains a minimum height legal stretching Q that can be routed using s tracks for any given s. This algorithm determines the values q_i^L and q_i^R for $i = 1, 2, \ldots, n$ in that order and employs the sufficient and necessary conditions of Lemma 1 to assure routability. The algorithm, called *MinHeight* is given in Figure 3.

Theorem 1 : *Let $Q = \{q_i^A | 1 \leq i \leq n, A \in \{L, R\}\}$ be the set of terminal positions computed by procedure MinHeight. Q satisfies the following properties:*

a) Q is a legal stretching of cells L and R.

b) Q can be routed using s tracks.

c) Let $S = \{s_i^A | 1 \leq i \leq n, A \in \{L, R\}\}$ be any s track routable legal stretching of cells L and R. $s_i^L \geq q_i^L$ and $s_i^R \geq q_i^R$, $1 \leq i \leq n$.

d) Q is a minimum height legal stretching that is s track routable.

Proof From the algorithm, it is clear that $q_{i+1}^A - q_i^A \geq \Delta_i^A$ $0 \leq i < n$. So, Q is a legal stretching. The second **for** loop ensures that the necessary and sufficient conditions for s track routability (Lemma 1) are satisfied. So, Q is routable using s tracks. (c) follows from the observation the procedure *MinHeight* places each terminal at the lowest grid point possible while maintaining a legal stretching that can be routed in s tracks. (d) follows from (c) and the observation that the height is $max\{q_n^L + \Delta_n^L, q_n^R + \Delta_n^R\}$. □

Let $Q(s)$ denote the s track stretching computed by procedure *MinHeight*. A minimum area stretching can be obtained in $O(n^2)$ time by computing $Q(j)$, $0 \leq$

$j \leq n$ and determining which of these gives the smallest area. Note that for any $Q(j)$ the area can be computed in $O(1)$ time using the area formula of Section 2. The actual computing time can be reduced somewhat by first determining the maximum number of tracks, s_{max}, that can be used. This is obtained in $O(n)$ time by determining the number of tracks needed when there is no stretching [Pin83, Mir87] and then computing $Q(j)$ only for j in the range $[0, s_{max}]$. Since $0 \leq s_{max} \leq n$, the overall complexity remains $O(n^2)$. Unfortunately, the s value that results in minimum area cannot be found by a binary search in the range $[0, n]$. The reason for this is that area is not a monotone function of s. For example, when $n = 10$, $w_L = 57$, $w_R = 61$, $c = 7$, $\Delta^L_{0:10} = (10, 1, 1, 1, 1, 1, 1, 1, 1, 1, 32)$ and $\Delta^R_{0:10} = (1, 2, 3, 4, 5, 5, 4, 3, 2, 1, 20)$, the minimum area for $s = 0, 1, 2, 3, 4, 5, 6(s_{max})$ is $8378, 9240, 9313, 9052, 8568, 8320, 8517$ respectively.

4 Algorithm For MAJMMW

Let Q be a minimum height s track routable legal stretching obtained using procedure *MinHeight*. We shall show first how to rearrange the terminals in Q so that the resulting stretching U satisfies the following properties:

$P1$: It is a minimum height s track routable legal stretching.

$P2$: The maximum wire length in it is minimum over all stretchings that satisfy $P1$.

We shall say that terminal i of $L(R)$ is *fixed* if moving it from its position in Q results in one or more of the following conditions:

$C1$: Q is no longer a legal stretching.

$C2$: Q is no longer s track routable.

$C3$: The height of Q increases.

$C4$: The maximum wire length is not reduced.

We now describe how to label some of the terminals in Q fixed. From Theorem 1(c), it follows that no terminal can be moved down from its position in Q without violating $C1$ or $C2$. So we need only be concerned with possible upward moves of terminals. Let us examine the terminals of L and R in the order $n, n-1, \ldots, 1$ and attempt to identify some as fixed. Since the height of Q is $max\{q^L_n + \Delta^L_n, q^R_n + \Delta^R_n\}$, one or both of the q^L_n and q^R_n is fixed (if $q^L_n + \Delta^L_n \geq q^R_n + \Delta^R_n$, then T^L_n is fixed as increasing q^L_n increases the height, similarly if $q^R_n + \Delta^R_n \geq q^L_n + \Delta^L_n$, then T^R_n is fixed).

Put either T^R_n or T^L_n or both, depending on which has (have) been labeled fixed, into a queue. This queue will contain only terminals labeled fixed that can lead us to other terminals that can be labeled fixed.

In each pass of our labeling procedure, we extract a terminal from the queue. Assume this is terminal T^L_i (the case T^R_i is similar). Let j be the smallest integer, $j \leq i$, such that $q^L_{a+1} - q^L_a = \Delta^L_a$, $j \leq a < i$ (note that $q^L_{n+1} = q^L_n + \Delta^L_n$). Figure 4 depicts the situation. Since q^L_i is fixed and the separation between q^L_{a+1} and q^L_a is the minimum possible, none of the terminals $T^L_a, j \leq a < i$ can be moved up without violating condition $C1$. As noted earlier, no terminal can be moved down. So, T^L_a, $j \leq a < i$ are all fixed terminals. From procedure *MinHeight*, it follows

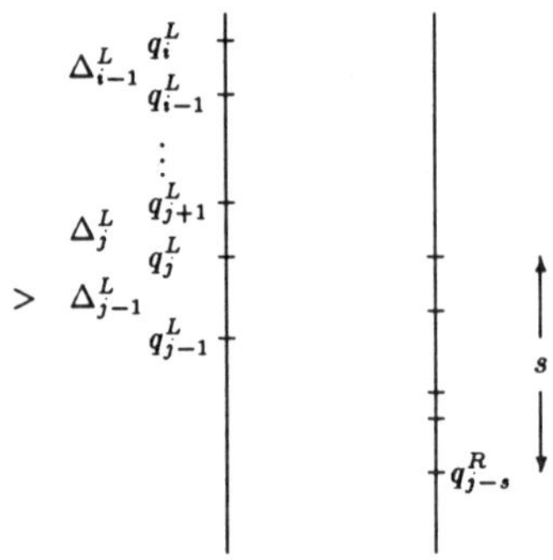

Figure 4: Relationship

that j is either 1 or greater than s. If $j = 1$, we extract another terminal from the queue and proceed as above. In case the queue is empty, the labeling procedure terminates. If $j > s$, then it follows that $q_j^L = q_{j-s}^R + s$. If terminal T_{j-s}^R is moved up, then T_j^L will also need to be moved up (Lemma 1). Since T_j^L is a fixed terminal, it cannot be moved. So, T_{j-s}^R is a fixed terminal. It is labeled as such and entered into the queue.

Furthermore, since T_j^L is fixed and $\Delta_b^L \geq 1$ for all b, the position s_a^L of T_a^L, $a \leq j$ satisfies the inequality

$$s_a^L \leq q_j^L - (j - a) = q_{j-s}^R + s + a - j$$

Also, since $\Delta_b^R \geq 1$ for all b, we get the following inequality for the position of T_a^R, $a \geq j - s$:

$$s_a^R \geq q_{j-s}^R + a - (j - s) = q_{j-s}^R + s + a - j$$

Combining the two inequalities yields:

$$s_a^L \leq s_a^R, j - s \leq a \leq j$$

Hence in every position of T_a^L and T_a^R that satisfies conditions $C1$ through $C3$, T_a^L is either below or at the same level as T_a^R, $j - s \leq a \leq j$. Hence, we cannot get a lower value for the maximum wire length by moving any of the terminals T_a^R, $j - s \leq a \leq j$ up from their positions in Q (recall these are their minimum positions). So, terminals T_a^R, $j - s \leq a \leq j$ are labeled fixed. Note that this does not restrict the upward movement of T_a^L, $j - s \leq a < j$. Any upward movement of these will reduce the length of the wires connected to them.

Following this labeling, we extract a fixed terminal (if any) from the queue and continue to identify other fixed terminals as above. A more formal description of the labeling procedure is given in Figure 5.

Lemma 2 *For each i, $1 \leq i \leq n$, at least one of T_i^L and T_i^R is labeled fixed by procedure FixedLabeling.*

Proof Easily seen by examining the procedure. □

```
procedure FixedLabeling(Q);
    /* label some terminals as fixed. The labeled terminals
       satisfy conditions C1 – C4 */
begin
    Initialize a queue to be empty;
    if q_n^L + Δ_n^L ≥ q_n^R + Δ_n^R then label T_n^L fixed and add to the queue;
    if q_n^R + Δ_n^R ≥ q_n^L + Δ_n^L then label T_n^R fixed and add to the queue;
    while queue not empty do begin
        delete a terminal, say T_i^A, from the queue;
        find least j such that q_{a+1}^A – q_a^A = Δ_a^A, j ≤ a < i;
        label T_a^A, j ≤ a < i, fixed;
        if j > s then begin
            let B = {L, R} – A;
            add T_{j-s}^B to the queue;
            label T_{j-s}^B, ..., T_j^B fixed;
        end;
    end;
end;
```

Figure 5: Procedure to label fixed terminals

Lemma 3 *There exists a minimum height s track routable legal stretching U which satisfies the properties:*

a) The maximum wire length in U is minimum over all such stretchings.

b) $u_i^A = q_i^A$ *for every* T_i^A *labeled fixed by procedure FixedLabeling (*u_i^A *is the position of* T_i^A *in U).*

Proof Follows from the definition of a fixed terminal and the correctness of procedure *FixedLabeling*. □

As a result of Lemma 3, we need to only consider repositioning the non-fixed terminals of Q so as to minimize the maximum wire length. For this, we divide the terminals into blocks. This is done by scanning terminal pairs (T_i^L, T_i^R) in the order $n, n-1, \ldots, 1$. The first block begins at the first pair (T_j^L, T_j^R) with the property that either T_j^R or T_j^L has not been labeled fixed. Assume that T_j^R has not been labeled fixed. The first block consists of the terminal pairs (T_k^L, T_k^R), $a \leq k \leq j$ where a is the largest integer less than j such that both T_a^L and T_a^R are labeled fixed. From the labeling procedure it follows that $T_{a+1}^L, \ldots, T_{j-1}^L$ are also labeled fixed. For the next block we find the largest integer b less than a such that one of T_b^L and T_b^R is not fixed. This next block begins here and includes all terminal pairs (T_k^L, T_k^R), $d \leq k \leq b$ where d is the largest integer less than b such that both T_d^L and T_d^R are labeled fixed. This process continues until all terminal pairs that have one non-fixed terminal have been assigned to a block. Figure 6 shows an example block decomposition. This has three blocks $B1$, $B2$ and $B3$. Note that when all terminals are labeled fixed, there are no blocks.

Consider a block B of terminals in which all the terminals on the right side (e.g., block $B2$ of Figure 6) are fixed. Let (T_k^L, T_k^R), $j \leq k < i$ be the terminal

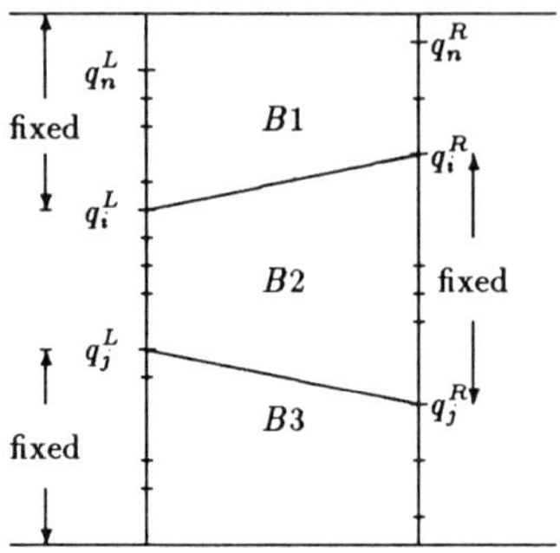

Figure 6: Block Decomposition

pairs in this block. From the terminal labeling procedure it follows that terminals T_i^L and T_i^R are fixed and $q_i^R \geq q_i^L$. Hence,

$$q_i^R \geq q_i^L \geq q_{i-s}^L + s$$

Now suppose that as a result of terminal relocation, terminals T_{i+t}^R and T_{i+t-s}^L where $i+t \leq n$, $i+t-s \geq 1$, and $0 \leq t \leq s$ are relocated to positions s_{i+t}^R and s_{i+t-s}^L respectively. Since a relocation can only move a terminal upwards (Theorem 1(c)), $s_{i+t}^R \geq q_{i+t}^R$ and $s_{i+t-s}^L \geq q_{i+t-s}^L$. So, we obtain:

$$s_{i+t}^R \geq q_{i+t}^R \geq q_i^R + \sum_{k=i}^{i+t-1} \Delta_k^R \geq q_i^R + t \geq q_i^L + t \geq s_{i+t-s}^L + s - t + t = s_{i+t-s}^L + s$$

Hence all legal relocations of terminals T_{i+t-s}^L and T_{i+t}^R for $i+t \leq n$, $i+t-s \geq 1$, $0 \leq t \leq s$ satisfy the necessary and sufficient conditions of Lemma 1 . Consequently non-fixed terminals in block B may be relocated without regard to terminals in other blocks. The same is true when the fixed terminals of B are on the left side of B. This development results in the following lemma.

Lemma 4 *Let S be an s track routable minimum height legal stretching obtained from Q by relocating the non-fixed terminals in each block so as to minimize the maximum wire length in each block, then S has the minimum maximum wire length over all s track routable minimum height stretchings.*

As a result of Lemma 4, we need only consider how to minimize the length of the longest wire in a single terminal block. We explicitly consider only the case when all terminals on the left side of the block are fixed. The case when all terminals on the right side are fixed is similar.

Figure 7 shows a block in which the left terminals are fixed. We are to reposition the right terminals so as to obtain an s track routable legal stretching which has minimum maximum wire length. Let the terminals of the block be $T_j^L, \ldots, T_i^L, T_j^R, \ldots, T_i^R$. Let q_a^A, $A \in \{L, R\}$, $j \leq a \leq i$ be the current position of T_a^A, By definition of a block, terminals T_a^R, $j < a \leq i$ are not fixed and may be

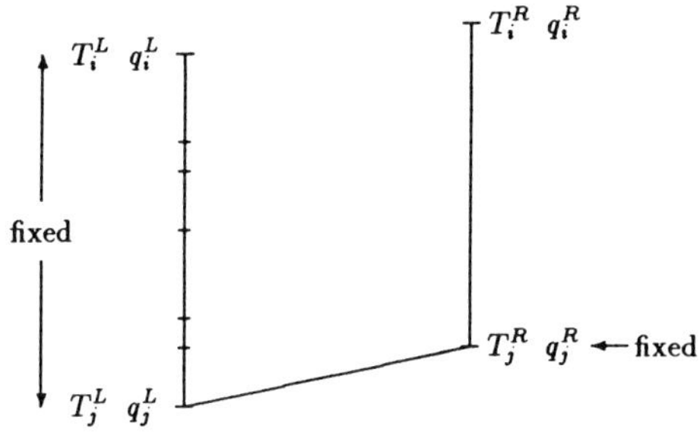

Figure 7: A block with left terminals fixed

moved upwards from their current locations. Our algorithm to relocate the $q_a^{R\prime}$s, $j < a \leq i$ will make two passes over these terminals. In the first, the terminals will be examined in the order $i, i-1, \ldots, j+1$. In this pass, called the *length determination* pass, we shall determine the minimum value, $maxL$, for the longest wire length. During this pass some of the right terminals of the block may be moved up. In the second pass, called the *relocation pass* the terminals are again examined in the order $i, i-1, \ldots, j+1$. In this pass the terminals are relocated, as necessary, so that the maximum wire length is $MaxL$.

Since the number of routing tracks is fixed at s, the horizontal distance between any pair (T_a^L, T_a^R) of terminals is fixed at $c(s+1)$ where c is a technology dependent parameter. The vertical distance is $|q_a^L - q_a^R|$. The length of the wire that connects terminals T_a^L and T_a^R is therefore $c(s+1) + |q_a^L - q_a^R|$. So, the wire length is minimized by minimizing $|q_a^L - q_a^R|$. Hence, in our future discussion, we shall refer to $|q_a^L - q_a^R|$ as the length of the wire between terminals T_a^L and T_a^R.

To accomplish the objectives of the first pass we shall employ two variables $MaxL$ and *high*. When terminal T_a^R, $j \leq a \leq i$, is being examined, $MaxL$ will be the length of the longest wire used so far in connecting the terminal pairs (T_b^L, T_b^R) $a < b \leq i$. When terminal T_a^R is relocated in pass one, its q_a^R value is changed. However, a further relocation may be necessary as a result of the relocation of a terminal that is examined later. All relocations increase the value of q_a^R. However, it will always be the case that $MaxL = \max_{a<b\leq i}\{|q_b^L - q_b^R|\}$ when terminal T_a^R is examined. When terminal T_a^R is being examined, *high*, will give the maximum possible value for q_{a+1}^R. If $high = q_{a+1}^R$, then T_{a+1}^R cannot be moved further up from its current location without violating either the s track routability requirement or the minimum terminal separation requirement. When T_{a+1}^R can be moved without causing such a violation (note that an upward movement in T_{a+1}^R may require an upward movement in terminals T_b^R, $a+1 < b \leq i$ in order to preserve feasibility), $high > T_{a+1}^R$.

The initial value of $MaxL$ is 0. If $i = n$, then we are dealing with the topmost block and q_{n+1}^R is $q_{n+1}^L = q_n^L + \Delta_n^L$. If $i < n$, then q_{i+1}^R is the already determined position of T_{i+1}^R (we assume blocks are examined top to bottom). The initial value of *high* corresponds to T_{i+1}^R. When the current block is being considered, T_{i+1}^R has been fixed at q_{i+1}^R. So, the initial value of *high* is q_{i+1}^R.

After the examination and possible relocation of T_a^R in the first pass, the following will be true:

$Q1$: $MaxL = \max_{a \le b \le i}\{|q_b^L - q_b^R|\}$

$Q2$: *high* is the maximum possible value for q_a^R

$Q3$: If $high > q_a^R$, then increasing q_a^R by δ units, $\delta \le high - q_a^R$, will require an increase in the length of the longest wire so far by δ units.

$Q1 - Q3$ are trivially seen to hold when $a = i + 1$ (i.e., they hold before we examine any of the terminals in the block). The procedure for the length determination pass, pass one, is given in Figure 8. This, of course, is only for the case of a block whose fixed terminals are on the left side. When terminal T_a^R is examined, we determine if $q_a^L + MaxL < q_a^R$. If so, then the length of the wire that connects T_a^L and T_a^R is already longer than $MaxL$ and any upward movement of T_a^R will only increase the wire length further (recall that T_a^R cannot be moved downward because of Theorem 1). In this case, we simply update $MaxL$ to be the length, $q_a^R - q_a^L$, of this wire. The highest value q_a^R can have without violating the s track routability constraint is $q_{a+s}^L - s$ (Lemma 1, in case $a + s > n$, set $q_{a+s}^L = \infty$). To satisfy the minimum separation constraint, we must have $q_a^R \le high - \Delta_a^R$ as by definition of *high*, $q_{a+1}^R \le high$. So, the maximum permissible value of q_a^R is $min\{high - \Delta_a^R, q_{a+s}^L - s\}$.

In case $q_a^L + MaxL \ge q_a^R$, then we move q_a^R as far up as possible without increasing the value of $MaxL$ and while preserving feasibility (first line of the else clause). Note that it is possible that $q_a^L - q_a^R > MaxL$ and it is not possible to move T_a^R far enough up to bring the length of this wire to a value $\le MaxL$. In any case, the highest value of q_a^R consistent with the length and feasibility requirements is given by $min\{q_a^L + MaxL, q_{a+1}^R - \Delta_a^R, q_{a+s}^L - s\}$. The first term ensures that T_a^R is not moved so far up as to increase $MaxL$ (if possible) and the remaining two terms ensure feasibility. Following this, it is possible that $q_a^L - q_a^R > MaxL$. Increasing q_a^R increases the $MaxL$ but decreases $q_a^L - q_a^R$. So, we increase q_a^R by as much as $\lceil (q_a^L - q_a^R - MaxL)/2 \rceil$ subject to the constraint that q_a^R must be no more than *high* or $q_{a+1}^R - \Delta_a^R$ (third statement of the else clause). The remaining two statements of the else clause of Figure 8 update *high* and $MaxL$ correctly. One may easily verify that properties $Q1 - Q3$ hold at the end of each iteration, provided they held at the start of the iteration.

Theorem 2 *Procedure LeftMaxL computes the minimum value possible for the maximum wire length in a block whose left terminals are fixed.*

Proof Follows from the observation that the value of $MaxL$ is increased only when there is no way to relocate terminals T_b^R, $a \le b \le i$ without such an increase. □

Once the value of $MaxL$ has been determined the final positions of T_a^R, $j < a \le i$ (note that T_j^R is a fixed terminal), can be obtained by moving each terminal T_b^R to the highest possible position consistent with feasibility and $MaxL$. This is simply done using the procedure *LeftRelocate* (Figure 9). Its correctness is immediate.

The corresponding procedures for a block with fixed terminals on the right side are similar. Each of the passes for a block takes time proportional to the number of terminals in the block. So, the time needed to transform a minimum height s track routable legal stretching as obtained by procedure *MinHeight* into a

```
procedure LeftMaxL(i,j);
    /* Determine the minimum maximum wire length for
       a block with fixed terminals on the left side. The
       terminals in the block are T_a^A, A ∈ {L, R}, j ≤ a ≤ i */
begin
    if i = n then q_{n+1}^R = MinHeight;
    high := q_{i+1}^R;
    MaxL := 0; /* length of longest wire */
    for a := i downto j + 1 do
        if q_a^L + MaxL < q_a^R
        then begin /* wire a is longer than MaxL */
            MaxL := q_a^R − q_a^L;
            high := min{high − Δ_a^R, q_{a+s}^L − s};
        end
        else begin /* wire a may be longer than MaxL */
            q_a^R := min{q_a^L + MaxL, q_{a+1}^R − Δ_a^R, q_{a+s}^L − s};
            high := min{high − Δ_a^R, q_{a+s}^L − s};
            if q_a^L − q_a^R > MaxL
               then q_a^R := min{high, q_{a+1}^R − Δ_a^R, q_a^R + ⌈(q_a^L − q_a^R − MaxL)/2⌉};
            MaxL := max{MaxL, q_a^L − q_a^R};
        end;
end;
```

Figure 8: Determine minimum maximum length

```
procedure LeftRelocate(i,j);
    /* Relocate the right terminals of a block whose left
       terminals are fixed. The maximum wire length is to be MaxL
       and the terminals are T_a^A, A ∈ {L, R}, j < a ≤ i} */
begin
    for a:=i downto j + 1 do
        q_a^R := min{q_a^L + MaxL, q_{a+1}^R − Δ_a^R, q_{a+s}^L − s}
end;
```

Figure 9: Relocate the right movable terminals

minimum height s track routable legal stretching that has the minimum maximum wire length is $O(n)$.

To use the algorithm outlined above to obtain a solution for **MAJMMW**, one would have to consider all values of s that result in a minimum area solution. For each of these s values, we will start with the solution obtained by procedure *MinHeight* and minimize the maximum wire length using the above procedure. The resulting solution with smallest maximum wire length is a solution to **MAJMMW**. Since the number of s values that could result in a minimum area solution is $O(n)$, the complexity of our procedure for **MAJMMW** is $O(n^2)$. In practice, however, we expect that usually only $O(1)$ s values will minimize the area so that after these have been determined (this takes $O(n^2)$ time as the minimum area algorithm needs to be run), an additional $O(n)$ time will suffice to minimize the length of the longest wire.

5 Algorithm For MAJMTW

Our strategy for **MAJMTW** is similar to that for **MAJMMW**. We consider all values of s that result in a solution to **MAJ**. For each of these we start with the solution obtained by the procedure *MinHeight* and obtain a solution with the same height that also minimizes the total wire length. Our algorithm for this transformation takes $O(n\log n)$ time. Since we may have up to n starting solutions to be transformed, the total transformation time becomes $O(n^2\log n)$. The transformed solution with least total wire length is the solution to **MAJMTW**. Due to space limitation the details are omitted and can be found in [LCS92]

6 Experimental Results

Our algorithms for **MAJ**, **MAJMMW**, and **MAJMTW** as well as the joining heuristic of [CD89] were programmed in C and the programs run on a SUN SPARCstation *SLC*. For test data, we used six pairs of cells named Bi, $i \in \{64, 128, 250, 500, 1000, 5000\}$. Cell pair Bi has i terminals in each of its cells. The number, s_{max}, of tracks needed to join the cells in a cell pair without any stretching is given in column 2 of Table 1. This table also gives the area of the layout obtained by our algorithm for **MAJ**, the heuristic of [CD89], and when joining is done by using only stretching to align all terminal pairs (also called pitchmatching) or by using only river routing (no stretching is done). The minimum area layout (this is obtained by our algorithm) uses up to 27% less area than the layouts obtained using [CD89]. The layouts of [CD89] were consistently better than when only pitchmatching or only river routing is used. Surprisingly, our optimal algorithm is considerably faster than the heuristic of [CD89]. The heuristic took between 1 times to 62 times as much time as did our optimal algorithm on the tested cell pairs.

Table 2 gives the length of the longest wire in the solution from our **MAJMMW** algorithm as well as the sum of the wire lengths in the solution from our **MAJMTW** algorithm. The corresponding figures for the heuristic of [CD89] and for the river routing solution are also given. Note that when pitchmatching is used, there are no vertical wire segments so the maximum and total wire lengths as measured by us are zero. Interestingly, the longest wire length and total wire length in our solutions are smaller than those in the layout of [CD89] even though

Tests cases	S_{max}	Our Algorithm			[CD]		Stretching	River-routing
		area	time	% improv.	area	time		
$B64$	16	11.7	0.01	7.1	12.6	0.01	12.5	17.9
$B128$	34	31.2	0.02	27.1	42.8	0.04	34.8	59.0
$B250$	41	86.9	0.03	3.8	90.3	0.15	109	99.0
$B500$	22	310	0.11	21.3	394	0.55	383	420
$B1000$	44	1104	0.21	8.8	1207	2.26	1397	1337
$B5000$	41	25000	1.04	2.8	25730	65.3	34492	26006

Times are in seconds.

Table 1: Area of join

Test cases	Our Algorithms						[CD]		River-routing	
	longest	opt	time	total	opt	time	long	total	long	total
$B64$	.022	.016	< .01	.85	.77	< .01	.023	1.03	.071	3.77
$B128$	.028	.017	< .01	1.67	1.54	< .01	.062	5.41	.143	15.3
$B250$	.031	.019	< .01	2.89	2.79	.02	.034	5.10	.068	13.3
$B500$	.035	.026	.01	8.83	8.33	.04	.135	53.7	.167	69.7
$B1000$	.045	.031	.02	20.5	19.9	.09	.094	75.1	.176	156
$B5000$	.045	.034	.09	111	110	.69	.087	355	.167	754

Times are in seconds.

Table 2: Longest wire length and total wire length of join

the layouts of [CD89] do not have minimum area.

7 Conclusion

We have shown that the minimum area join of two compacted cells can be obtained by a fast polynomial time algorithm. This is true even when we wish to minimize not just the layout area but also the length of the longest wire and the total wire length (subject to area being minimum). Experimental results indicate that our algorithms not only obtain significantly better layouts than those obtained by the heuristic of [CD89] but they also take less computer time.

8 Acknowledgement

This research was supported, in part, by the National Science Foundation under grant MIP86-17374. The first author's research was also supported, in part, by the ACM SIGDA and IEEE DATC scholarship.

References

[Boy88] D. BOYER, *Symbolic Layout Compaction Review*, in ACM/IEEE Design Automation Conference, June 1988, pp. 383–389.

[CD89] G. CHENG AND A. DESPAIN, *Fang: A Joiner for Compacted Cells*, in VLSI 89, 1989, pp. 455–463.

[LCS92] A. LIM, S. CHENG, AND S. SAHNI, *Optimal Joining of Compacted Cells*, IEEE Transactions on Computer, (1992). To appear.

[LP83] C. LEISERSON AND R. PINTER, *Optimal placement for river routing*, SIAM Journal on Computing, (1983), pp. 447–462.

[Mir87] A. MIRZAIAN, *River Routing in VLSI*, Journal of Computer and System Sciences, (1987), pp. 43–54.

[Pin82] R. PINTER, *On routing two point nets across a channel*, in ACM/IEEE Design Automation Conference, 1982, pp. 899–902.

[Pin83] ——, *River-Routing: Methodology and Analysis*, in Third Caltech Conference on VLSI, March 1983.

[Wes81] N. WESTE, *Virtual Grid Symbolic Layout*, in ACM/IEEE Design Automation Conference, June 1981, pp. 225–233.

A New Method to Compute Prime and Essential Prime Implicants of Boolean Functions

O. Coudert and **J. C. Madre**

Bull Corporate Research Center
Rue Jean Jaurès
78340 Les Clayes-sous-bois, FRANCE
coudert@crg.bull.fr, madre@crg.bull.fr

Abstract

Computing the prime implicants and the essential prime implicants of Boolean functions is a problem that has applications in several areas of computer science, for instance it is critical in circuit synthesis and optimization, but also in automated reasoning. In this paper we propose a new method to compute *implicitly* the sets of prime implicants and of essential prime implicants of incompletely specified Boolean functions. This method allows us to handle functions that have sets of prime implicants and of essential prime implicants several orders of magnitude larger than those of the functions that could be handled by existing methods. The key idea that makes these computations possible is to manipulate sets of prime and of essential prime implicants denoted by their *meta-products*, which are characteristic functions. These functions are represented with binary decision diagrams that are a very compact canonical graph representation of Boolean functions and that support very efficiently all the operations needed here.

1 Introduction

In the worst case, the number of prime implicants and essential prime implicants of a Boolean function is exponential with respect to the number of variables of this function [1]. Thus all techniques that perform a prime by prime computation are by nature limited to functions with relatively limited numbers of variables and of prime implicants. For instance the Espresso minimizer [1, 9] is practically limited to functions with less than 20,000 primes, and the reasoning maintenance system described in [7], which is based on a similar technology, has the same limitations.

We have introduced recently [3] a technique for verifying sequential circuits that is not limited by the number of states of these circuits. The key idea that makes this verification technique possible is to represent subsets of $\{0,1\}^{\aleph}$ with their characteristic functions. These Boolean functions are denoted by binary decision diagrams (BDD) that are a very compact graph representation of propositional formulas. This representation allows us to operate on these sets without considering their elements individually. In this paper we show that this idea can be used with very good results to implement an *implicit* prime and essential prime implicant computation procedure.

This paper is divided in 7 parts. Section 2 presents the problem that will be treated here and the elementary concepts that will be used to solve it. Section 3 introduces the canonical representation, which we call *meta-product*, of sets of products. Section 4 explains how meta-products can be represented with binary decision diagrams and gives the functions *ProductCost* and *LiteralCost* that allow us to evaluate in linear time on meta-products the standard cost functions associated with sum of products. Section 5 shows how meta-products are used to compute implicitly the sets of prime and of essential prime implicants of Boolean functions with care sets. Section 6 gives experimental results. Finally Section 7 discusses this method and the experimental results obtained with it.

2 Definitions

We consider here propositional formulas built out of the finite set of propositional variables $\{x_1, \ldots, x_n\}$. A *literal* is a propositional formula of the form (x_k) or $(\overline{x_k})$. A *product* is a formula of the form $(l_1 \ldots l_k)$, $k \leq n$, whose literals are built out of distinct variables. When $k = n$, such a product is called a *minterm.* The product p *contains* the product p' if and only if the formula $(p' \Rightarrow p)$ is a tautology. The product p *strictly contains* the product p' if p contains p', and p' does not contain p.

A *sum of products* is a formula of the form $(p_1 + \ldots + p_m)$, where $p_1, \ldots, p_m$ are distinct products. Any propositional formula f can be rewritten into a formula expressed as a sum of products. However this rewriting process is not canonical, which means that there exist different formulas written as sums of products that are equivalent. There are two standard cost functions for formulas expressed as sums of products [6, 9]. The first cost function of the formula $f = (p_1 + \ldots + p_m)$, noted *ProductCost*(f), is the number of products in this sum, and the second cost function, noted *LiteralCost*(f), is defined as *LiteralCost*$(f) = \sum_{k=1}^{k=m}$ *LiteralCost*(p_k), where *LiteralCost*(p_k) is the number of literals that occur in the product p_k.

Any Boolean function from the set $\{0,1\}^n$ into the set $\{0,1\}$ can be represented by a propositional formula built out of the finite set of propositional variables $\{x_1, \ldots, x_n\}$. Conversely any propositional formula built out of the finite set of propositional variables $\{x_1, \ldots, x_n\}$ denotes a unique Boolean function from the set $\{0,1\}^n$ into the set $\{0,1\}$. In the sequel we will use the same symbol to represent a propositional formula and the Boolean function it denotes.

Any subset S of $\{0,1\}^n$ can be represented by a unique Boolean function

χ_S from the set $\{0,1\}^n$ into $\{0,1\}$, called its *characteristic function*, that is defined in the following way:

$$\chi_S(x) = 1 \qquad \text{if and only if} \qquad x \in S.$$

Conversely, any Boolean function f from $\{0,1\}^n$ into $\{0,1\}$ denotes a unique subset S_f of $\{0,1\}^n$ made of all the elements x of $\{0,1\}^n$ such that: $f(x) = 1$. For instance the characteristic function of the subset $\{[01010], [11010], [01011], [11011]\}$ of $\{0,1\}^5$ can be represented by the product $(x_2\overline{x_3}x_4)$. Characteristic functions are a very interesting implicit representation of Boolean sets because there is a direct correspondence between the set operators and the logical operators, for instance union corresponds with disjunction, intersection with conjunction, and complementation with negation.

A Boolean function f_C with a care set C is denoted by a couple (C, f), where C is a subset of $\{0,1\}^n$ and f is a Boolean function from $\{0,1\}^n$ into $\{0,1\}$, and is defined by the following equation:

$$\begin{array}{lcll} f_C(x) & = & f(x) & \text{if } x \in C, \\ f_C(x) & = & * & \text{if } x \notin C, \end{array}$$

where the symbol "*" can be either 0 or 1. A product p built out of the variables $\{x_1, \ldots, x_n\}$ is a *cube* of the function f_C if and only if the formula $(\chi_C \wedge p)$ is not an antilogy, and the formula $(\chi_C \Rightarrow (p \Rightarrow f))$ is a tautology. This expresses that the intersection between the set S_p denoted by p and the care set C is not empty, and that for any element of this intersection, the function f evaluates to 1. The product p is a *prime implicant* of f_C if and only if it is a cube of f_C, and there is no cube of f_C different from p that contains p. Finally, the product p is an *essential prime implicant* of f_C if and only if it is a prime implicant of f_C and the Boolean function denoted by the sum of all the primes of f_C that are different from p is not equal to f on the care set C.

Prime implicants and essential prime implicants play a very important part in the minimal cover problem [9]. The set of prime implicants of the Boolean function f_C is the set of products that can be used to build a minimal cover of f_C. Amongst all these prime implicants the essential ones are critical since they are prime implicants that occur in any minimal cover of f_C. In particular the cost of any minimal cover of f_C is at least equal to the cost associated with its essential prime implicants.

In the case where the care set of the function f_C is equal to the complete input set $\{0,1\}^n$, the function f_C becomes a standard Boolean function. For this reason, we will in the following consider only functions with a care set, Boolean functions being just a particular case of such functions.

In this paper we will also use *quantified propositional formulas* that are formulas of the form $(Q_1x_1 \ldots Q_kx_k\ f)$ where $Q_1, \ldots, Q_k$ are either the universal or the existential quantifier, $x_1, \ldots, x_k$ are propositional variables, and f is a propositional formula [3]. For instance the formula $(\forall x_1\ \exists x_2\ (x_1 \vee (\neg x_2 \wedge x_3)))$ is such a quantified propositional formula. Any quantified propositional formula can be rewritten into a propositional formula using the following rewriting rules:

$$\begin{array}{rcl} (\forall x\ f) & \to & f[x \leftarrow 0] \wedge f[x \leftarrow 1], \\ (\exists x\ f) & \to & f[x \leftarrow 0] \vee f[x \leftarrow 1], \end{array}$$

if the variable x is free in the formula f. For instance, the quantifier free form of the quantified formula given above is (x_3). Adding quantifiers to Propositional Logic does not increase its expressive power, but it allows us to write very concisely expressions whose quantifier-free forms have exponential sizes.

3 Meta-Products

There are 3^n elements in the set P_n of products that can be built out of the set of variables $\{x_1, \ldots, x_n\}$. Indeed there is a one to one mapping between this set and the set of strings $\{x_1, \overline{x_1}, \varepsilon\} \times \ldots \times \{x_n, \overline{x_n}, \varepsilon\}$, where ε is the empty string. In order to represent any set of products with a Boolean function it is thus sufficient to use $\lceil \log_2(3^n) \rceil$ Boolean variables. However, though this number is theoretically sufficient, it is not the most interesting from the computing point of view. In this section we propose a representation of subsets of P_n that uses more Boolean variables (i.e. $2n$) but that has very good properties regarding the implicit prime implicant and the essential prime implicant computation technique presented in this paper.

3.1 Definition

In order to represent sets of products built out of the set of variables $\{x_1, \ldots, x_n\}$, we introduce 2 finite sets of variables $\{o_1, \ldots, o_n\}$ and $\{s_1, \ldots, s_n\}$. We will call the variables $o_1, \ldots, o_n$ the *occurrence* variables and the variables $s_1, \ldots, s_n$ the *sign* variables. We define the mapping σ from the set $\{0,1\}^n \times \{0,1\}^n$ into the set P_n in the following way:

$$\sigma(o, s) = l_1 \ldots l_n,$$

where l_k is the empty string ε if $o_k = 0$, l_k is the string $\overline{x_k}$ if $o_k = 1$ and $s_k = 0$, and finally l_k is the string x_k if $o_k = 1$ and $s_k = 1$. For instance, with $n = 4$, we have $\sigma([0111], [1101]) = (x_2 \overline{x_3} x_4)$.

The mapping σ defined above can not be a one to one mapping from the set $\{0,1\}^n \times \{0,1\}^n$ into the set P_n because it maps 4^n elements onto 3^n elements. There are two ways to get a canonical representation of the elements of P_n using the sets of variables $\{o_1, \ldots, o_n\}$ and $\{s_1, \ldots, s_n\}$. The first way is to choose, amongst all the couples that denote the same product, a particular one that will be taken as reference. The second way, which leads to a simpler and better implementation, is to consider the canonical representation of a product to be the set of all the couples that denote this product. For instance, for $n = 4$, the canonical set representation of the product $(x_2\overline{x_3}x_4)$ is the set $\{([0111],[0101]),([0111],[1101])\}$. Since any subset of $\{0,1\}^n \times \{0,1\}^n$ has a unique characteristic function from the set $\{0,1\}^n \times \{0,1\}^n$ into the set $\{0,1\}$, we get the following canonical functional representation of sets of products.

Definition 3.1 *Let* $P = \{p_1, \ldots, p_m\}$ *be a subset of products of* P_n. *Its* meta-product $\mathcal{P}$ *is the characteristic function of the following subset of* $\{0,1\}^n \times \{0,1\}^n$*:*

$$\left(\bigcup_{k=1}^{k=m} \sigma^{-1}(p_k)\right)$$

3.2 Properties

Meta-products have immediate properties that are direct consequences of the fact that they are characteristic functions. For any meta-products $\mathcal{P}$ and $\mathcal{P}'$, the function $(\mathcal{P} \vee \mathcal{P}')$ is the meta-product of the union of the sets of products P and P' denoted by $\mathcal{P}$ and $\mathcal{P}'$ respectively ; $(\mathcal{P} \wedge \mathcal{P}')$ is the meta-product of the intersection of these sets ; $(\neg\mathcal{P})$ is the meta-product of $(P_n \setminus P)$; the set P is included in the set P' if and only if the function $(\mathcal{P} \Rightarrow \mathcal{P}')$ is a tautology.

The product p associated with the couple (o, s) denotes a unique subset of the set $\{0,1\}^n$, whose characteristic function, noted $\mathcal{X}(o, s)$, is defined by the equation:

$$\mathcal{X}(o,s) = \lambda x. \left(\bigwedge_{k=1}^{k=n} (o_k \Rightarrow (x_k \Leftrightarrow s_k))\right)$$

Two couples (o, s) and (o', s') of $\{0,1\}^n \times \{0,1\}^n$ denote the same product if and only if the functions $\mathcal{X}(o, s)$ and $\mathcal{X}(o', s')$ are identical, which is the case if and only if the formula $Equal((o,s),(o',s'))$ is a tautology, where $Equal$ is defined by the equation:

$$Equal((o,s),(o',s')) = \left(\bigwedge_{k=1}^{k=n} (o_k \Leftrightarrow o'_k) \wedge (o_k \Rightarrow (s_k \Leftrightarrow s'_k))\right)$$

This formula expresses that a variable occurs in the product denoted by the couple (o, s) if and only if it occurs in the product denoted by the couple (o', s'), and in this case it occurs with the same sign in both products.

The product p denoted by the couple (o, s) contains the product p' denoted by the couple (o', s') if and only if the formula $(\mathcal{X}(o', s') \Rightarrow \mathcal{X}(o, s))$ is a tautology. This is the case if and only the formula $Contain((o,s),(o',s'))$ is a tautology, where *Contain* is defined by the equation:

$$Contain((o,s),(o',s')) \quad = \quad \left(\bigwedge_{k=1}^{k=n} o_k \Rightarrow (o'_k \wedge (s_k \Leftrightarrow s'_k))\right)$$

This formula expresses that if a variable occurs in the product p, it also occurs in the product p', and it occurs with the same sign in both products.

The following theorem, whose proof can be found in [4], gives a very easy way to compute the Boolean function that is equal to the sum of all the products belonging to a set of products denoted by a meta-product. This theorem is critical in the implicit essential prime implicant computation method that will be introduced in Section 3.4.

Theorem 3.1 *Let $P = \{p_1, \ldots, p_m\}$ be a set of products built out of the variables $\{x_1, \ldots, x_n\}$, $\mathcal{P}$ its meta-product, and f the Boolean function denoted by the sum of these products. We have the identity:*

$$f \quad = \quad \lambda s.(\exists o\ \mathcal{P}(o,s))$$

3.3 Relatively Prime Implicants

The product p of the set of products $P = \{p_1, \ldots, p_m\}$ is a *relatively* prime implicant of P if and only if there is no product p' in the set P in which p is strictly contained. For example, in the set $P = \{x_1\overline{x_2},\ \overline{x_1}x_2,\ \overline{x_1}x_2\overline{x_3}\}$, the products $(x_1\overline{x_2})$ and $(\overline{x_1}x_2)$ are relatively prime implicants, but $(\overline{x_1}x_2\overline{x_3})$ is not, because it is contained in the product $(\overline{x_1}x_2)$. The following theorem [4] makes explicit the relation that exists between the meta-product of the set P and the meta-product of the subset of relatively prime implicants of P.

Theorem 3.2 *Consider a set of products P and its meta-product $\mathcal{P}$. The meta-product $\mathcal{I}(\mathcal{P})$ of the subset of relatively prime implicants of P is defined by the equation:*

$$\begin{aligned}\mathcal{I}(\mathcal{P}) \quad = \quad & \lambda o.\lambda s.(\mathcal{P}(o,s) \wedge \\ & \quad (\forall o'\ \forall s'\ (\mathcal{P}(o',s') \wedge Contain((o',s'),(o,s))) \Rightarrow \\ & \quad\quad Equal((o',s'),(o,s)))),\end{aligned}$$

that can be simplified into:

$$\mathcal{I}(\mathcal{P}) \quad = \quad \lambda o.\lambda s.(\mathcal{P}(o,s) \wedge (\forall o'\ \mathcal{P}(o \wedge o', s) \Rightarrow (o \leq o'))),$$

where $(o \wedge o') = (o_1 \wedge o'_1, \ldots, o_n \wedge o'_n)$, and $(o \leq o') = \left(\bigwedge_{k=1}^{k=n} (o_k \Rightarrow o'_k)\right)$.

3.4 Relatively Essential Prime Implicants

The product p of the set of products $P = \{p_1, \ldots, p_m\}$ is a *relatively essential* prime implicant of P if and only if the Boolean function that is the sum of all the products of the set $P \setminus \{p\}$ is different from the Boolean function that is the sum of all the products of the set P. The following theorem [4] states that the meta-product of the subset of relatively essential prime implicants of the set of products P can be computed from the meta-product $\mathcal{P}$ of the set P.

Theorem 3.3 *Consider a set of products P and its meta-product $\mathcal{P}$. The meta-product $\mathcal{E}(\mathcal{P})$ of the subset of relatively essential prime implicants of P is defined by the equation:*

$$\begin{aligned}\mathcal{E}(\mathcal{P}) \quad = \quad & \lambda o.\lambda s.(\mathcal{P}(o,s) \wedge \\ & \qquad (\exists s' \; (\exists o' \; \mathcal{P}(o',s')) \oplus \\ & \qquad\qquad (\exists o' \; \mathcal{P}(o',s') \wedge \neg Equal((o',s'),(o,s))))).\end{aligned}$$

4 Representing Meta-Products with BDD's

Binary decision diagrams are a very compact canonical graph representation of Boolean functions that has been introduced in 1986 by R. E. Bryant [2]. In addition with being canonical, the BDD of a function is in most cases smaller than any sum of products that denotes this function. Combining two BDD's G_1 and G_2 using the usual operators such as the conjunction or the disjunction is done in $O(|G_1| \times |G_2|)$, where $|G|$ is the number of vertices of the graph G. This means that the algorithms manipulating BDD's have a much more predictable behavior than the algorithms manipulating sums of products, because these basic operations are exponential in the worst case on this representation. Binary decision diagrams also are a good representation of subsets of $\{0,1\}^n$ because there is no direct correspondence between the number of elements in such a subset and the size of the BDD that denotes this subset, so very large subsets of $\{0,1\}^n$ can be represented with small BDD's.

4.1 The Variable Ordering Problem

Except for the special class of symmetric functions, the size of the BDD of a Boolean function heavily depends on the input variable ordering used to build this BDD. There exist functions from $\{0,1\}^n$ into $\{0,1\}$ whose BDD has a polynomial size (with respect to n) for the best input variable ordering and an exponential size for the worst ordering [2]. This section shows that there exists a good variable ordering of the variables $\{o_1, \ldots, o_n\}$ and $\{s_1, \ldots, s_n\}$ used to built BDD's of meta-products.

The equations that define the meta-products of relatively prime and relatively essential prime implicants use the terms $Equal((o,s),(o',s'))$,

Contain$((o, s), (o', s'))$, and $\chi(o, s)(x)$. There exist several variable orderings that minimize the sizes of the BDD's of these expressions. For the term *Equal*$((o, s), (o', s'))$, these variables orderings have the following property:

$$\begin{array}{ccc} \{o_{\pi(1)}, o'_{\pi(1)}\} & < & \{s_{\pi(1)}, s'_{\pi(1)}\} < \\ \{o_{\pi(2)}, o'_{\pi(2)}\} & < & \{s_{\pi(2)}, s'_{\pi(2)}\} < \\ & \vdots & \\ \{o_{\pi(n)}, o'_{\pi(n)}\} & < & \{s_{\pi(n)}, s'_{\pi(n)}\}, \end{array}$$

where π is a permutation of the integers $\{1, \ldots, n\}$. This relation states that the variables o_k and s_k must be clustered and that the different clusters thus formed can be permutated. The BDD of the term *Equal*$((o, s), (o', s'))$, shown in Figure 1, has $6n$ vertices for any variable ordering compatible with the relation given above.

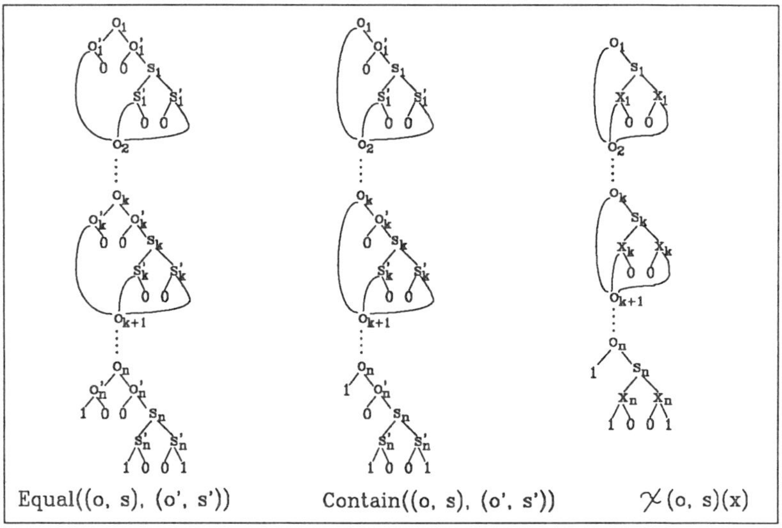

Figure 1. Minimal BDD's of the terms *Equal*$((o, s), (o', s'))$, *Contain*$((o, s), (o', s'))$, and $\chi(o, s)(x)$.

The different variable orderings that minimize the BDD of the term *Contain*$((o, s), (o', s'))$ are defined by the relation:

$$\begin{array}{ccccc} o_{\pi(1)} & < & o'_{\pi(1)} & < & \{s_{\pi(1)}, s'_{\pi(1)}\} < \\ o_{\pi(2)} & < & o'_{\pi(2)} & < & \{s_{\pi(2)}, s'_{\pi(2)}\} < \\ & & & \vdots & \\ o_{\pi(n)} & < & o'_{\pi(n)} & < & \{s_{\pi(n)}, s'_{\pi(n)}\}, \end{array}$$

where π is a permutation of the integers $\{1, \ldots, n\}$. This relation states that the variables o_k, s_k, o'_k, and s'_k must be clustered, and that the different clusters thus formed can be permutated. The BDD of the term *Contain*$((o,s),(o',s'))$, shown in Figure 1, has $5n$ vertices for any variable ordering compatible with the relation given above. These variable orderings are compatible with the variable orderings that minimize the BDD of the expression *Equal*$((o,s),(o',s'))$.

Finally, the different variable orderings that minimize the BDD of the term $\mathcal{X}(o,s)(x)$ are defined by the relation:

$$\begin{array}{lcl} o_{\pi(1)} & < & \{s_{\pi(1)}, x_{\pi(1)}\} < \\ o_{\pi(2)} & < & \{s_{\pi(2)}, x_{\pi(2)}\} < \\ & \vdots & \\ o_{\pi(n)} & < & \{s_{\pi(n)}, x_{\pi(n)}\}, \end{array}$$

where π is a permutation of the integers $\{1, \ldots, n\}$. This relation states that the variable o_k must be put before the variables s_k and x_k, and that the different clusters thus formed can be permutated. The BDD of the term $\mathcal{X}(o,s)(x)$, shown in Figure 1, has $4n$ vertices for any variable ordering compatible with the relation given above.

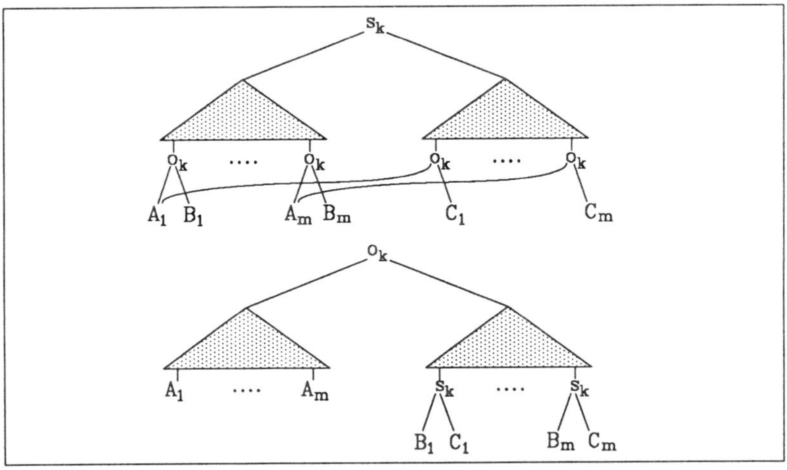

Figure 2. Swapping the variables o_k and s_k in BDD's of meta-products.

The different variable orderings given above are compatible, and it can be shown that they are also good for representing meta-products. Figure 2 shows that the variable s_k should not be put before the variable o_k because in many cases it will lead to a larger BDD. It can also be shown that the interleaving defined above makes some useful computations, for instance counting the number of products in the set denoted by a meta-product (see Section 4.2), linear with respect to the size of the BDD of this meta-product.

In the following we will consider that the BDD's of meta-products are built with the ordering:

$$o_1 < s_1 < o_2 < s_2 < \ldots < o_n < s_n$$

4.2 Product and Literal Costs

Though there is no direct correspondence between the number of elements in the subset S_f of $\{0,1\}^n$ denoted by the function f and the size of the BDD of f, counting this number of elements can be done in a time linear with respect to the number of vertices in this BDD. Since several couples (o, s) can denote the same product, the number of elements in the subset $S_{\mathcal{P}}$ of $\{0,1\}^n \times \{0,1\}^n$ denoted by the *characteristic function* $\mathcal{P}$ is not equal to the number of products in the subset of P_n denoted by the *meta-product* $\mathcal{P}$. We show here that the two cost functions on sums of products defined in Section 2 can be evaluated in times linear with respect to the numbers of vertices of the BDD's of their meta-products.

Consider the meta-product $\mathcal{P}$ built out of the set of variables $\{o_1, \ldots, o_n, s_1, \ldots, s_n\}$, and P the set of products it denotes. Shannon expansion of $\mathcal{P}$ with respect to the variable o_k provides us with the couple of functions $(\mathcal{P}_{\overline{o_k}}, \mathcal{P}_{o_k})$. The function $(\overline{o_k} \wedge \mathcal{P}_{\overline{o_k}})$ denotes the subset of P of products that do not contain the variable x_k. The function $(o_k \wedge \mathcal{P}_{o_k})$ denotes the subset of P of products that contain the variable x_k. This set is the union of two disjoint sets that are the subsets of P of products that contain the literal $(\overline{x_k})$ and the literal (x_k) respectively. These sets are denoted by the functions $(o_k \wedge \overline{s_k} \wedge \mathcal{P}_{o_k\overline{s_k}})$ and $(o_k \wedge s_k \wedge \mathcal{P}_{o_k s_k})$ respectively, where the couple of functions $(\mathcal{P}_{o_k\overline{s_k}}, \mathcal{P}_{o_k s_k})$ is Shannon expansion of the function $\mathcal{P}_{o_k}$ with respect to the variable s_k. We thus have the equality:

$$\begin{aligned} \mathit{ProductCost}(\mathcal{P}) &= \mathit{ProductCost}(\mathcal{P}_{\overline{o_k}}) + \\ &\quad \mathit{ProductCost}(\mathcal{P}_{o_k\overline{s_k}}) + \\ &\quad \mathit{ProductCost}(\mathcal{P}_{o_k s_k}), \end{aligned}$$

from which it is immediate to derive a product counting function that has a complexity linear with respect to the number of vertices in the BDD of the meta-product $\mathcal{P}$.

The literal cost of the meta-product $\mathcal{P}$ can also be evaluated with a complexity linear with respect to the number of vertices in the BDD of the meta-product $\mathcal{P}$. The literal cost associated with the subset of P of products containing the literal $(\overline{x_k})$ is the cost of the products in the set denoted by $\mathcal{P}_{o_k\overline{s_k}}$ plus the number of elements in this set, because the literal $(\overline{x_k})$ will be added to each element of this set. The same can be said of the set of products that contain the literal (x_k). We thus have the equality:

$$\begin{aligned} \mathit{LiteralCost}(\mathcal{P}) &= \mathit{LiteralCost}(\mathcal{P}_{\overline{o_k}}) + \\ &\quad \mathit{LiteralCost}(\mathcal{P}_{o_k\overline{s_k}}) + \mathit{ProductCost}(\mathcal{P}_{o_k\overline{s_k}}) + \\ &\quad \mathit{LiteralCost}(\mathcal{P}_{o_k s_k}) + \mathit{ProductCost}(\mathcal{P}_{o_k s_k}), \end{aligned}$$

from which it is also immediate to derive a literal counting function that has a complexity linear with respect to the number of vertices in the BDD of the meta-product $\mathcal{P}$.

5 Cubes, Primes and Essential Primes of Boolean Functions

The usual Boolean operators in addition with the quantified variable elimination are sufficient to compute the meta-products that denote the sets of prime implicants and of essential prime implicants of Boolean functions as well as Boolean functions with a care set. This computation is done in two main steps. The first step consists in computing the meta-product of the set of cubes of this function, and the second step consists in extracting the meta-products of the set of prime implicants and of essential prime implicants from this meta-product using the equations given in Section 3.

5.1 Cubes

The following theorem states the meta-product $\mathit{Cube}(f_C)$ of the set of cubes of a Boolean function f_C with the care set C can be computed using the usual Boolean operators in addition with the quantified variable elimination.

Theorem 5.1 *Let f_C be a Boolean function from the set $\{0,1\}^n$ into the set $\{0,1\}$ with the care set C. The meta-product $\mathit{Cube}(f_C)$, which denotes the set of all products built out of the set of variables $\{x_1,\ldots,x_n\}$ that are cubes of the function f_C, is a Boolean function from the set $\{0,1\}^n \times \{0,1\}^n$ into $\{0,1\}$ defined by the equation:*

$$\begin{aligned}\mathit{Cube}(f_C) \quad = \quad & \lambda o.\lambda s.((\exists x\ \mathcal{X}(o,s)(x) \wedge \chi_C(x)) \wedge \\ & (\forall x\ \chi_C(x) \Rightarrow (\mathcal{X}(o,s)(x) \Rightarrow f(x)))).\end{aligned}$$

In the case where the care set of the function f_C is the complete input set, this equation can be simplified into:

$$\mathit{Cube}(f_{\{0,1\}^n}) \quad = \quad \lambda o.\lambda s.(\forall x\ \mathcal{X}(o,s)(x) \Rightarrow f(x)).$$

5.2 Prime Implicants and Essential Prime Implicants

Since the meta-product $\mathit{Cube}(f_C)$ of a function f_C represents all the products built out of the set of variables $\{x_1,\ldots,x_n\}$ that are cubes of this function, computing the meta-product of all the products that are prime implicants of f_C comes down to computing the relatively prime implicants of this set of products.

Theorem 5.2 *Let f_C be a Boolean function from the set $\{0,1\}^n$ into the set $\{0,1\}$ with the care set C, and Cube(f_C) be the meta-product of the set of cubes of f_C. The meta-product $\mathcal{P}rime(f_C)$, which is the meta-product of the sets of products built out the set of variables $\{x_1,\ldots,x_n\}$ that are prime implicants of the function f_C, is equal to:*

$$\begin{aligned} \mathcal{P}rime(f_C) &= \mathcal{I}(Cube(f_C)) \\ &= \lambda o.\lambda s.(Cube(o,s) \wedge (\forall o'\ Cube(o \wedge o', s) \Rightarrow (o \leq o'))) \end{aligned}$$

Theorem 5.3 *Let f_C be a Boolean function from the set $\{0,1\}^n$ into the set $\{0,1\}$ with the care set C, Cube(f_C) be the meta-product of the set of cubes of f_C, and $\mathcal{P}rime(f_C)$ be the meta-product of the set of prime implicants of f_C. The meta-product $\mathcal{E}ssential(f_C)$, which is the meta-product of the sets of products built out the set of variables $\{x_1,\ldots,x_n\}$ that are essential prime implicants of the function f_C, is defined by the equation:*

$$\begin{aligned} \mathcal{E}ssential(f_C) &= \mathcal{E}(Cube(f_C)) \\ &= \lambda o.\lambda s.(Cube(o,s) \wedge \\ &\qquad (\exists s'\ f(s') \oplus \\ &\qquad\qquad (\exists o'\ Cube(o',s') \wedge \neg Equal((o',s'),(o,s))))) \\ &= \mathcal{E}(\mathcal{P}rime(f_C)) \\ &= \lambda o.\lambda s.(\mathcal{P}rime(o,s) \wedge \\ &\qquad (\exists s'\ f(s') \oplus \\ &\qquad\qquad (\exists o'\ \mathcal{P}rime(o',s') \wedge \neg Equal((o',s'),(o,s))))) \end{aligned}$$

The equations above show that there are two ways to compute the meta-product $\mathcal{E}ssential(f_C)$, either directly from the meta-product $Cube(f_C)$, or in two steps, by first extracting the meta-product $\mathcal{P}rime(f_C)$ from the meta-product $Cube(f_C)$, and then by extracting the meta-product $\mathcal{E}ssential(f_C)$ from $\mathcal{P}rime(f_C)$. Both ways use only the standard BDD operations such as the conjunction and the disjunction, in addition with the quantified variables elimination which itself can be implemented using these operators.

6 Experimental Results

We have used the method presented in this paper to compute the sets of prime and of essential primes implicants of all the output functions of some combinational and sequential circuits. Table 1 gives the characteristics of the circuits. Except for the entries *indust*, *math*, and *random*, **#Ins** is the number of input variables, **#Funs** is the number of output functions,

Name	#Ins #Files	#Funs	#Primes	#Essentials
math	23	128	8616	1937
indust	111	2412	87797	17284
random	11	189	118141	169
cbp16	32	17	655287	655287
cbp32	64	33	42949672960	42949672960
dsip	452	421	22850	2223
mm10	43	40	10015327	127139
$mm20_{no\ outputs}$	83	60	5243035	5242976
pewd	123	213	11442	1119
s1423	91	79	469397	36226
prod	20	33	24831	1950
clm3	46	32	55386	782
clm1	46	115	15219	862
bred	49	1	27778	27778
add3	21	11	77008	199
add4	29	12	688167	247
addsub	31	15	408190	102413
rip04	8	5	75	75
rip08	16	9	1499	1499
mul06	12	12	5430	765
mul07	14	14	28972	1551
mul08	16	16	152051	3879

Table 1. Characteristics of Examples.

#Primes is equal to $\sum_f ProductCost(\mathcal{P}rime(f))$ for all output functions f, and **#Essentials** is equal to $\sum_f ProductCost(\mathcal{E}ssential(f))$. The entries *indust*, *math*, and *random* correspond to the directories of the MCNC 2-level minimization benchmark. Note that the output functions of these circuits have care sets. For these entries, column **#Files** is the number of circuits in this directory, and **#Funs**, **#Primes**, and **#Essentials** are the sums of the figures obtained for each circuit of the directory. All the circuits of Table 1 put together have 3867 output functions.

Each file has first been translated from its original format into an intermediate form in which Boolean functions are represented with a textual form of BDD's. The CPU times needed to perform these translations are not included in the results presented here. The variable ordering used to build these BDD's, computed using the heuristics described in [8], is the same for all functions of a file, and the occurrence and sign variables are ordered according to this input variable ordering.

Table 2 gives the CPU times in seconds and the memory sizes in Mbytes needed to perform the computations on a SUN Sparc2 workstation. Column $\textbf{Time}_\textbf{P}$ is the CPU time needed to compute the meta-product of the set of prime implicants (Theorem 5.2) of each output function of the circuit,

Name	Primes		Essentials			
	$\mathbf{Time_p}$	$\mathbf{Mem_p}$	$\mathbf{Time_{ec}}$	$\mathbf{Mem_{ec}}$	$\mathbf{Time_{ep}}$	$\mathbf{Mem_{ep}}$
math	17.6	1.2	50.5	2.1	**32.5**	**1.7**
indust	289	1.8	1273	4.3	**639**	**3.5**
random	511	2.4	1427	4.3	**892**	**2.7**
*cbp*16	2.5	0.8	**4.3**	1.1	5.4	1.1
*cbp*32	12.3	1.4	**22.5**	2.0	30.5	**1.9**
dsip	21.9	1.0	71.8	1.5	**62.9**	**1.3**
*mm*10	598	4.0	5694	10.8	**884**	**5.3**
*mm*20	10.7	1.3	**23.5**	2.0	24.2	**1.9**
pewd	15.1	1.0	46.5	1.5	**39.6**	**1.3**
*s*1423	154	1.8	891	3.8	**347**	**3.0**
prod	14.2	1.1	45.7	1.8	**34.2**	**1.5**
*clm*3	32.1	1.6	128	3.2	**69.2**	**2.2**
*clm*1	47.4	2.9	358	9.4	**93.5**	**4.3**
bred	1.1	0.5	**2.3**	0.9	2.5	0.9
*add*3	11.6	1.3	36.1	1.7	**30.3**	**1.4**
*add*4	18.5	1.3	55.0	2.0	**43.4**	**1.5**
addsub	3.2	0.9	**6.4**	1.3	7.7	1.3
*rip*04	0.1	0.1	0.2	0.1	0.2	0.1
*rip*08	0.5	0.2	**0.8**	**0.4**	0.9	0.5
*mul*06	23.8	1.4	108	3.9	**48.2**	**1.8**
*mul*07	138	3.9	932	18.2	**275**	**5.5**
*mul*08	1001	15.6	–	> 24	**1903**	**22.1**

Table 2. Experimental Results.

and $\mathbf{Mem_p}$ is the memory size needed to make this computation. Columns $\mathbf{Time_{ec}}$ and $\mathbf{Mem_{ec}}$ are the CPU time and memory size to compute the meta-product of the set of essential prime implicants of each function, by first computing the meta-product of the set of cubes of this function (Theorem 5.1), and then using this meta-product to compute the requested meta-product (Theorem 5.3). Finally, columns $\mathbf{Time_{ep}}$ and $\mathbf{Mem_{ep}}$ are the CPU time and memory size to compute the meta-product of the set of essential prime implicants of each function, by first computing the meta-product of the set of cubes of this function, then extracting the meta-product of the set of prime implicants of the function (Theorem 5.2), and finally using this meta-product to compute the result (Theorem 5.3).

These experimental results show that the method presented here can be applied to Boolean functions with very large number of primes. For all but small examples, the essential prime implicant computation that uses the intermediate meta-product $\mathcal{Prime}$ is more efficient that the computation using the meta-product *Cube*. This comes from the fact that for all large examples, the BDD of the meta-product $\mathcal{Prime}$ is much smaller than the BDD of *Cube* and that computation costs are directly related to the sizes of these BDD's. This is particularly the case for the multipliers *mul*07 and *mul*08, and the *mm*10 circuit. Moreover the results show that there is *no relation* between the CPU times and the number of primes in the sets denoted by the meta-products computed here.

7 Conclusion

In this paper we have presented a method to compute in an implicit way the sets of cubes, of prime implicants, and of essential prime implicants of Boolean functions with care sets. The key ideas used in this method are to represent products built out of the set of variables $\{x_1, \ldots, x_n\}$ with formulas, called *meta-products*, built out of 2 sets of variables $\{o_1, \ldots, o_n\}$ (the occurrence variables) and $\{s_1, \ldots, s_n\}$ (the sign variables), and to denote these formulas with binary decision diagrams.

Though this method allows us to deal with functions with very large numbers of prime implicants and of essential prime implicants, there are functions for which the memory size needed to store intermediate results is too large to make these computations possible. In order to decrease this memory size, we have developed an *incremental* implicit prime computation method that performs a depth-first traversal of the BDD of the function under consideration and computes the result bottom-up. This method [5], which will be presented in a forthcoming paper, allows us to handle functions that cannot be handled by the method presented here.

It has been shown [10, 11] that computing the prime implicants and essential primes implicants of a *multiple-valued* input and output function comes down to computing the prime implicants and essential primes implicants of a multiple-valued input Boolean function. There are several ways to extend the method presented here in order to deal with such Boolean functions, that correspond to different ways to encode the multiple-valued input variables of the function into Boolean variables. With these different encodings correspond different meta-product representations and their associated calculi. These different techniques will also be presented in a forthcoming paper.

Acknowledgments

The authors would like to thank Henri Fraisse who introduced the meta-product representation, and Gary Hachtel and Bill Lin for many fruitful discussions.

References

[1] Brayton, R. E., Hachtel, G. D., McMullen, C. T., Sangiovanni-Vincentelli, A. L., *Logic Minimization Algorithms for VLSI Synthesis*,

Kluwer Academic Publishers, 1984.

[2] Bryant, R. E., "Graph-based Algorithms for Boolean Functions Manipulation", *IEEE Trans. on Computers*, Vol C35, 1986.

[3] Coudert, O., Madre, J. C., "A Unified Framework for the Formal Verification of Sequential Circuits", in *Proc. of ICCAD'90*, Santa-Clara, November 1990.

[4] Coudert, O., Madre, J. C., *A New Method to Compute Prime and Essential Prime Implicants of Boolean Functions*, Bull Research Report N°91028, November 1991.

[5] Coudert, O., Madre, J. C., *Implicit and Incremental Computation of Prime and Essential Prime Implicants of Boolean Functions*, Bull Research Report N°91033, November 1991.

[6] Hong, S. J., Muroga, S., "Absolute Minimization of Completely Specified Switching Functions", *IEEE Trans. on Computers*, pp. 53–65, Vol 40, 1991.

[7] Madre, J. C., Coudert, O., "A Logically Complete Reasoning Maintenance System Based on a Logical Constrain Solver", in *Proc. of International Join Conference on Artificial Intelligence (IJCAI)*, Sydney, Australia, August 1991.

[8] Minato, S., Ishiura, N., Yajima, S., "Fast Tautology Checking Using Shared Binary Decision Diagrams - Experimental Results", in *Proc. of the Workshop on Applied Formal Methods for Correct VLSI Design*, L. Claesen Ed., North Holland, 1989.

[9] Rudell, R. L., *Multiple-Valued Logic Minimization for PLA Synthesis*, Technical Report M86/65, UCB, 1986.

[10] Rudell, R. L., Sangiovanni-Vincentelli, A. L., "Multiple Valued Minimization for PLA Optimization", *IEEE Trans. on CAD*, Vol 6, No 5, September 1987.

[11] Sasao, T., "An Application of Multiple-Valued Logic to a Design of Programmable Logic Arrays", in *Proc. of 8th International Symposium on Multiple Valued Logic*, 1978.

Equation-Free Synthesis of High-Performance Linear Analog Circuits

Emil S. Ochotta, Rob A. Rutenbar and **L. Richard Carley**
Department of Electrical and Computer Engineering
Carnegie Mellon University
Pittsburgh, Pennsylvania 15213

Abstract

We describe a new synthesis strategy for cell-level linear analog circuits. An evolving class of "equation-based" synthesis tools has offered some hope of automatic synthesis in this application: translation from an unsized circuit topology to device sizes and a dc bias point. The term "equation-free" is meant to convey our strong belief that an ideal synthesis scheme, in contrast to most existing approaches, should not rely on designers to derive, code, manipulate, or guide the symbolic pruning of performance equations for the circuits they wish to synthesize. Our synthesis strategy relies on four key ideas. First, to avoid accuracy problems we use industrial device models extracted from detailed circuit simulators. Second, to avoid robustness problems, we use a novel formulation of simulated annealing on continuous variables for numerical optimization. Third, we employ a "relaxed" representation of circuit behavior in the annealer that does not require circuits to be feasible, *i.e.*, to satisfy Kirchoff's laws, at all times. Finally, to automate fully the solution process, we rely on Asymptotic Waveform Evaluation techniques, augmented with some simple symbolic analysis, to evaluate the behavior of each circuit visited by the annealer. These ideas have been implemented in a pair of synthesis tools called ASTRX and OBLX. Experimental results demonstrate that it is possible to synthesize new circuits from nothing more than a netlist, a set of device models, and a set of linear performance targets.

1 Introduction

By the end of the 1990s, most new application-specific integrated circuit (ASIC) designs will incorporate both digital and analog functionality. This is an inevitable consequence of increases in the density and performance of ASIC technologies, which allow circuit boards full of expensive discrete components to be implemented on a single, less expensive IC. Most critical applications for such ASICs–consumer products, telecommunications, network control, speech recognition, machine vision, data conversion, etc.–rely on a core of analog circuitry. Unfortunately, this trend toward mixed-signal design is impeded by a troubling bottleneck in the design process for these chips: the continuing lack of mature analog CAD tools.

In the digital world, the pairing of logic synthesis tools [2] with semi-custom layout techniques [12] now allows relatively high-level descriptions of circuit function to be reduced reliably and efficiently to silicon. We have as yet no tools of comparable maturity and robustness for automating the analog circuit design process. Attempts to mimic successful digital approaches, such as hand-crafted standard cell libraries, provide some relief but fall short of a comprehensive solution. The problem is that a typical analog cell, *e.g.*, an operational amplifier, may have 20 unique performance specifications. An attempt to cover each "corner" of the design space for this amplifier would result in about one million different cells. This is clearly an impractical solution, especially since libraries must frequently be remapped to track evolving fabrication technologies. In contrast, a digital library that offers a modest number of cell types, each available across a small range of power, speed and area trade-offs, can cover essentially all design situations. As a result, the digital cells in today's mixed-signal ASICs are typically selected from a cell library, while many of the necessary analog cells still require time-intensive, full-custom design. Moreover, much of this custom design is routine, diverting skilled analog designers away from more interesting circuit- and system-level problems.

This paper offers a new approach to the problem of analog synthesis, *i.e.*, translating a specification into a circuit schematic with sized devices. The scope of this work is synthesis for *cell-level* (10 to 50 devices) *linear* circuits: starting from a transistor schematic, we seek both to design a dc bias point and size all devices to meet performance targets such as gain, bandwidth and phase margin.

To motivate our new synthesis strategy, we begin by summarizing previous approaches. These can be partitioned into two major synthesis strategies: *simulation-based* and *equation-based.* The earliest attempts at analog synthesis were simulation-based. These rely on iterative optimization techniques, usually a sophisticated form of gradient-descent, wrapped around a detailed circuit simulator that evaluates circuit performance [6][17]. Such techniques have proved successful primarily for optimization, *e.g.*, tuning a good circuit design for enhanced performance or remapping it into a slightly different technology.

An alternative approach, equation-based synthesis, replaces the circuit simulator with a set of simplified behavioral equations. The onus is now on the designer to construct the relevant equations for the circuit to be synthesized, and simplify and arrange them in a form amenable to automatic solution. Several strategies have been tried. Notable here are the OASYS system [7], which organized the equations hierarchically (*e.g.*, for cells, subcircuits, devices) and used backtracking heuristics to deal with the under-constrained nature of the design equations; IDAC [3], which used heuristics to find an initial solution to a flat equation representation, then a specially-tuned simulator for performance evaluation; and OPASYN [11], which relied on numerical optimization of simplified equations. Recent approaches have focused on symbolic analysis techniques to ease the burden of acquiring the requisite equations. ARIADNE [5] is a good example. It incorporates tools to analyze linear circuits, produce symbolic frequency-domain transfer functions, manipulate these symbolic forms (*e.g.*, it is frequently necessary to separate the independent and dependent variables by numerically inverting the expressions), and solve them numerically.

These prototype synthesis systems, and equation- and simulation-based synthesis strategies in general, have several drawbacks which we can categorize as follows:

- **Accuracy:** Early equation-based approaches relied heavily on simplifications to device models. The trend toward smaller devices invalidates the use of many of these simplifications. For example, in a 3μ MOS process, $I_{DS} = K'W/2L(V_{GS} - V_T)^2$ is a workable model of the current-voltage relationship for a device, and can be inverted to allow either voltage or current as the independent variable. But this is a grossly inaccurate model for a device with a submicron channel length. The need to support complex device models and circuit design goals that exploit the limits of these models is fundamentally at odds with equation-based strategies that rely on simplified models.

- **Robustness:** Existing simulation-based approaches are notably sensitive to the starting circuit solution, often failing to converge if started too far from a "good" answer. This is due in part to the reliance on gradient techniques, which easily become trapped in local minima. The manual design work necessary to obtain a good starting point for simulation-based tools in effect disqualifies them as true synthesis tools, which start with an unsized circuit topology. Despite this well-known starting point problem, gradient descent algorithms are still the method of choice for these approaches because they tend to minimize the number of invocations of the detailed circuit simulator. It is this large time penalty for each circuit evaluation that ultimately limits the exploration of the reachable design space. At the other extreme, the accuracy/speed trade-offs that enable some equation-based strategies to explore more of the design space are ultimately fragile when asked to design circuits whose performance relies on effects that have been simplified out of the underlying equations.

- **Automation:** Equation-based strategies must acquire their equations somehow. This is an arduous process if done manually, and in our experience usually deters designers. Symbolic analysis methods are an obvious alternative, but the state of the art here is still insufficient for practical synthesis. If realistic device models are used, they can generate symbolic expressions of huge complexity, *e.g.*, 10,000 terms for 10 devices. If the device models themselves are pruned before symbolic analysis, the resulting expressions lack accuracy for high-performance designs. If the original expressions are pruned, the result is usually faithful only to *some* performance concerns. For example, ISAAC [5] prunes terms in the transfer function $H(s)$ to keep its coefficients faithful to (within a specified error bound of) the original expression. But such pruning can destroy both magnitude and especially the phase information, distorting measures such as phase margin on which the circuit's performance may critically depend.

This paper suggests an alternative synthesis strategy devised to handle the problems enumerated above. We refer to this as *equation-free* synthesis primarily to distinguish it from equation-*based* synthesis in general, and symbolic schemes in particular. The central attribute of equation-free synthesis is that it does not burden designers with any need to derive, manipulate, invert, or guide the symbolic pruning of any equations for the circuits they seek to synthesize. Starting with only a fixed circuit topology, we can synthesize a dc bias point and device sizes to meet linear performance constraints (small-signal ac constraints) while optimizing figures of merit such as active device area. To our knowledge this is the only synthesis system that can start from an arbitrary circuit topology and size it to meet linear performance specifications *without* designer-supplied equations.

The paper is organized as follows. Section 2 outlines the components of this synthesis strategy. Section 3 then formulates the problem more precisely, paying particular attention to the structure of the annealing optimizer and its cooperation with the AWE circuit evaluator. Section 4 presents experimental results. Finally, Section 5 offers some concluding remarks.

2 Equation-Free Synthesis: Strategy

Our circuit synthesis strategy relies on four key ideas. First, to avoid accuracy problems we use industrial device models extracted from detailed circuit simulators. Second, to avoid robustness problems, we use a novel formulation of simulated annealing on continuous variables for numerical optimization; annealing is particularly resistant to the problems of local minima and starting-point sensitivity that plague simulation-based schemes. Third, following [14], we employ a "relaxed" representation of circuit behavior in the annealer that does not require circuits to be feasible, *i.e.*, to satisfy Kirchoff's laws, at all times. This relaxation allows us to use complex device models without a time consuming dc-solve after every annealing perturbation to the circuit. Finally, to automate fully the solution process, we rely primarily on AWE techniques [18], augmented with some simple, automatically-generated symbolic equations, to evaluate the behavior of each circuit visited by the annealer. AWE has the significant advantage that it can analyze general RLC models (in our case, small-signal models) robustly, accurately and quickly. We describe each of these four ideas in more detail below.

There is, unfortunately, no longer any alternative to using industrial-strength device models in a practical circuit synthesis system. High-performance circuits rely on the device performance characteristics achievable in aggressive technologies. Simplified models of the current-voltage relationships at the terminals of these devices fail to capture many of these important characteristics. The real problem is not that such models are difficult to come by–indeed, they form the core of any detailed circuit simulator. Rather, the problem is that we cannot arbitrarily invert the terminal relationships of these models and choose which variables are independent and which dependent. This is perhaps the most critical simplification employed in equation-based approaches, because it often enables one to solve for the dc bias point of the circuit analytically. Freed from the need to solve for dc correctness numerically, many equation-based tools have thus been able to employ aggressive iterative optimization techniques that examine many possible circuits.

In [14] Maulik and Carley suggested an alternative formulation of the synthesis problem that resolves this trade-off between model accuracy and the inefficiency of the dc bias solution. The idea is to use device models extracted from a circuit simulator, and *encapsulate* these with a layer of insulation that allows them to be invoked by a synthesis tool. Supplied with terminal voltages and device parameters such as width/length, the encapsulated model returns both the terminal currents, and the complete small-signal model. To deal with the inefficiency of solving for dc biasing, [14] *relaxes* the requirement that each circuit visited during iterative optimization be dc-correct. Just as we must formulate optimization goals such as meeting gain or bandwidth constraints, we now formulate dc-correctness as yet another goal to meet. Of course, the idea of relaxing the dc constraints in this manner is not new, *e.g.*, an analogous formulation for microwave circuits is discussed in [19]; however, it has been controversial [13]. The problem is to guarantee convergence to a realizable circuit. Simulation-based techniques resolve this by solving for dc for each candidate circuit. Maulik and Carley resolved this problem by casting dc and performance goals as constraint functions, then using sequential quadratic programming techniques (SQP) to solve the resulting constrained optimization problem. The technique first finds the region of dc-feasible solutions and then searches along the boundary of this region to optimize the other performance goals. Using encapsulated device models to generate all the relevant circuit quantities, this approach has been quite successful in converging to accurate solutions. However, it still requires a good starting point, and difficult circuit-level (but not device-level) equation extraction and simplification.

We resolve the model-versus-dc problem similarly. However, we relax the constraint on dc-correctness even further, insisting only that our circuit be physically realizable at the *end* of the optimization process. In other words, we allow both dc goals and performance goals to evolve slowly over the course of synthesis. This becomes a reasonable strategy because we choose simulated annealing as the optimization algorithm [10]. Annealing is empirically robust on many problems with complex non-convex cost surfaces. Annealing's hill-climbing abilities have encouraged experimentation with a variety of problem formulations with the same flavor as relaxed dc-correctness, *i.e.*, particular terms of the cost function to be minimized correspond to penalties that quantify the distance from the current solution to a feasible one. The twin goals of annealing are then to both drive these specific penalty terms to zero and to minimize the remaining terms. Further, because annealing starts with an essentially randomized solution (during the high-temperature regime of the gradual cooling process) it also avoids completely the starting point sensitivity problems prevalent with gradient-descent techniques. Our focus is thus on the proper design of the annealer's cost function, move generation mechanism, and cooling schedule to support circuit synthesis.

The issue yet to be resolved is how we evaluate a circuit quickly (our annealer will need to visit many circuit configurations), and without requiring the designer to supply equations. Our solution is to employ asymptotic waveform evaluation (AWE) techniques. AWE is a robust, efficient approach to analysis of arbitrary linear RLC circuits. By matching the initial boundary conditions and the first $2q-1$ moments of the actual RLC circuit transient response to a reduced q-pole model, AWE can predict circuit response using a reduced complexity model valid for frequencies below a specified limit [18]. For each circuit visited during annealing–physically realizable or not–we use encapsulated device models to supply AWE with an "accurate" small-signal model. AWE analysis is extremely fast here, allowing a complete re-evaluation for each circuit visited. The cell-level analog circuits we target have 10-50 devices, and produce small-signal models with a few hundred elements. AWE reduces such a small-signal model to an accurate low-order transfer function in 10 to 100 ms on a fast workstation. This transfer function can then be used to calculate ac performance. AWE can also generate other information that we are beginning to exploit: for example, the LU factors of the small signal matrix computed during AWE's analysis give us the dc voltage gradient. In practice, we rely on AWE to handle ac constraints such as gain and bandwidth. The dc constraints are formulated on the currents violating Kirchoff's current law (KCL) using simple automatically-generated symbolic equations. Previous synthesis strategies fall short of handling both sets of constraints automatically. These evaluation techniques liberate us from the need to extract from the designer *any* essential performance equations for linear circuit synthesis.

3 Equation-Free Synthesis: Formulation

This section will describe our equation-free synthesis formulation in detail. Given an unsized analog circuit topology and a set of linear performance specifications, the path to a completed, synthesized circuit comprises these steps:

1. *Analysis:* From the input circuit netlist, we automatically derive the independent and dependent variables with respect to numerical optimization, and symbolic equations for some aspects of the dc bias point calculation.
2. *Code Generation:* Based on this analysis, we automatically generate the code for the annealer's cost function that evaluates circuit performance after each move. Aspects of dc biasing, as well as specific ac (linear small-signal) performance targets, and actual optimization goals (*e.g.*, minimize area), appear in this cost function.
3. *Solution:* This cost function evaluation code is compiled and linked to a "core" annealing engine that numerically solves for the relevant device sizes and dc bias point to meet specifications.

We refer to the first two steps as *compilation*, and the last step simply as *annealing*. The overall flow is reminiscent of [5], except that we automate the synthesis of the dc bias point (rather than ask the designer to add the relevant equations), use AWE to evaluate ac performance (rather than pruned symbolic equations), and use an efficient, aggressive annealing strategy (to evaluate 10^5 rather than 10^3 circuit candidates over a typical synthesis run).

The first part of compilation identifies the dependent and independent variables for subsequent annealing. The input format is essentially that used by HSPICE [15], which defines a circuit topology while allowing component values and device sizes to be specified as variables. Using this technique, matched devices are simply devices that use all the same variables. Obviously, variables in the input file become independent variables to be manipulated by the annealer. Because of our relaxed dc formulation, the annealer also manipulates node voltages as independent variables in order to satisfy KCL. Since the node voltages automatically form a basis set, Kirchoff's voltage law (KVL) is always satisfied. As the annealer perturbs these voltages, calls to the device models transform the perturbations into changes in the currents at each node of the circuit. To satisfy KCL, an associated penalty term, p, is included in the cost function. Whenever KCL is not satisfied, $\sum_{\text{nodes}} (\sum_{\text{branches } b} I_b) = p > 0$, and the value of p can be used to produce the appropriate penalty. These equations, $\sum_{\text{branches } b} I_b = 0$, are extracted automatically during the compilation process and the code necessary to produce the KCL penalty is automatically generated. It is worth noting that the compiler recognizes cases where zero dc current flows and eliminates node voltages that are not truly independent, *e.g.*, the middle node of a series RC connection. Simple heuristics like this permit substantial reduction in the dimensionality of the problem we hand off to the annealer.

These voltage perturbations (to the operating point of each device) also change the small-signal model we must use in analyzing overall circuit response. Again, the compiler generates the code to incrementally update the small-signal model of the overall circuit, the input to the AWE evaluator. The results passed back from each call to AWE are similarly embedded as terms in the annealer's cost function. To summarize, the compiler generates all the code necessary for the annealer to: manipulate a basis set of bias voltages; invoke device models to evaluate new bias currents and small-signal models; invoke AWE to analyze the overall behavior of the circuit; and evaluate the complex multiple-component cost function that defines the quality of each circuit configuration visited.

The next phase of the synthesis is the annealing. To describe the annealing formulation, we must describe its four principal components: the *representation* of the synthesis problem manipulated by the annealer; the *moves* used to transform one circuit configuration into another; the *cost function* that evaluates the quality of each visited circuit configuration; and the *control mechanisms* that direct the overall cooling process.

The problem representation is straightforward. The annealer manipulates the values of the independent variables determined during compilation. These come in several classes: passive component values, active device sizes, and operating point voltages. The interesting point to note is that we do not represent all these as continuous variables. The operating point voltage values must clearly be continuous to determine an accurate bias point. Highly nonlinear devices can be extremely sensitive to very small changes in their operating point, and we are aware of no voltage discretization scheme that is viable for analog applications. Device sizes, however, can reasonably be regarded as discrete quantities. The underlying limitations are either lithographic (how accurately we can etch a device) or reflect the need to restrict designs to a known set of carefully qualified device sizes for which accurate models are available. Moreover, there is considerable advantage to be had from properly discretizing device sizes: the coarser the discretization, the smaller the space of reachable sizes that must be explored. Because small changes in device sizes make proportionally less difference on larger devices, we use a logarithmically spaced grid.

The more interesting problem is clearly the design of the *move-set*–the set of allowable perturbations–for the continuous variables. The basic problem is the need for an efficient mechanism to generate each perturbation $\Delta \underline{x}$ to the current n-dimensional real-valued solution, $\underline{x} \in R^n$. For discrete variables, there is always a smallest allowable move, an *atomic* perturbation, and the choice is what larger moves should be included in the move-set for efficiency. For example, for the device size variables, we could rely only on moves that changed a variable by 0.1μ, though this would be remarkably inefficient. The problem is more acute for the continuous variables: what is the right *smallest* move? To synthesize an accurate dc bias point, we may need to explore across a voltage range of several volts, with an accuracy of a few tens of nanovolts. But it is unlikely we can afford the time required to search for a feasible bias point if we restrict ourselves solely to moves of 10 nV in magnitude. Hence we need some large moves for exploration efficiency. The question then becomes: what is the right *mix* of move sizes for the continuous variables?

We are aware of several attempts at solving this problem, with two of particular note: the *topography-following* approach of Vanderbilt and Louie [21] and the *tree-annealing* approach of Bilbro [1]. Vanderbilt and Louis focus on the second moments of the random walk formed by a sequence of accepted annealing moves generated on an n-dimensional cost surface. They show that the covariance matrix generated by these moves can be used to gauge the shape of the "valley" currently being explored. In contrast, Bilbro implicitly builds a model of the n-dimensional cost surface in the form of an n-dimensional tree. The distance from root to leaf is deeper in the regions where more successful solutions have been found. The tree is incrementally refined after each successful move, and is used to bias the generation of random moves toward more promising regions of the space. Although both methods show promise, there is a lack of data for problems with large dimensionality, *i.e.*, more than 20 or 30 variables. There is also concern about the numerical properties of the matrix manipulations used in the Vanderbilt and Louis topography-following scheme, and concern about the storage requirements of the tree in Bilbro's tree annealing, which grows inexorably with each accepted move.

Because of these concerns we have adapted a method from Hustin [9] developed for discrete annealing problems. Each variable v_i manipulated by the annealer has a range of possible values $v_i \in [v_{min}, v_{max}]$. We then define a set of moves for this variable, each of which has a range within these bounds. For example, a voltage might take values between 0 and 10 V. We might then define separate voltage moves with range 100 nV, 1 mV, 10 mV, 100 mV, and 1 V. The idea is that during annealing, we will randomly select not just which variable to perturb, but which move-size with which to bound it. If we have 10 variables each with 5 different move sizes, we regard these as a palette of 50 unique move *types* from which to select. Once selected, we generate Δv_i as a uniform random number in the appropriately specified range, then compute $v_i^{new} = v_i^{old} + \Delta v_i$.

To select which type of move, we track how each move has contributed to the overall cost function at each temperature, following [9]. Suppose there are M unique move types. Suppose the mth move type is attempted K times at this temperature. Let $\phi_{ij} = 1$ if the jth attempt at the ith move type was accepted; $\phi_{ij} = 0$ otherwise. Let $\Delta cost_{ij}$ be the change in the cost function due to the jth attempt at the ith move type. Then we can define a quality factor

$$Q_m = (1/K) \sum_{n=1}^{K} |\Delta \text{cost}_{mn}| \cdot \phi_{mn} \tag{1}$$

for the mth move type that increases if this type of move is accepted frequently, or if it changes the value of the overall cost function appreciably. Note that $f_m = Q_m / \sum_{i=1}^{M} Q_i$ can then be regarded as the fraction of moves that should be dedicated to type m moves at the next temperature in the cooling schedule. Using this scheme, we automatically select the size of each move at each temperature. In particular, the moves devoted to the continuous operating point voltage variables automatically restrict themselves to reasonable ranges as the annealing proceeds.

In addition to these purely random, *undirected* moves, the LU factored small-signal matrix generated as a by-product of the AWE analysis provides a convenient method to augment the annealer's move-set with *directed* moves that follow the dc gradient. Similar to the theoretically-convergent continuous annealing strategy in [4], where successively smaller amounts of noise are used to perturb gradient-based moves, we incorporate gradient-based voltage steps as part of our palette of unique move types. Using the KCL-violating currents calculated at each node, $\Delta \underline{i}$, and the factored small-signal matrix, Y^{-1}, we can calculate the voltage steps as:

$$\Delta \underline{v} = \alpha (Y^{-1} \Delta \underline{i}) \tag{2}$$

where α is a scaling factor that bounds the range of the move. Since the factored matrix is already available, this calculation is very fast. We can also use this as the first step of a Newton-Raphson iterative dc solve. In practice, we augment our move-set with a collection of gradient steps of various sizes, full Newton-Raphson solves, and combinations of perturbations of other variables followed by gradient voltage steps, allowing the quality factors to determine the most appropriate perturbation–purely random, gradient-directed, or a combination–at a given point in the optimization process.

The heart of the annealer is the cost function used to evaluate each visited circuit configuration $\underline{x}$. Our cost function has the form:

$$\text{cost}(\underline{x}) = \underbrace{w_o C_{obj}}_{objective} + \underbrace{w_p C_{perf} + w_s C_{sat} + w_n C_{strinv} + w_k C_{KCL}}_{penalty\ terms} \tag{3}$$

where the w_i are appropriately chosen weights. Observe that there are two distinct kinds of terms: the *objective* term is to be minimized, while the *penalty* terms are to be driven to zero.

C_{obj} is the simplest term. It merely evaluates figures of merit for the circuit to be optimized. For example, to optimize for (active device) area and unity gain frequency (*UGF*), this term would be formulated as:

$$C_{obj} = \left(W_{Area} \sum_{devices\ d} W_d L_d \right) + W_{UGF} / (UGF) \tag{4}$$

where the W_i weigh the relative importance of area and unity gain frequency. In contrast, the penalty terms measure the distance from the current solution to a desired target performance. C_{perf} sums a penalty for each separate linear performance constraint, *i.e.*, the constraints evaluated by AWE. Recall that each move propagates though the device models to update our small-signal circuit model for AWE to analyze. AWE returns dc gain, UGF, and phase margin information. Hence, C_{obj} includes terms such as $(\text{gain}(\underline{x}) - G)$ to penalize the difference between the AWE-derived gain of the current circuit configuration $\underline{x}$ and the desired specification G. To be precise, it includes these terms just if the performance specification is not met: we attach no penalty to exceeding a performance specification. In this example, the gain penalty component of C_{perf} drops to zero and remains there when $\text{gain}(\underline{x}) > G$. The penalty terms C_{sat} and C_{strinv} are similarly formulated in terms of device operating point voltages to ensure that all MOS devices remain in the saturation region, and operate in strong inversion, respectively. (Note this is a very conservative choice on our part. There is, in fact, no compelling reason to demand strong inversion if one is using device models that are accurate under conditions of weak inversion.)

The most unusual component of the cost function is the C_{KCL} penalty term. If we were to require that KCL be satisfied after every move, we would have to resort to essentially a simulation-based synthesis scheme that completely solves for dc-correctness after every move. Instead, we relax this requirement, evaluate the KCL violation and penalize it. This term is formulated:

$$\sum_{\text{nodes}} \left(\sum_{\text{branches } b} \frac{I_b}{|I_b|} \right) \tag{5}$$

where the $|I_b|$ factor simply normalizes the sum at each node to avoid numerical problems with very small currents.

It is worth considering a seeming inconsistency in our overall synthesis formulation. We clearly cannot guarantee that each annealing move visits a physical (KCL correct) circuit. Nevertheless, we evaluate this probably nonphysical circuit using detailed device models and highly accurate AWE techniques. Exactly what are we evaluating here? The perhaps surprising answer is that there is considerable useful performance information in these nonphysical circuits. Although KCL may not be satisfied by these intermediate circuits during the annealing process, these circuits can be pictured as dc-correct circuits with an extra current source at each node sinking the KCL violating currents. The annealer can still learn much from these circuits. For example, if we need to achieve more gain in our circuit, we probably need to increase the g_m on some critical device. We do not need a precise dc bias point to know that we must either increase that device's width, its current, or both. We can successfully make coarse decisions such as this even in the presence of these extra current sources. During the annealing process, as the magnitudes of these current sources are forced to zero, the interactions between devices are more precisely modelled and our decisions become more accurate.

To achieve this, we cool the problem carefully, and choose the weights on each term of the cost function dynamically. We employ a standard adaptive cooling schedule to determine the initial temperature, temperature decrement rule, equilibrium conditions, and freezing criterion [8]. More interesting is the determination of the weights. We believe it is impossible to pick the weights w_i statically, independent of the problem being solved. Hence, we use a novel adaptive weighting scheme that adjusts weights dynamically during annealing. The two basic goals are to keep the magnitude of each term of the cost function roughly equivalent, and to keep the circuits considered near the feasible region, *i.e.*, the penalty terms near zero, to increase the amount of information obtained from each perturbation. For example, we attempt to strike a balance between the critical C_{KCL} and C_{obj} terms. If we weight KCL too heavily, we freeze too quickly into a region of feasible circuits with poor performance (under-optimized C_{obj} term). If instead we weight KCL too lightly, we escape the region of KCL-valid circuits and freeze in a region where the small signal models indicate high-performance, but, unfortunately, these models prove overly optimistic when the resultant circuit is simulated. We employ a novel feedback mechanism that modulates the weights so that each penalty term *tracks* the C_{obj} term. Interestingly, the scheme is essentially a variant of a PI controller from classical linear control theory. We begin by annealing at the initial hot temperature to sample the cost surface; we use the results to set the initial weights to render each cost term approximately equal. We then start the real annealing, and update the weights after every temperature. The weight update rule has three regimes. In the *attack* phase, the corresponding penalty term is larger than the C_{obj} term, and we carefully increase its weight to increase the importance of reducing it. At *crossover*, the C_{obj} term suddenly dominates this penalty term, and we relax the penalty to give the annealer freedom to explore the cost surface vigorously, to minimize C_{obj}. In the *realization* phase, the C_{obj} term has reached its minimum, and the minimization process can reduce only the cost of the penalty terms, eventually forcing them to zero. Roughly speaking, the attack phase directs the solution toward regions of almost-physical circuits that almost meet performance constraints; crossover defines the point at which we focus primarily on the minimizing the objective term; and the realization phase forces the remaining penalty terms to zero.

4 Experimental Results and Analysis

We have implemented this equation-free synthesis strategy in a pair of tools called ASTRX and OBLX. ASTRX is the compiler that analyzes an unsized analog circuit and generates C code to evaluate the cost function and manipulate the relevant design variables. OBLX is the annealing engine that actually solves all the circuit constraints generated by ASTRX. All our results were run on a DECstation 5000 workstation.

Figure 1 shows three circuit topologies used to exercise ASTRX. *DiffPair* is a simple MOS differential pair with an active load and represents a toy example. *2Stage* is a standard two-stage operational amplifier configuration. *FoldCas* is a novel high-performance folded cascode operational amplifier recently published in [16]; this is a particularly nice synthesis example because it is not a textbook design. Figure 1 also illustrates the results of the compilation process for the small *DiffPair* circuit.

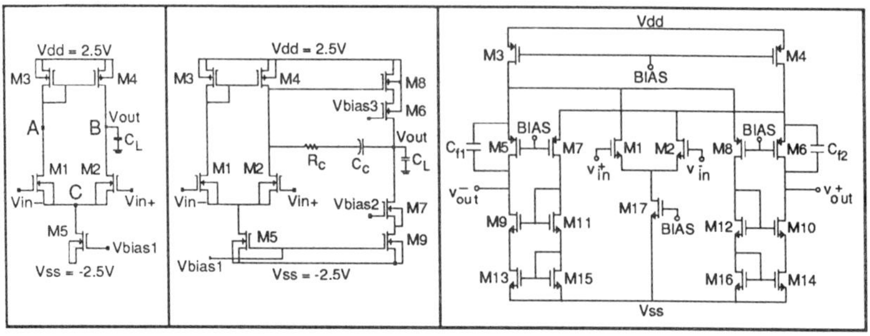

Ckt: ***DiffPair*** **Ckt:** ***2Stage*** **Ckt:** ***FoldCas***

Independent vars: $W_2, L_2, W_4, L_4, W_5, L_5, V_{bias1}, V_A, V_B, V_C$

KCL: $I_{ds1} + I_{ds2} - I_{ds5} = 0,\ I_{ds3} - I_{ds1} = 0,\ I_{ds4} - I_{ds2} = 0$

Small-signal circuit elements: 48

ASTRX Analysis Summary for Circuit *DiffPair*

Figure 1: Three Test Circuits and Example ASTRX Compilation

Specifications	*DiffPair* Proc: 2μ; Opt: UGF	*2Stage* Proc:2μ; Opt: Area	*2Stage* Proc: 1.2μ; Opt: Area
	Nominal: Predict/Simul	Nominal: Predict/Simul	Nominal: Predict/Simul
Gain (dB)	40: 39.5 / 39.5	86: 85 / 86	86: 86 / 88
UGF (MHz)	10: 38 / 39	10: 16 / 16	10: 33 / 32
Phase (deg)	45: 45 / 48	45: 45 / 45	45: 45 / 42
Area ($10^3\mu^2$)	18.3	4.5	1.6
Moves (10^3)	67	438	276
CPU (hrs)	0.75	11.5	4.7

Table 1: OBLX Synthesis Results for Circuits *DiffPair*, *2Stage*

FoldCas Specification	Proc: 2μ Optimize: area	Proc: 1.2μ Optimize: area	Proc: 1.2μ Opt: area, UGF	Proc: 1.2μ Opt: area, UGF
	Nominal: Predict/Simul	Nominal: Predict/Simul	Nominal: Predict/Simul	Nominal: Predict/Simul
Gain (dB)	83: 83 / 79	83: 84 / 80	83: 83 / 80	60: 67 / 69
UGF (MHz)	90: 92 / 96	90: 104 / 104	90: 132 / 134	90: 251 / 241
Phase (deg)	45: 92 / 92	45: 89 / 89	45: 88 / 88	45: 89 / 89
Area ($10^3\mu^2$)	43	12.2	22	9.6
Moves (10^3)	460	-	753	304
CPU (hrs)	7.2	5.7	9.8	5.5

Table 2: OBLX Synthesis Results for Circuit *FoldCas*

Tables 1 and 2 show final synthesis results generated by OBLX for the three circuits, in both 1.2μ and 2μ processes, using BSIM device models [20], for several different specifications. These examples were run directly through ASTRX and OBLX with no designer intervention. These results were generated using an early prototype of the tools that did not incorporate directed (dc gradient) moves– which accounts for the CPU time. The critical observations are that we are quite successful meeting both nominal specifications and optimization goals, and we achieve very close agreement between the result predicted during synthesis, and the actual simulation result (via HSPICE [15]).

Nominal Specification	Trial 1 Predict/Simul	Trial 2 Predict/Simul	Trial 3 Predict/Simul	Trial 4 Predict/Simul
Gain (dB): 83	84 / 86	84 / 80	86 / 87	84 / 84
UGF (MHz): 90	120 / 121	237 / 239	99 / 92	145 / 146
Phase (deg): 45	90 / 90	86 / 86	91 / 91	88 / 88
Area ($10^3\mu^2$)	7.0	9.1	5.5	6.6
Moves (10^3)	60	72	60	63
CPU (hrs)	1.0	1.2	1.0	1.0

Table 3: OBLX Results for Circuit *FoldCas* with Gradient Moves. Four Consecutive Trials; 1.2μ Process; Optimize Area & UGF

The result of adding directed gradient-based moves is illustrated in Table 3. For this table, the annealer automatically generated by ASTRX for the *FoldCas* circuit was augmented with a prototype version of the dc gradient-based moveset. Of particular note in Table 3 are the new run times: these have been reduced by a factor of nearly 10 over similar results in Table 2, with comparable performance. Table 3 was generated from four consecutive runs of OBLX using different random number seeds and represents typical results rather than a carefully selected set of "best" results. For these results, the final value of the objective cost had a standard deviation of less than 20% of the mean. However, note that as we should expect, OBLX was free to trade-off these performance constraints against each other since the objective function was the simple weighted sum of active area and (inverse) unity gain frequency. This accounts for the four very different circuits (area/bandwidth tradeoffs) OBLX found to meet these loose specifications.

Figure 2 offers some additional insight into the synthesis process for the *FoldCas* circuit. The evolution of a few critical components of the cost function is plotted as a function of cooling steps (*i.e.*, time) for one run. Note that the C_{KCL} term initially increases, until the automatic weight adjustment forces it to begin tracking C_{obj}. After the crossover point, around step 35, C_{obj} starts to decline appreciably as we explore for maximum performance in this region of almost-physical circuits. After C_{obj} finally freezes (around step 70), the remaining penalty terms quickly decline toward zero, and we reach a feasible, optimized circuit.

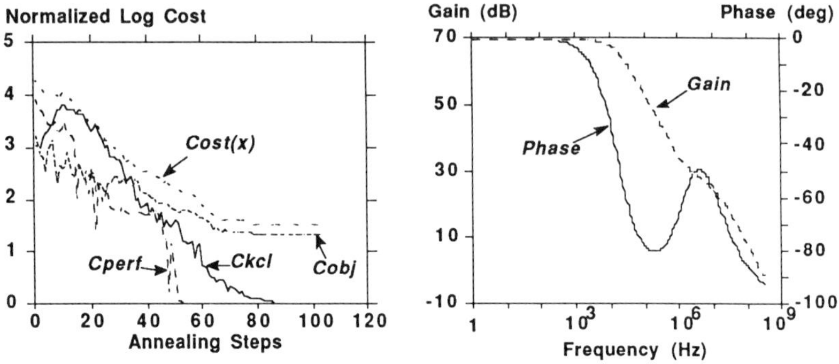

Figure 2: Annealing Dynamics (left), Frequency Response (right) for Synthesized Circuit *FoldCas*

The frequency response for a synthesized instance of *FoldCas* in Figure 2 clearly illustrates the advantages of using AWE in the inner loop of our annealer. AWE builds a new low-order approximation of the transfer function *H(s)* for *each* circuit visited during annealing. This allows us to handle complex custom circuits with unusual frequency response characteristics that would be challenging to model accurately, *a priori*, by means of a simple symbolic expression. For example, the *FoldCas* circuit is actually designed so that *six* poles contribute to its phase margin.

5 Conclusions

We have described a new synthesis strategy for linear analog circuits of cell-level complexity. The strategy relies on a relaxed formulation of dc biasing, annealing optimization, and AWE analysis of each circuit configuration visited during annealing. Preliminary results are encouraging: we can synthesize a new circuit, in a few hours, from nothing more than a netlist, a set of device models, and a set of linear performance specifications.

Our immediate focus is on establishing the robustness of our (evolving) tools across a broader spectrum of circuit benchmarks, and tuning for efficiency. Our longer term work focuses on attacking a broader set of critical performance specifications, notably transient specifications such as slew rate, and nonlinear constraints such as distortion. We are studying mechanisms for converging to multiple dc bias points simultaneously, in effect, synthesizing the circuit to respond to multiple input stimuli, from which we hope to approximate these nonlinearities and transients.

Acknowledgments

P. C. Maulik and Katsu Nakamura provided benchmark circuits, and insight on their behavior and synthesis formulation. Ron Rohrer, John Lee, and Richard

Trihy provided assistance with taming a version of AWE for our use in OBLX. We gratefully acknowledge their generous help. This work was funded by the National Science Foundation under grant MIP-8657369, the CMU-NSF Engineering Design Research Center, and the Semiconductor Research Corporation.

References

[1] G.L. Bilbro and W.E. Snyder, "Optimization of functions with many minima," *IEEE Trans. Systems, Man, Cyber.*, in press.

[2] R.K. Brayton, A.L. Sangiovanni-Vincentelli and G.D. Hachtel, "Multi-level logic synthesis," *Proc. IEEE*, vol. 78, no. 2, Feb. 1990.

[3] M. Degrauwe *et al.*, "Towards an analog system design environment," *IEEE JSSC*, vol. sc-24, no. 3, June 1989.

[4] S.B. Gelfand and S.K. Mitter, "Simulated Annealing Type Algorithms for Multivariate Optimization," *Algorithmica (1991), 6: 419-436.*

[5] G. Gielen, *et al.*, "Analog circuit design optimization based on symbolic simulation and simulated annealing," *IEEE JSSC*, vol. sc-25, no. 3, June 1990.

[6] G.D. Hachtel and R.A. Rohrer, "Techniques for the Optimal Design and Synthesis of Switching Circuits," *Proc. IEEE*, vol 55, No. 11, Nov. 1967.

[7] R. Harjani, R.A. Rutenbar and L.R. Carley, "OASYS: a framework for analog circuit synthesis," *IEEE Trans. CAD*, vol. 8, no. 12, Dec. 1989.

[8] M. Huang and F. Romeo and A. Sangiovanni-Vincentelli, "An efficient general cooling schedule for simulated annealing," *Proc. IEEE ICCAD*, Nov. 1986.

[9] S. Hustin and A. Sangiovanni-Vincentelli, "TIM, a new standard cell placement program based on the simulated annealing algorithm", presented at IEEE Physical Design Workshop on Placement and Floorplanning, Hilton Head, SC, April 1987.

[10] S. Kirkpatrick, C.D. Gelatt, M.P. Vecchi, "Optimization by simulated annealing," *Science*, vol. 220, no. 4598, 13 May 183.

[11] H.Y. Koh, C.H. Sequin, and P.R. Gray, "OPASYN: a compiler for MOS operational amplifiers," *IEEE Trans. CAD*, vol. 9, no. 2, Feb. 1990.

[12] E.S. Kuh and T. Ohtsuki, "Recent advances in VLSI layout," *Proc. IEEE*, vol. 78, no. 2, Feb. 1990.

[13] K.S. Kundert and A. Sangiovanni-Vincentelli, "Reply to: Comments on 'Simulation of nonlinear circuits in the frequency domain,'" *IEEE TCAD*, vol. 8, no. 8, Aug. 1989.

[14] P.C. Maulik and L.R. Carley, "Automating analog circuit design using constrained optimization techniques," *Proc. ACM/IEEE ICCAD*, Nov. 1991.

[15] Metasoft Corp. HSPICE manual, 1990.

[16] K. Nakamura and L.R. Carley, "A current-based positive-feedback technique for efficient cascode bootstrapping," *Proc. VLSI Circuits Symposium*, June 1991.

[17] W. Nye, *et al.*, "DELIGHT.SPICE: an optimization-based system for the design of intergrated circuits," *IEEE Trans. CAD*, vol. 7, no. 4, April 1988.

[18] L.T. Pillage and R.A. Rohrer, "Asymptotic waveform evaluation for timing analysis," *IEEE Trans. CAD*, vol. 9, no. 4, April 1990.

[19] G.W. Rhyne and M.B. Steer, "Comments on 'Simulation of nonlinear circuits in the frequency domain,'" *IEEE Trans. CAD*, vol. 8, no. 8, Aug. 1989.

[20] B. Sheu, *et al.*, "BSIM: Berkeley short-channel IGFET model for MOS transistors," IEEE JSSC, vol. sc-22, no. 4, Aug. 1987.

[21] D. Vanderbilt and G. Louie, "A Monte Carlo simulated annealing approach to optimization over continuous variables," *J. Comput. Phys.*, vol. 36, 1984.

Circuit Technology

A CMOS Parallel Adder Using Wave Pipelining

David Fan, C. Thomas Gray, William Farlow, Thomas Hughes
Wentai Liu, Ralph K. Cavin
Department of Electrical and Computer Engineering
North Carolina State University
Raleigh, NC 27695-7911, USA
fan@ececroc.ece.ncsu.edu
wentai@mcnc.org

Abstract

Wave pipelining (also known as maximal rate pipelining) is a timing methodology used in digital systems to increase the number of effective pipelined stages without increasing the number of physical registers in the pipeline. Using this technique, new data are applied to the inputs of a combinational block before the previous outputs are available thus effectively pipelining the combinational logic. Achieving a high degree of wave pipelining in CMOS technology requires careful study of delay balancing technique involving circuit design, layout method, and testing structure. We have designed a 16-bit parallel adder using wave pipelining. It is implemented with MOSIS 2 micron technology and initial test results of fabricated devices show more than six times speed up over non-pipelined operation.

1 Introduction

Wave pipelining (or maximal rate pipelining) is a timing methodology that can be used to pipeline combinational logic in digital systems without the use of physical registers. With conventional logic design, combinational logic between registers is allowed to fully stabilize before new input data is applied. In contrast, wave pipelined systems apply new inputs to combinational logic before the previous results are available. In this manner, the combinational logic is effectively pipelined with multiple coherent "waves" of data present at various stages in the logic. The clock period of such a system is bounded by the difference of the maximum and minimum delay through the logic. Thus, by balancing delay differences, the system can be clocked at a frequency much higher than that possible with a conventional design. For this reason, wave pipelining is especially suited for regular and parallel structures such as the parallel adder presented in this paper.

Previous design of wave pipelining systems has included the design of a floating point unit[1], an experimental computer[2], population counter[3], and FIFO system[4]. While the work described in [1] and [3] were based on bipolar ECL technologies, our work concentrates on the study of wave pipelining techniques in CMOS. In particular, this paper presents a CMOS parallel adder using wave pipelining.

In this paper, we address the various sources of delay imbalance in the parallel adder at the gate level, the circuit level, and at the detailed layout level. We also present the design of an on-chip test circuit for high speed testing the adder without the requirement for high speed input/output.

We first review the theoretical background of wave pipelining technique. Next, we present a modified parallel computation tree structure. We then describe the circuit structures and the basic circuit cell designs and explain our layout design approach through a discussion our balancing methodology and consideration of noise prevention at high frequencies. Next, we show a high speed circuit test methodology and the test circuits. Finally, we analyze the simulated performance of the wave pipelined adder layouts and point out their limitations.

2 Wave Pipelining Concepts and Timing Constraints

In order to fully demonstrate the design consideration and constraints of a wave pipelined system, it is necessary to understand the wave pipelining technique from a theoretical point of view. Previous work in this area includes [5, 6, 7, 8, 9, 3, 10]. In this section, we present highlights and main results from this work. For a more detailed study in this area, the reader is referred to [5].

In Figure 2.a, a conceptual view of wave pipelining is presented. As is shown in the block diagram, data enters the combinational logic through an input register and leaves through an output register. The output register is activated by the input clock delayed through a delay unit (Δ). The logic depth diagram[9] shows the flow of data through the combinational logic block versus time. After new data is latched into the input register, the data begins to flow through the combinational logic block. Due to path delays differences, process/temperature variation, etc., data will emerge from the combinational logic at different times. The minimal and maximal delay due to these variations are represented by t_{min} and t_{max} and are shown in the figure. The shaded area between the t_{max} and t_{min} lines depict regions of uncertainty in the data value. As the diagram indicates, outputs must be strobed outside these cones of uncertainty to insure proper operation.

As can be seen in the diagram, proper clocking can be insured only when data uncertainty cones do not overlap. Thus output data must be sampled after the latest data has arrived and before the earliest data from the next clock period has arrived. This gives rise to a set of constraints on the clock period (T_{clk}) represented by the constraint space diagram in Figure 2.b.

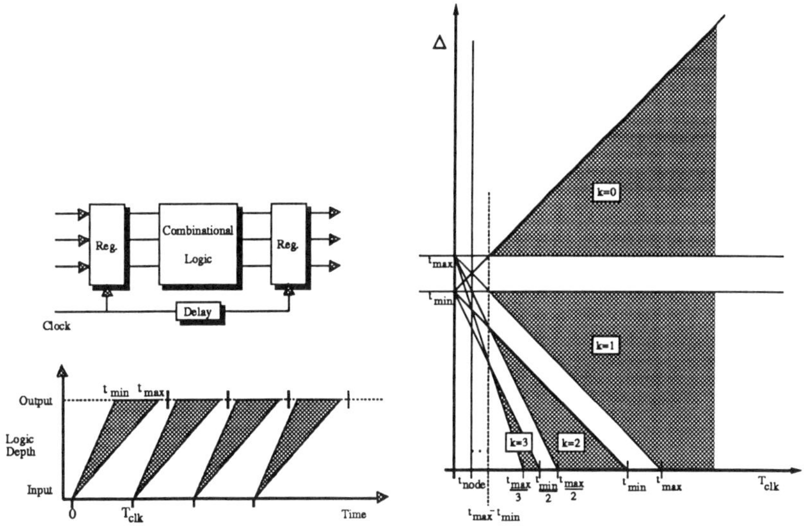

a. Wave Pipelined Operation **b.** Timing Constraint Space

Figure 1: Timing of Wave Pipelined Systems

It is obvious from figure 2 that the factor limiting reduction of T_{clk}, is the difference in the maximum and minimum delays (t_{max} -t_{min}). This is shown in the following equation:

$$T_{clk} >= (t_{max} - t_{min}) + \text{unintentional skew} + \text{setup/hold time} \qquad (1)$$

Refer to [5] for a mathematical derivation of this constraint.

From the above discussion, it is evident that the major objective in the design of maximum speed wave pipelined circuits, is the minimization of variation in delay within the circuit. These delay variations can occur from several different sources including delay differences due to path lengths, data dependency, static process and environmental variations, and dynamic process and environmental variations. Each of these sources of variation as they relate to the adder will be dealt with in subsequent sections of this paper.

The first source, path delay variation, is dealt with at the architectural, gate level, and circuit level to minimize differences in delays through various circuit path from input register to output register. The second source refers to the differences in gate delay for different input patterns. While this variation is minimal in ECL technologies[11], it is of much concern in CMOS technologies where standard static CMOS gates inherently have a factor of two difference in delay for different input patterns. In section 4, we introduce an alternate biased CMOS cross-coupled NAND gate that has much less data dependency.

The last two sources for delay variation arise from process variation and environmental variations. These sources are categorized into two major groups: *static variation* – deviation from nominal design delay (fixed at run-time) and *dynamic operational variation* – variation in delay at run-time.

The static component refers to the inevitable processing variations that occur between chips of different fabrication runs, between chips at different locations on the same wafer, and between different devices on the same chip. Slight variations on channel lengths of fabricated devices can produce variations in nominal delay of the logic gates which could cause phase misalignment at a latch. Also included in the static component are variations in operating environment such as ambient temperature and power supply. The static variations that have the most effect on a single chip system and thus for our adder are differences in p-channel devices versus n-channel devices after fabrication and random process variation distributed across the chip causing differences in path delays. Our biased CMOS gates can compensate for some variation in p-channel and n-channel devices by use of a bias voltage described in section 4.

The dynamic component is caused by the variation in gate delay induced by changes in the operating environment of the device. This would include changes in temperature due to changes in power dissipation which in turn produces dynamic variations in gate delay. For example, the amount of power will vary depending on whether the data causes many devices to switch or few devices to switch thus effecting data path delays. The dynamic component also includes the effect of random noise on signal and power lines. These variations must be dealt with through careful layout and detailed balancing as demonstrated in section 5.

3 Architecture for High Speed Addition

In this section, we present a modified parallel computation adder structure. It is well known[12] that binary addition can be transformed into a parallel computation by introducing an associative operator $\circ$. Brent and Kung[12] have proposed a VLSI design for fast addition based on this parallel approach which we have modified to save area and power. The new architecture contains half of the logic depth of the original Brent/Kung design although at a cost of larger drivers to drive higher fanout. The adder architecture and the modified carry computation structure are shown in Figures 2 and 3 respectively. Figure 3 also shows the internal logic of the four basic computational cells present in the carry structure.

This tree structure propagate term, p, for each bit position i, then the carry c for each bit obeys the following recurrence relation:

$$g_i = a_i b_i \tag{2}$$

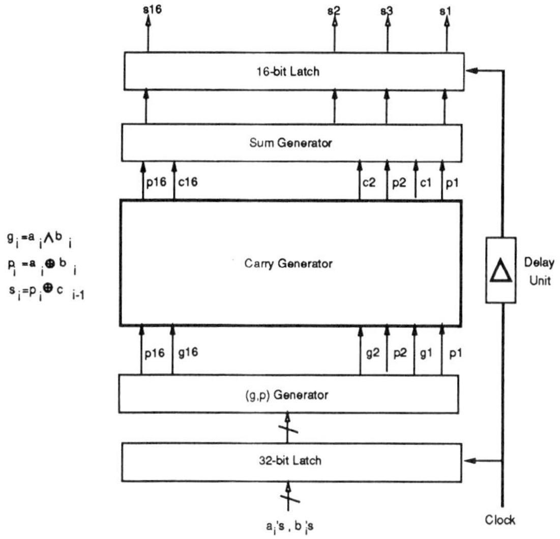

Figure 2: Block Diagram of Parallel Adder

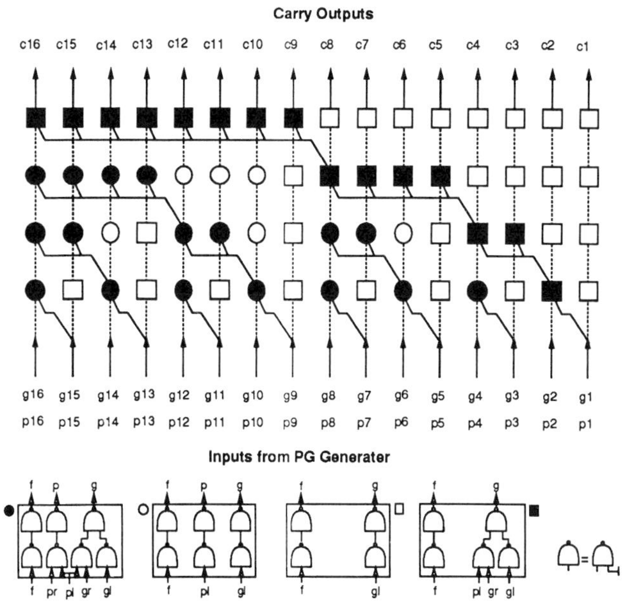

Figure 3: Computation of all the carries for n = 16

$$p_i = a_i \oplus b_i \tag{3}$$
$$c_i = G_i, \quad for\ i\ =\ 1, 2, ..., n \tag{4}$$

where

$$(G_i, P_i) = (g_i, p_i) \quad for\ i\ =\ 1 \tag{5}$$
$$(G_i, P_i) = (g_i, p_i) \circ (g_{i-1}, p_{i-1}) \quad for\ i\ >\ 1 \tag{6}$$

and $\circ$ is a concatenation operator defined as follows:

$$(g_l, p_l) \circ (g_r, p_r) = (g_l + p_l g_r, p_l p_r) \tag{7}$$

After the carry bit c is computed, the sum bit s is

$$s_i = p_i \oplus c_{i-1} \tag{8}$$
$$s_1 = p_1 \tag{9}$$

To implement the functions defined above, three circuit blocks shown in Figure 2 are required. The first circuit (labeled (g,p) generator) takes adder inputs a and b and generates the initial p and g in equations 2 and 3. The initial p and g are then fed into the carry generator to compute c (equation 4) and delay the initial p to be used in the final sum (equation 8). The third block is a sum circuit, consisting of a row of exclusive OR gates, which combine the c and p to produce the final sum.

The parallel carry computation is implemented with four basic cell types: black circle, black square, white circle, and white square cells. The terms "black" and "white" cells come from Brent and Kung and are shown in Figure 3. The black cell performs the associative concatenation and is the "computation" cell, and the white cells only copy inputs to outputs and are "padding elements". The f paths in the black and white cells are used to delay the initial p to be used by the SUM unit.

Based on our implementation, the black and white cells are of two categories, circle and square. The circle cell shown in Figure 3 produces p, g, and f, while the square cell need only produce g and f. Padding elements of NAND gates with one input tied to Vdd are used to balance path delay. Since every cell in the network uses the same gate, this approach provides an almost balanced delay path from adder inputs to outputs. These balanced paths through the computation and padding networks are essential to adder clock frequency maximization of wave pipelined circuits.

4 Components for Adder Circuit

The ideal gate for wave pipelining should have no delay variations on different input patterns thereby not contributing to the difference in maximal and minimal delays through the logic. Although CMOS technology is very attractive for this application in terms of cost and integration capabilities, conventional static CMOS has severe data dependent delay variation[13, 14] due to parallel transistor paths and, to a lesser extent, serial transistor paths. In addition, static CMOS has high input capacitance, large feedthrough current, and high dynamic temperature variation.

To overcome these problems, we use a biased CMOS gate. As shown in Figure 4, the pull-up paths of conventional static CMOS have been replaced by a single PMOS transistor with its gate connected to a bias voltage. This arrangement eliminates variation in delay associated with static CMOS parallel paths and give post process control of the rise-time versus fall-time of the gate to compensate for process variations. For the adder, we use a cross-coupled biased CMOS NAND gate as shown in Figure 4. This also eliminates some of the delay variation caused by the serial path. Simulation results show that the cross-coupled structure has substantially less delay variation than a standard static CMOS NAND gate. Since the number of transistors used to make a NAND gate has gone from four to five, there is a slight area penalty for using the new NAND gate.

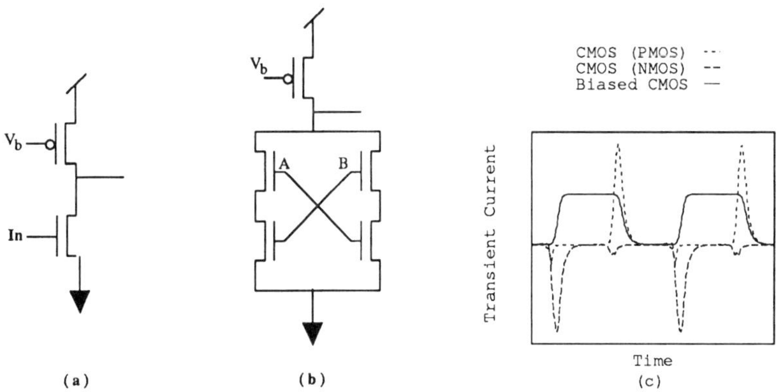

Figure 4: Components for adder circuit: (a) biased inverter (b) biased NAND gate (c) Transient current waveforms at 200 Mb/s

In addition to less delay variation and post-process control, the biased CMOS NAND has lower input capacitance since its inputs are connected only to NMOS transistors and, from simulation results, Figure 4c shows that the waveform of the charging/discharging current is smoother than in the static CMOS. This indicates that biased CMOS has less switching noise.

5 Layout Design

In the previous sections, we presented the design of the adder from a gate level and presented a circuit implementation of the basic gates. In this section we present the layout of the adder by describing the detailed physical balancing process, and by considering how cross talk and power supply noise

Table 1: Layout Statistics

Transistors (adder)	6067
Transistors (adder & tester)	15056
Adder Size	3508μm$\times$3782μm
Die Size	4600μm$\times$6800μm
Total IO Pins	106
Power/Ground Pins (adder)	16
Power/Ground Pins (tester)	27
Process	MOSIS 2μm PWELL
Package	108 pin PGA

affect the operation of the adder. The layout design is done using the MAGIC layout system and some layout statistics are summary in table 1.

5.1 Detail Balancing Methodology

As mentioned in section 2, the operating frequency of a wave pipelined circuit can be maximized by balancing path delay differences due to variation in path lengths, data dependencies in the gates, differences path loading, process variations, and environmental variations. The first two categories were discused in sections 3 and 4. At the architectural level path lengths are equalized by the addition of "white" cells on slack paths and, at the circuit level, biased CMOS NAND gates are used to minimize differences in delay due to data dependency in the gates. The last three categories of variation will be compensated for or minimized at the layout level in a detailed balancing process. Ideally, this consists of balancing each path such that paths delays are equalized over all variations in process and environment.

As is evident in Figure 3, even though paths are balanced such that each path from input to output consists of the same number of gates, substantial differences are present in the fanout of different gates and the amount of capacitive load due to interconnect. Thus, in order to balance delays, logic paths must be balanced at the detailed layout level.

In order to drive interconnects, we use four biased CMOS inverters as buffer between all computational cells. The widths and lengths of PMOS and NMOS devices are made equal to match the transfer characteristics of the cross-coupled NAND. The most area efficient way to compensate for difference fanouts and different loading capacitance at different bit position would be to minimize routing capacitance as much as possible and size drivers at different bit positions such that delays are equalized. While delays could be equalized using this method on a nominal process, fabrication variations on driver strength (e.g. due to variation in channel widths/length) versus capacitance values (due to variation in oxide thicknesses or width of interconnect) could cause paths balanced at design time to become unbalanced after fabrication. In short, this is due to the fact that process variations of the transistors will not in general track process variations on the interconnect. Thus, in order to design conservatively and provide for the maximal process tracking and thus minimal differences in path delays, we use the same buffers at each bit position at each level in the adder and balance routing such that each driver has the same loading capacitance.

Furthermore, to provide maximal tolerance of process variation in the loading capacitance, each routed line at a particular level in the adder has the same length and width of polysilicon, metal 1, metal 2, and number of vias. To equalize fanout, transistors with source and drain grounded are added to paths not having maximum fanout. Also additional coupling capacitance is added such that each interconnect line has the same amount of overlap, sidewall, and fringe capacitance. This included metal1 to metal2, metal1 to metal1, metal1 to poly, metal to substrate, etc. capacitances. Detailed balancing was done by hand using a geometrical approach. Thus, we did not have to rely on extremely accurate capacitance extraction and process parameter information to verify our balancing.

The area penalty for this type of balancing is substantial – approximately 40 percent. However, it provides for maximal tolerance of process variation and should provide information on the upper bound on clock frequency for wave pipelined design. It should be noted that with better information on process parameter distributions, better process control, and more exact parasitic extraction, it may be possible to relax the balancing requirements and trade off process invariance for chip area.

5.2 Reducing Cross Talk

The potential of cross talk can affect the operation of the adder in two ways. First coupling between signal lines can effectively change propagation times through the interconnect and, second, coupling between signal lines and the routed bias line (for the biased CMOS circuits) can induce variations in bias voltage applied to the gates and thus affect the delays of these circuits.

We minimize coupling between signal lines by avoiding routing signal lines in parallel in close proximity as much as possible and by minimizing the number of times a signal line crosses another signal line. Where we must route two signals in parallel for long distances, we separate the lines by placing a ground line between.

In order to minimize induced noise on the bias line (Vb), we placing a shield plane above and below the bias line. The Vb line is routed in metal 1 with metal 2 above and polysilicon below electrically connected to a quiet shield line. This forms a shield wall to protect the Vb line from other noise sources. Figure 5 shows the cross-section of this shielded structure.

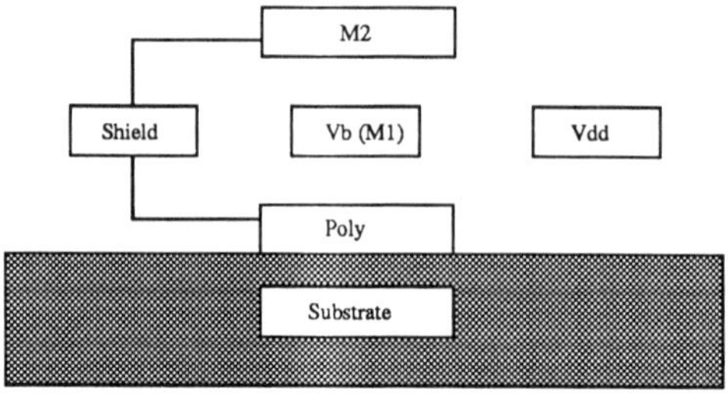

Figure 5: A conceptual cross-section of the cross talk prevention structure

5.3 Reducing Power Supply level Fluctuations

In addition to dynamic variations due to signal line cross talk and bias voltage variation, dynamic power supply level fluctuations due to simultaneous switching need to be taken into consideration in delay calculations. The worst case on-chip power supply noise is generated at the edge of a clock cycle when many latches and logic gates are simultaneously activated thus increasing chip power supply current.

Simultaneous switching noise can affect the circuit in two ways. First, simultaneous switching noise may increase delays since current spikes on power supply lines will induce voltage drops along the lines thus reducing the output current of all active drivers. Second, the noise pulse may cause logic malfunction.

For these reasons, we have inserted many decoupling capacitors between power and ground pairs. These nonswitching capacitors can share their charge with the switching capacitors and prevent on-chip power levels from collapsing. Also, to minimize the effective inductance of the return path connection off-chip and the resulting noise, many power/ground pins are supplied.

Lastly, to minimize power level fluctuation induced by the large clock buffer of the test circuitry (described in the next section), the power supply for the test circuitry is independent of the power supply for the adder circuitry and the adder is electrically isolated from the surrounding test circuitry with a substrate guard ring that encloses the adder. In this way, the adder power supply and substrate are effectively isolated from the effects of the test circuitry.

6 High Speed Testing of the Wave Pipelined Adder

While slow speed testing will find gross faults in the adder, wave pipelining requires high speed testing/tuning to assure full speed functionality. To find the two sided timing constraints, the adder is run at speed with worst case vectors while the skew between the input and output clocks is varied. Due to a lack of very high speed test equipment and the difficulties of high speed I/O in 2 micron CMOS we chose to move much of the testing capabilities to the chip. To accomplish this we implemented an 8 to 1 multiplexer for each input bit and 1 to 8 demultiplexer for each output of the adder. The input of the mux and the output of the demux are RAMs that allow the data I/O of the chip to be asynchronous. Thus, the chip only requires two high speed clock inputs and no high speed outputs. Figure 6 conceptually shows the main blocks of the chip. We choose the ratio 8 to 1 so that we can have enough independent data to fill the pipeline before repeating the same input sequence. As long as the adders speed up is less than eight it is possible to fill the pipeline with noncyclic data.

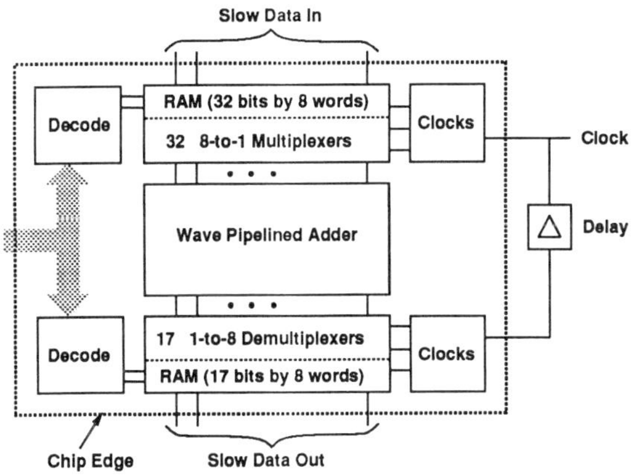

Figure 6: Block Diagram of the Chip

6.1 Clock generation

The clock generation is accomplished with a tree of toggle flip flops. The main cell takes a single phase clock and produces a delayed version of itself along with two divide-by-two clocks lined up with the edges of delayed clock. One clock at half frequency aligns with the positive edge of the delayed clock and the other half frequency clock aligns with the negative edge. This is accomplished by balancing delays. The clocks, c1, c2_1, c2_2, c4_1, c4_2, c4_3, and c4_4 of Figure 7.a are generated by the circuit shown in Figure 7.b This figure also shows how these signals are used by the multiplexer to generate high speed serial data. Note how d0 is passed to the output. From this timing diagram it is clear that there is about one pulse width of c1 skew tolerance between c2_2 and c4_1 in both directions. However there is no skew tolerance between c2_2 and c1 in the c1-late direction. We introduced a programmable delay line to compensate for the lack of skew tolerance in the highest speed clock. Figure 7.b shows how C1 is affected by the programmable delay line.

Two clock inputs are provided to the chip so the skew between the adder input and output can be set off chip. Differential inputs are provided for the two clocks. Duty cycle can be controlled by adjusting the reference voltage input of the differential amplifier or by adjusting the edge placement of the clock generated off chip.

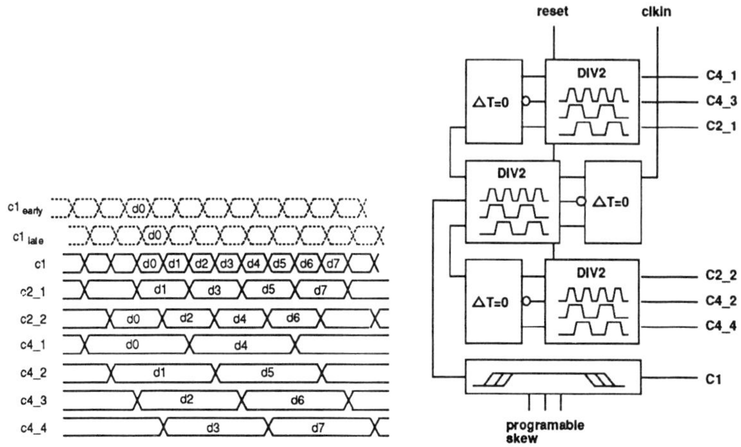

a. Timing Diagram of Clocks **b.** Clock Generation

Figure 7: Timing of the Multiplexer Circuits

6.2 8 to 1 Multiplexer

The multiplexer, shown in Figure 8, accepts eight parallel bits from a RAM and time multiplexes them into a serial data stream. One bit is output for each transition of the clock thus achieving a bit rate twice the clock frequency. That is, for a 100MHz clock, we have 200Mbit/s data. The circuit operates by steering data through a tree structure of 2 to 1 multiplexers. The control consists of the 7 clocks generated on chip.

6.3 1 to 8 Demultiplexer

The demultiplexer (Figure 8) is almost the same circuit As the multiplexer; of course the data direction is reversed. A less obvious difference is the RAM cell. The 8 outputs of demultiplexer are skewed in time by the pulse width of c1(PW_{c1}). This requires that the simple RAM cell of the multiplexer be replaced by a clocked register that allows only latched data to be written into the cell. Since the output of the multiplexer is periodic with period

$8 \cdot PW_{c1}$, every eighth cycle the same data gets written to a given word of the demultiplexer's RAM. A global line labeled latch in Figure 8 allows data acquisition to be started or stopped at any time. The pull-down output of the RAM avoids the possibility of flipping the data during a read. The same clock generation circuitry is used for controlling the demultiplexer.

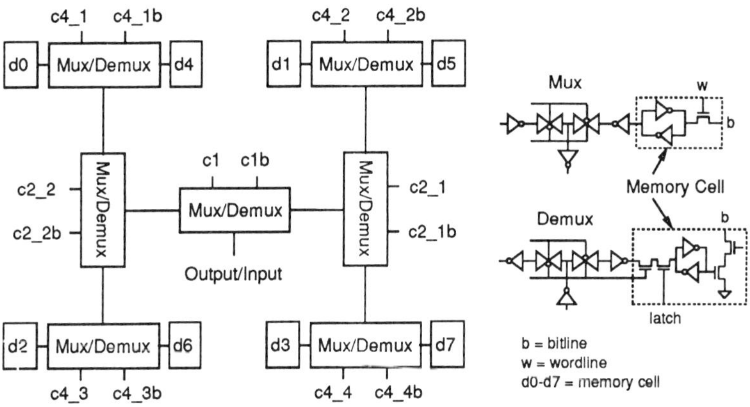

Figure 8: Multiplexer/Demultiplexer Block Diagram

7 Simulation and Test Results

In order to verify our design, we have performed intensive circuit level simulations using the SPICE-like simulator CAzM from MCNC[15]. Our simulations have included maximum frequency simulations across worst/best case process parameters, sensitivity analysis to determine sensitivity to different process parameters, and an analysis of transient power consumption. In this section we present a summary of the results obtained from these simulations. A more detailed presentation of simulation results can be found in [16].

Since we have done detailed balancing of logic paths, routing capacitance, and fanout at each level in the adder the major source of variation in delay is due to the data dependency of the cross-coupled NAND. For a 16 bit adder there exist 2^{32} input patterns. Therefore it is necessary to determine which input vectors cause worst case data dependency variation in delay at the outputs of the adder. By identifying which data patterns give worst case delay variation at individual gates and backtracking to primary inputs, worst case vectors were determined and used in all simulations to determine maximum operating frequency.

Figure 9 shows results of simulations using worst case vectors and worst/best case process parameter models. Maximum frequencies are plotted versus V_{bias} at 100°C (a), supply voltage at 100°C (b), and temperature at V_{bias} =

0.3V and supply voltage = 5V (c). The test chips are expected to fall within the envelope of worst case conditions.

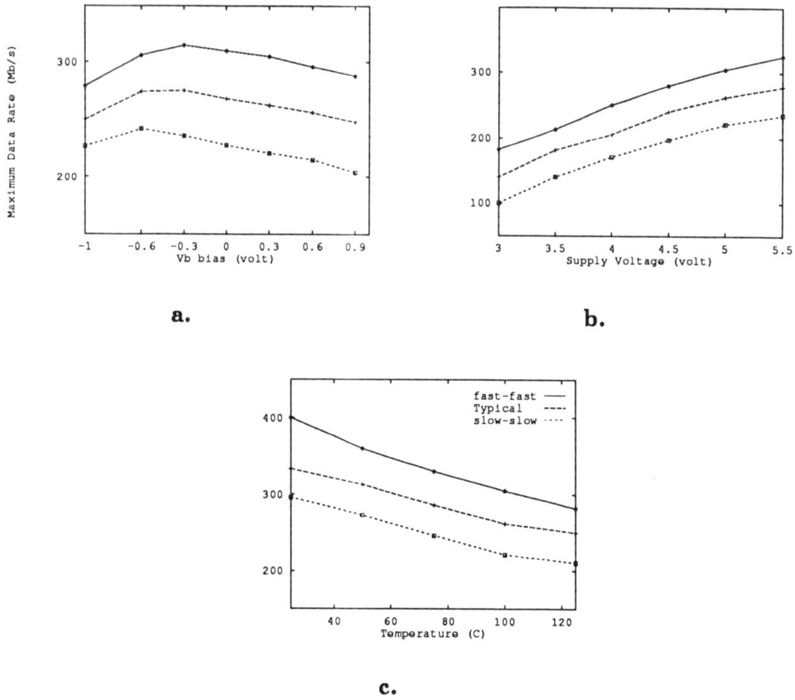

Figure 9: Maximum data rates versus a. Vb biases, b. temperatures, c. supply voltages

Figure 10 depicts overlayed output waveforms for all 16 outputs. The transition (at time 25) from the first vector to the second vector represents the lastest rise time of an output and the transition (at time 30) represents the earliest fall time. The worst case output delay variation is 1.2 ns.

In order to find the parameters most sensitive to this wave pipelining design, we performed a serial sensitivity analysis. Three figures of merit were measured from simulation data: t_{max} - t_{min} , power, and t_{max} . We have found T_{ox}, Δ L, Δ W, and K_p to be the most sensitive set for the three figures of merit.

It is well known that interwafer delay variations are large[17] due to variations in these sensitive parameters. However, Glasser[18] notes that T_{ox}, V_t, Δ L, and Δ W can be treated as constant within a single chip for modern CMOS technology. This implies that process/device variation among PMOS (or NMOS) devices on-chip are relatively small although potentially different

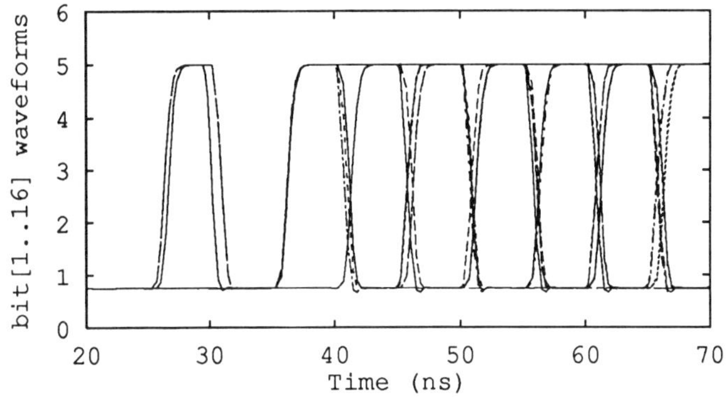

Figure 10: Typical overlapped 16 bit waveforms from the adder output

from nominal values. This makes our post fabrication tuning technique feasible in that we can adjust the bias to compensate for differences in relative strengths of PMOS and NMOS devices.

An analysis of the transient power consumption is required to understand the potential on-chip temperature variation. Applying random test vectors, simulations show almost constant power dissipation with the observed maximum variation in power being around five percent. Since, the transient time between power fluctuations is much shorter than the thermal time constant of modern IC packages, we expect that the chip temperature will remain relatively constant thereby not contributing to further variation in circuit parameters.

By performing sensitivity analysis, worst, best and nominal case transient simulation, and random input sequence simulation we have centered the design in a safe operating region, verified the operating and expected maximal operational data rate, and have verified the detailed balancing of interconnect and fanout.

We have received 20 chips back from MOSIS and have begun testing. From initial test data using a HP 82000 tester, we have measured computation rates in excess of 250 million operations per second yielding a speedup over non-pipelined operation of greater than six. Currently we are completing full functional tests and characterizing delay variation, jitter, and maximum operating speed of all 20 chips.

8 Conclusion

Using the wave pipelining concept, we have designed a 16 bit adder that is capable of a speedup of more than six relative to a non-pipelined design but has no internal physical registers. This lack of internal registers can greatly

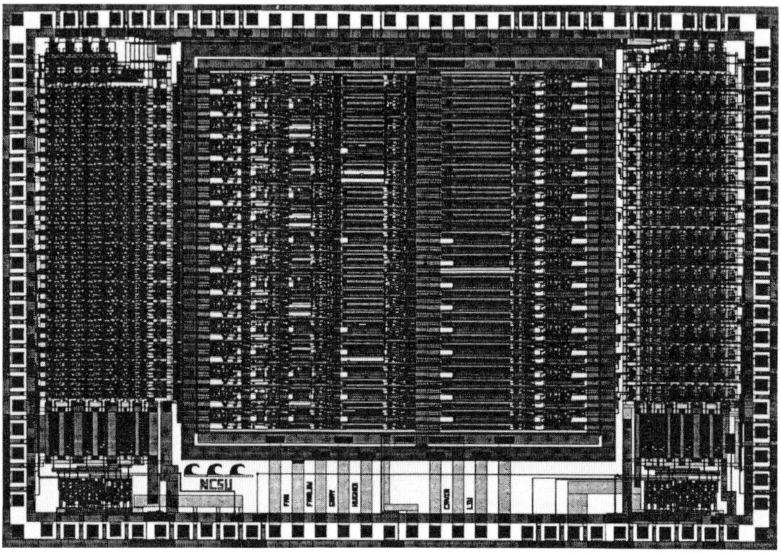

Figure 11: Layout of Parallel Adder

simplify clock distribution and skew problems while also reducing pipeline latency. To achieve this amount of wave pipelining we have developed a biased-CMOS cross-coupled NAND gate with minimal data dependence, and we have performed a custom layout (Figure 11) using geometrical balancing to assure process insensitivity. Through extensive circuit level simulation, we have verified both the functionality and the maximal operational speed of the adder. By placing the high speed tester circuits on chip, we maximized our chances of being able to thoroughly test the adder at speed by avoiding the difficulty of very high speed chip I/O. We submitted our adder design for fabrication by MOSIS and have received 20 chips back for testing. Initial test results indicate speeds in excess of 250 million additions per second with a speedup over non-pipelined operation of six or more.

Throughout the design process, it was apparent that better tools for design and analysis of wave pipelined systems and high speed systems in general are a necessity. For example, better extraction tools allowing for extraction of distributed RC nets for power, clock, and signal nets would aid in more accurate balancing of routing and potentially less conservative power and clock routing. Also, tools to find paths with worst case data dependent variation are needed. Concerning process and design, area penalties could be much less significant with tighter fabrication process control by relaxing balancing constraints and additional work is needed with high-speed I/O. In addition, more work is needed for yield optimization of wave pipelined circuits under the constraint of minimum delay variation. In conclusion, with commonly available tools, we have found wave pipeline design to be challenging but manageable. However, we expect wave pipelined design to become much easier and worth the additional design effort as more sophisticated CAD tools are developed.

Acknowledgements

David Fan was supported by Analog Devices Fellowship. Thomas Gray and Thomas Hughes were supported by Semiconductor Research Corp. fellowships. William Farlow was supported by MCNC Center for Microelectronics. Prof. Wentai Liu received support from NSF under NSF-MIP-882142. The authors would also like to thank Carson Stuart, Wayne Dettloff, Subhash Roy, and Fred Heaton of MCNC for participation in a design review of the adder chip and MCNC for the use of the HP tester.

References

[1] S. Anderson, J. Earle, R. Goldschmidt, and D. Powers. The IBM system/360 model 91 floating point execution unit. *IBM Journal of Research and Development*, pages 34–53, January 1967.

[2] L. Qi and X. Peisu. The design and implementation of a very fast experimental pipelining computer. *J. of Computer Science and Technology*, 3(1):1–6, 1988.

[3] D. Wong. *Techniques for Designing High-Performance Digital Circuits using Wave Pipelining.* PhD thesis, Stanford University, September 1991.

[4] C.T. Gray, T. Hughes, D. Fan, G. Moyer, W. Liu, and R. Cavin. A high speed CMOS FIFO using wave pipelining. Technical Report NCSU-VLSI-91-01, North Carolina State University, 1991.

[5] T. Gray, T. Hughes, S. Arora W. Farlow, D. Fan, W. Liu, and R. Cavin. Theoretical and practical issues in CMOS wave pipelining. Technical Report NCSU-VLSI-91-02, North Carolina State University, 1991.

[6] L. W. Cotten. Maximum-rate pipeline systems. In *1969 AFIPS Proc. Spring Joint Computer Conf.*, pages 581–586, 1969.

[7] P. Kogge. *The Architecture of Pipelined Computers.* McGraw-Hill, 1981.

[8] B. Fawcett. Maximal clocking rates for pipelined digital systems. Master's thesis, University of Illinois, 1976.

[9] B. Ekroot. *Optimization of Pipelined Processors by Insertion of Combinational Logic Delay*. PhD thesis, Stanford University, 1987.

[10] F. Klass and J. M. Mulder. CMOS implementation of wave pipelining. Technical Report 1-68340-44(1990)02, Delft University of Technology, December 1990.

[11] D. Wong, G. De Micheli, and M. Flynn. Designing high-performance digital circuits using wave pipelining. In *Proc. of VLSI '89*, Munich, W. Germany, August 1989.

[12] R. P. Brent and H. T. Kung. A regular layout for parallel adders. *IEEE Trans. on Computer*, C-31(3), March 1982.

[13] D. Wong, G. DeMicheli, and M. Flynn. Inserting active delay elements to achieve wave pipelining. Technical Report CSL-TR-89-386, Computer System Laboratory, Stanford University, 1989.

[14] Sanjay Arora. Design of a pseudo NMOS wave pipelined adder. Master's thesis, North Carolina State University, 1991.

[15] D. Rose D. Erdman and G. Nifong. CAzM: Circuit analyzer with macromodeling user's guide. Technical report, MCNC, June 1990.

[16] D. Fan, C. T. Gray, W. J. Farlow, T. Hughes, W. Liu, and R. K. Cavin. A CMOS parallel adder using wave pipelining. Technical Report NCSU-μSYS-91-03, North Carolina State University, 1991. In Preparation.

[17] M. G. Buehler, B. R. Blaes, H. R. Sayah, and U. Lieneweg. Parameter distributions for complex VLSI circuits. In *Proc. of VLSI '89*, Munich, W. Germany, August 1989.

[18] L. A. Glasser and D. W. Dobberpuhl. *The Design and Analysis of VLSI Circuits*. Addison-Wesley, 1985.

Using Qualified Clocks in the NORA Clocking Methodology to Implement a Systolic Queue Design

Susan R. Dickey
Richard Kenner
Ultracomputer Research Laboratory
Courant Institute of Mathematical Sciences
New York University
715 Broadway, 10th Floor
New York, NY 10003
dickey@nyu.edu, kenner@nyu.edu

Abstract

A systolic queue design has been implemented with thc thc NORA clocking methodology using qualified clocks. NORA allows the use of compact CMOS circuits with high tolerance for clock skew. Qualified clocks provide a natural way to implement local data movement in a systolic design. Their use with NORA, however, involves certain complications.

A circuit to produce a qualified clock for use in the NORA methodology is presented, and its maintenance of NORA assumptions is proved. Charge-sharing and noise problems which can arise are detailed. The performance of chips we have fabricated using this methodology is described.

1 Introduction

A variant of the systolic queue designs developed by Guibas and Liang [6] has been used to implement a combining processor-to-memory interconnection network for the NYU Ultracomputer [4]. Our first implementation [2] was in nMOS, using dynamic shift registers controlled by qualified clocks with two-phase clocking as presented in Mead and Conway [10]. When we moved to CMOS to obtain a low-power implementation, we wanted to keep the compactness of this dynamic shift register implementation.

In this introduction we first summarize the characteristics of this variant of systolic queue design that are relevant to implementation with qualified clocks and give a description of the original nMOS implementation. Then we describe the use of clocked CMOS (C^2MOS) latches to convert Mead and Conway's dynamic shift register implementation to CMOS with the NORA (No Race) clocking methodology [3].

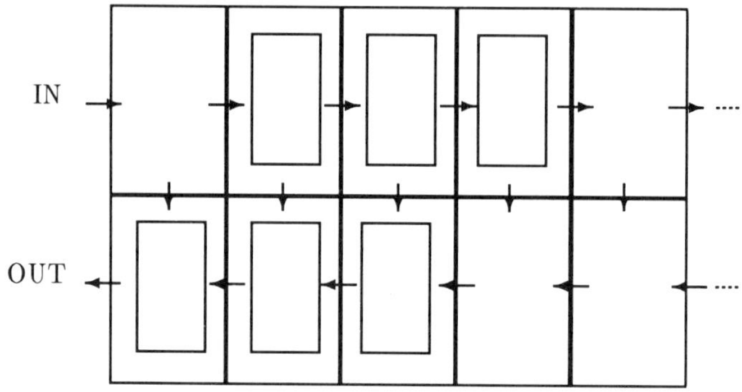

Figure 1: A systolic queue design.

1.1 Systolic Queue Design

The systolic queue design we have implemented is essentially the folded shift register shown in Figure 1. Items enter the IN row at the left and move right until the slot underneath, in the OUT row, is empty. As items are deleted from the front of the queue, items in the OUT row move to the left.

Guibas and Liang's design [6] is fully systolic, according to Leiserson's definition [9], requiring no global signals except a clock. In their design, as an item is deleted from the front of the OUT row, the cell the item has vacated is left "indisposed" instead of empty. Any item immediately to the right in the OUT row senses that it may move left on the next cycle, but an item in the IN row directly above will not move down. In general the OUT row will consist of a string of occupied and indisposed cells, terminated by a string of empty cells to the right. An item can be deleted or inserted every second cycle. Special cells must be used to handle the end conditions at the right of each row.

Our design is not fully systolic. We distribute global information about whether the queue is blocked to each cell in order to be able to delete an item the cycle immediately after it is inserted. If the queue is not blocked, a deletion will always occur and all items in the OUT row simultaneously move left. To handle the problem of finite queue size, we also distribute global information about whether the queue is full. This information determines whether the items in the IN row move right or retain their position.

Both designs, however, use the same local movements of data, transferring items either horizontally to neighboring cells within a row or vertically from the IN row to the OUT row. In a VLSI implementation, the control logic will usually not be repeated for each bit of a 32-bit data item. Instead, the information that controls data movement will be distributed as signals "locally global" to an array of 32 data cells. An identical implementation of

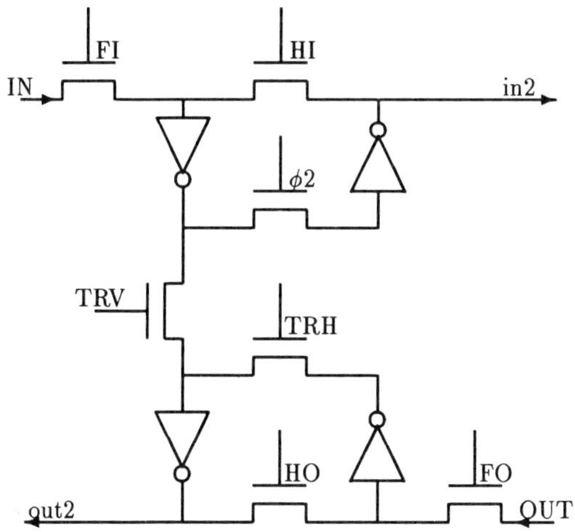

Figure 2: An nMOS implementation of the data cell of a systolic queue.

the single-bit data cell of the queue can be used for our design and for the fully systolic design, even though the control logic is different.

1.2 nMOS Implementation

In our original nMOS implementation, we controlled movement in the single-bit data cell with qualified clocks derived from two-phase non-overlapping clock signals, as is done in Mead and Conway's stack design [10]. Data movement within a row (IN or OUT) occurs on phase 1 (when clock signal $\phi 1$ is high); data transfer from IN to OUT occurs on phase 2 (when clock signal $\phi 2$ is high). Pass transistors control the horizontal and vertical data movement from cell to cell in the data path (Figure 2). During the periods when both clocks are low, all pass transistors are turned off. The decision as to which direction data will move in a phase is computed in the previous phase; thus data movement and control computations are completely overlapped.

In Figure 2, IN and OUT are input signals from the adjacent cells; in2 and out2 are output signals. The qualified clocks FI ("Flow In") and HI ("Hold In") are low whenever $\phi 1$ is low, and exactly one of them is high whenever $\phi 1$ is high. When FI is asserted, items in the IN row move to the right; when HI is asserted, items in the IN row stay in place. FO and HO are defined similarly for the OUT row. TRV ("Transfer Vertical") and TRH ("Transfer Horizontal") are a pair of $\phi 2$-qualified clocks controlling whether out2 is set from the IN or the OUT row; the pass transistor with gate $\phi 2$ is used to transfer the phase 1 value to in2.

A direct mapping of this design to CMOS transforms the pass transistors

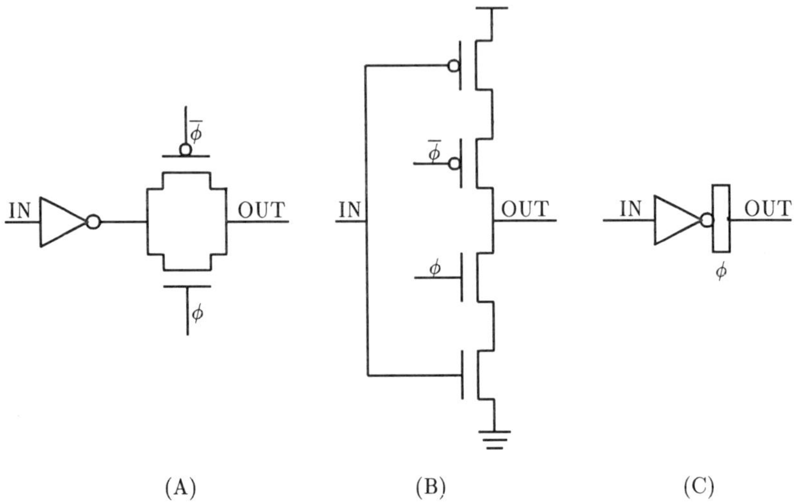

Figure 3: (A) Inverter and transmission gate. (B) C^2MOS latch. (C) Notation for C^2MOS latch.

to transmission gates, as in Figure 3(A). However, four basic clocks ($\phi 1$, $\overline{\phi 1}$, $\phi 2$ and $\overline{\phi 2}$) must be provided in order to have non-overlapping clock phases. As in nMOS, the non-overlap time must be guaranteed in the presence of clock skew, so the requirement of non-overlapping clock phases can result in considerable dead time in each clock cycle. In CMOS this problem is compounded by having four clock signals to distribute.

1.3 NORA Methodology

The C^2MOS latch (see Figure 3) can be used to replace an inverter followed by a pass transistor in the nMOS implementation. The key difference between the C^2MOS latch and the inverter followed by a transmission gate is that with the latch only one clock controls each transition of the output node, i.e., in Figure 3(B), OUT can be driven low only if ϕ goes high, and it can be driven high only if $\overline{\phi}$ goes low. This is true for every latch: the output of a latch can only be driven low when some clock is driven high and vice versa.

The NORA methodology described in [3] uses this latch to construct two-phase pipelined circuits that use only two basic clocks, ϕ and $\overline{\phi}$. In this methodology, *phase 1* corresponds to the time when ϕ is high and $\overline{\phi}$ is low, and *phase 2* corresponds to the time when $\overline{\phi}$ is high and ϕ is low. Therefore, in a standard NORA system, a phase 1 latch has ϕ on its N transistor and

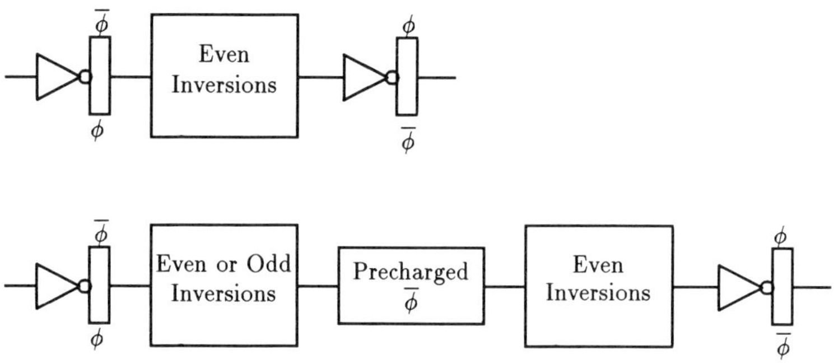

Figure 4: Summary of NORA inversion parity restrictions.

$\overline{\phi}$ on its P transistor; a phase 2 latch has the reverse. Restrictions on the parity of inversions between latches prevent data from flowing through a pair of latches during overlap periods when both ϕ and $\overline{\phi}$ have the same value.

The key to a properly-functioning NORA circuit is what happens during this overlap period. Consider the trivial circuit of one latch driving another latch and the case when both clocks are low. In that case, all the N clock transistors are off and all the P clock transistors are on; there is no path from ground to any output. During this overlap, the only transition that the output of the first latch can make is to go from low to high. That transition will not affect the output of the second latch because only an N transistor can be turned on and the path from the output to ground has been turned off by the low clock.

Generalizing, it can be seen that a signal that is an even number of inversions from the output of a latch on one clock phase can be safely connected to the input of a latch on the opposite clock phase. Similarly, a signal that is an even number of inversions from the output of an n-type precharged gate controlled by $\overline{\phi}$ cannot cause a transition on the output of a phase 2 latch. The use of precharged gates allows inverting as well as non-inverting logic between latches.

Goncalves and DeMan [3] developed rules for both n-type and p-type precharged logic blocks, giving added flexibility. As pointed out in Shoji [11] and discussed in Section 3.1, this kind of circuit is extremely sensitive to noise. We use only n-type precharged logic, connecting successive n-type precharged gates with static inverters, in the style of Domino CMOS [8].

Figure 4 summarizes the NORA inversion parity rules developed in [3] that are used in our design style. There must be an even number of inverting gates between two C^2MOS latches on different phases if only static gates are used. If either a single precharged gate or a series of precharged gates connected by Domino-style static inverters is used, there must be an even number of inverting gates after the last precharged gate.

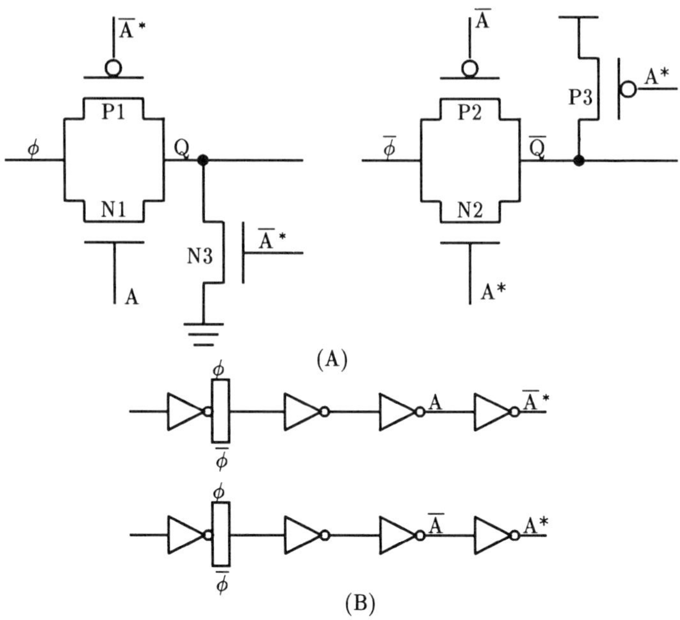

Figure 5: (A) Qualified clock circuits. (B) Parity from a latch of clock qualifier signals in NORA.

2 Qualified Clocking in NORA

In this section we present a circuit commonly used for clock qualification in CMOS logic and show how to modify it to work in a NORA environment.

2.1 Basic Clock Qualification Circuit

A circuit commonly used for producing qualified clocks is shown in Figure 5(A). ϕ and $\overline{\phi}$ are a pair of clocks denoting a particular clock phase. A, $\overline{\text{A}}$, A*, and $\overline{\text{A}}^*$ are the qualification inputs, with the latter two logically equivalent to the former two. Q and $\overline{\text{Q}}$ are the output qualified clocks, which are active only during the clock phase controlled by ϕ and when A is active. Q will be connected to the gate of an n-transistor in a latch and $\overline{\text{Q}}$ will be connected to the gate of a p-transistor.

This circuit is often used in preference to static logic to reduce the propagation delay from the clock to the qualified clock. Such a delay is a component of the clock skew of the system, which should be reduced as much as possible.

The operation of this circuit in an environment with non-overlapping clock phases is straightforward. In such an environment, A and $\overline{\text{A}}$ are assumed to be stable during the clock phase that is being qualified. Therefore, Q and $\overline{\text{Q}}$ are always deasserted by their respective input clocks being deasserted. Thus, the two transistors N3 and P3 serve no logical function. When this circuit is used with non-overlapping clocks, these transistors are included as bleeders to ensure that Q and $\overline{\text{Q}}$ remain quiet, especially if there is a long period between assertions of A.

2.2 Controlling Clock Qualification in NORA

The same clock qualification circuit can be used in NORA CMOS. In that case, the signals ϕ and $\overline{\phi}$ represent the two clocks used in a NORA design rather than a pair of clocks representing a single clock phase. A correct NORA clock qualification circuit must produce logically correct Q and $\overline{\text{Q}}$ signals during the two valid clock phases. During the overlap periods, it must not allow Q or $\overline{\text{Q}}$ to change from an inactive to an active state; i.e., it must produce behavior on Q and $\overline{\text{Q}}$ that is no worse than that of ϕ and $\overline{\phi}$.

For NORA logic, it is important to distinguish signals that are logically identical, but are a different parity (odd or even) of inversions away from a latch, or are produced from the output of precharged gates. Our convention is to mark signals that are an odd number of inversions from a latch with a "*"; signals not so marked are an even number of inversions away from a latch. Although it is possible to use precharged gates between the $\overline{\phi}$ latch and one or more of the qualifying signals, doing so would introduce delays which would defeat much of the advantage of using a clock-qualification scheme in the first place. Therefore, only fully-static logic is used between the latch and the clock qualifier circuit. We show below that the signals with parity as illustrated in Figure 5(B) are the ones required for correct operation of the clock qualification circuit.

To study the operation of this circuit in a NORA environment, we must consider all four possible states of the clock: phase 1, phase 2 and the two clock overlap phases when ϕ and $\overline{\phi}$ are both one or both zero. Instead of analyzing transitions of clock lines themselves, we assume that all signals have a positive propagation delay from a clock and analyze the transitions that data signals make during the state that the clock has entered. In addition, we assume that the rise and fall times of all clocks are zero and each clock reaches all gates to which it is connected at the same time (zero clock skew).

1. **Phase one ($\phi = 1$ and $\overline{\phi} = 0$).** In this case, the clock transistors in the latches that hold the qualification signals are all off so these lines cannot change. Therefore, the circuit will always function correctly during this phase, regardless of the connections used for the qualification signals.

2. **Phase two ($\phi = 0$ and $\overline{\phi} = 1$).** During this regime, the control signals A, $\overline{\text{A}}$, A*, and $\overline{\text{A}}$* are changing. Q and $\overline{\text{Q}}$ must both be de-

asserted, either directly or through the bleeders, for the entire phase. Considering the circuit that produces Q, this means that either N1 or N3 must be on at all times. The simplest way to guarantee this is for A and $\overline{A}^*$ to be on opposite sides of an inverter. The similar requirement for $\overline{Q}$ is satisfied if $\overline{A}$ and A^* are on opposite sides of an inverter.

3. **Overlap high ($\phi = \overline{\phi} = 1$).** As in phase 2, when $\overline{\phi}$ is high, $\overline{Q}$ will be driven high (and thus inactive). Therefore we need only enumerate the ways in which Q might be high (and thus active) and verify that none of them can cause difficulties. The only possibilities for Q to become high after being low are for the gate of N1 (A) to go high or the gate of P1 ($\overline{A}^*$) to go low. Since A must not go high, it must be an even number of inversions from a latch. Similarly, $\overline{A}^*$ must not go low and thus must be an odd number of inversions from a latch.

   Suppose Q is initially high and during the overlap period A goes low and $\overline{A}^*$ goes high. That action will turn off N1 and P1 and thus isolate ϕ and Q. So when ϕ later goes low, Q will not change. Thus, N3 must serve the purpose of bringing Q low.

   Logically, the signal on the gate of N3 must be the inverse of A. If it is an even number of inversions from a latch, it will only turn on when $\overline{\phi}$ goes down. Since there will be a (possibly substantial) propagation delay to the gate of N3, Q will not go down until significantly after $\overline{\phi}$ goes down. During this delay, data can flow through two consecutive latches on $\overline{\phi}$ and Q. If, instead, the signal on the gate of N3 is an odd number of inversions from a latch, as shown, Q will actually be turned off during the overlap period.

   To make A an even number of stages from a latch and $\overline{A}^*$ an odd number, we could generate either signal from the other. However, if $\overline{A}^*$ goes high first while A is still high (and hence N1 is still on), we have created a path from ground to ϕ while ϕ is high. To avoid this, the buffering circuits that create the controlling signals should tap A from an earlier point in the chain than $\overline{A}^*$.

4. **Overlap low ($\phi = \overline{\phi} = 0$).** This case is symmetric to the overlap high case; the relevant transistors are the ones driving $\overline{Q}$: N2, P2, and P3. Using an analysis like that in case 3, we find that the gates of N2 and P3 should be an odd number of inversions from a latch and the gate of P2 should be an even number of inversions from a latch.

2.3 Effects of Qualified Clocks on Precharged Logic

NORA circuits that contain qualified clocks may also contain precharged logic. If a precharged gate uses a qualified clock, any latches fed by that gate must have clocks that are at least as qualified as that of the gate. Otherwise, the latch would read the precharged data on the cycles when

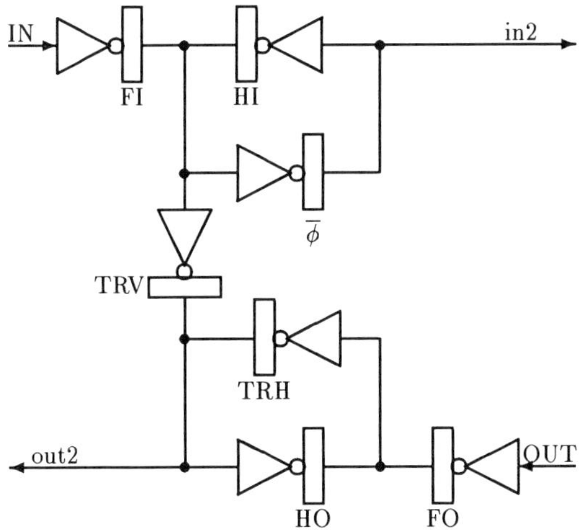

Figure 6: Data cell of a systolic queue with C^2MOS latches.

it was enabled but the clock used for the precharged gate was not. This behavior does not make logical sense.

The converse situation, where a precharged gate feeds a latch clocked by a more-qualified clock than the gate, is logically correct and does not violate any NORA rules. However, such use is discouraged because of the skew between the clock on the gate and the latch.

If a precharged gate feeds multiple latches qualified in different ways, either the gate must be duplicated, each with a different clock, or the gate's clock must have a qualification that is at least the union of all the clocks of the latches. The former method is preferred.

3 Analog Issues in Qualified NORA

The preceding sections concentrated on the logical correctness of qualified clocking in the NORA methodology. In particular, our discussions, as well as those in Goncalves and DeMan [3], assume that signals are either high or low at all times. In practice, of course, there is no such thing as a digital signal: all signals have non-zero rise and fall times and often are capacitively coupled to other signals. The analog behavior of signals in real circuits result in additional design constraints. We discuss these constraints below. Some of these issues are not specific to the addition of qualified clocking to NORA, but are included below for completeness.

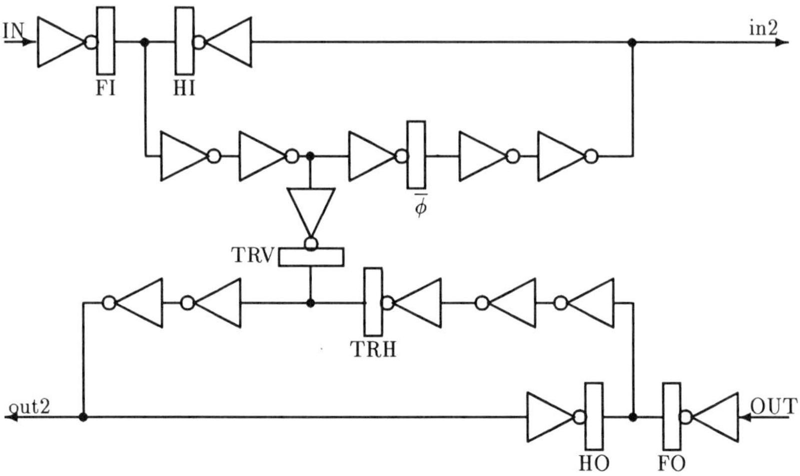

Figure 7: Revised systolic queue data cell.

3.1 Noise Problems of Dynamic Logic

It has long been understood that precharged gates have low noise immunity compared to static logic because the threshold of a precharged gate during the evaluation phase is the threshold of a single transistor rather than half the supply voltage. The noise immunity is typically reduced by a factor of three. In addition, a precharged gate is intrinsically noisy because the output is normally not actively held high (some designs use small bleeder p-transistors for this purpose) and thus capacitive coupling effects may slightly lower the voltage on the output.

If we connect the output of an n-type precharged gate to the input of a p-type precharged gate, we are connecting an intrinsically noisy signal to the input of a gate with a low noise immunity. This makes the circuit extremely sensitive to noise. (This effect is discussed in [11].) It is for this reason that we do not use both the n- and p-blocks of Goncalves and DeMan [3], but only n-type precharged gates (we could also use exclusively p-type gates, but n-transistors are normally faster). The inputs of precharged gates should not be directly connected to the outputs of latches for the same reason.

We fabricated our first NORA queue design in 1987; the schematic of the basic cell is shown Figure 6. Although our chip tolerated overlaps of up to 15 ns with both clocks either high or low and worked correctly at 11 MHz in a test environment, it was highly sensitive to different values of the supply voltage and clock timings. We were unable to get the part to operate in our system at a clock frequency above 4 MHz.

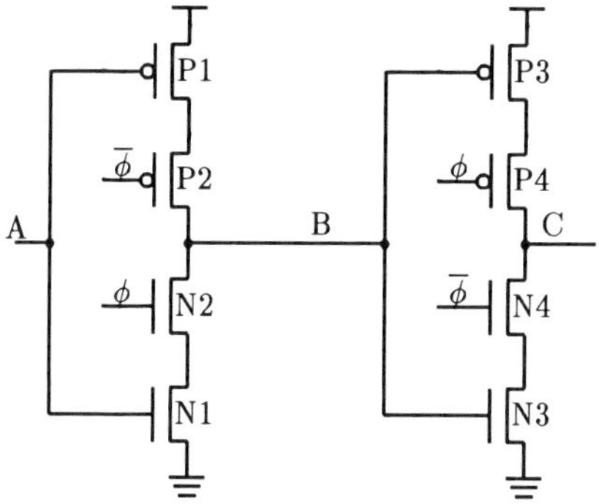

Figure 8: A chain of two C²MOS latches.

After analysis of the situation and consultation with Masakazu Shoji [11] and others, we realized that, when used in a NORA circuit, C²MOS latches also have a low noise immunity during the clock overlap period and that the same rules should be applied to such latches as to precharged gates. We therefore revised our design to include at least one static gate between any two dynamic gates (whether latches or precharged gates). The revised design is shown in Figure 7. A more complicated combining version of this design (described in Section 4) functions correctly in our system at much higher clock rates.

3.2 Clock Rise and Fall Times and Clock Skews

Another potential problem with NORA logic occurs when the clock skews or the rise or fall times of the clocks are comparable to the propagation delay of the logic. For NORA rules to work, one assumes that a single clock has the same value everywhere in the circuit; that is, if ϕ is low on the n-transistor of one latch it is also low on the p-transistor of the next latch.

Consider a chain of two C²MOS latches, shown in Figure 8. Suppose the rise time of ϕ is so long that it stays at half the supply voltage for a significant period of time. During that time, N2 and P4 are both weakly on. If A changes from low to high, B will start going low, which, because P4 is on, will cause C to start going low, thus allowing a signal to propagate through latches on both clock phases, which is not permitted. This effect is enhanced by the issues discussed in the previous section.

Fortunately, the remedies discussed in the previous section are also helpful here. In addition to decreasing the noise sensitivity of the latches, adding a pair of inverters between the two latches increases the delay from A to C and therefore increases the length of rise-time that is required to cause a malfunction.

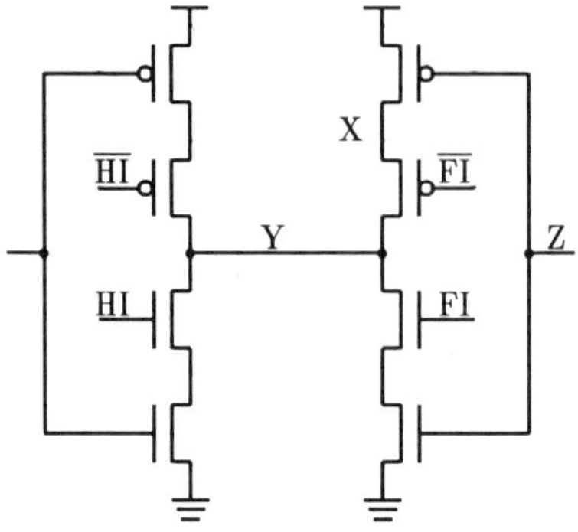

Figure 9: Charge-sharing in qualified C^2MOS latches.

Even with these inverters, however, care should be taken to size the transistors in the qualification logic to provide as short rise and fall times as practical. In [12] it is claimed that SPICE simulations show that NORA can tolerate clock rise times as long as 20ns in the technology being modeled. Our experiences lead us to believe that this is overly liberal; we try to obtain rise times at least as short as the propagation delay of an inverter driving an identical inverter.

Similar problems can exist if there is a significant propagation delay between a qualified clock and its non-qualified form when the former is controlling a latch whose output feeds a latch controlled by the latter. Adding a pair of inverters between the latches minimizes this problem, but we also try to ensure that the delay through the qualification transistors is as small as practical and arrange for the bleeder transistor to turn on as quickly as possible.

3.3 Charge Sharing

Consider the two C^2MOS latches shown in Figure 9, which is a detailed view of a part of our basic data cell, and suppose that during a previous clock cycle Y became high by the action of the HI and $\overline{\text{HI}}$ qualified clocks, but that the internal node X is at some intermediate voltage because Y was low the last time that $\overline{\text{FI}}$ was active. If Z is high when $\overline{\text{FI}}$ goes low again, X shares charge with Y, which may not be driven at the time. Similar problems exist on the N side of the latch.

In the specific circuit shown, this charge-sharing is highly unlikely to cause a failure, especially if Y is only connected to static logic, as we indicated in the previous sections. X will not reach a very low voltage due to the poor ability of P transistors to pass a low logic level. Also, the capacitance of Y will normally be significantly higher than that of X (if everything is of

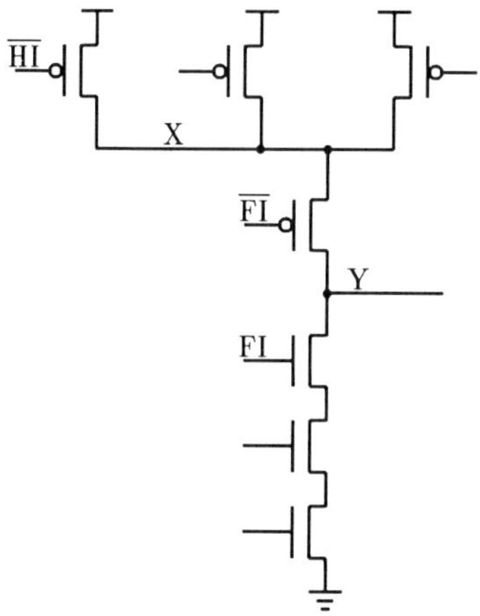

Figure 10: Two-input NAND C^2MOS latch, with charge-sharing protection.

minimum size, the capacitance of Y, not including its load, will be three times that of X; larger if parasitics are considered).

However, if a latch is implementing a more complex function than an inverter and is driving a smaller gate, there is the potential for this charge sharing to be significant. Therefore, in that case we add a transistor to precharge X when $\overline{HI}$ is asserted. An example of this technique used in a 2-input NAND latch is shown in Figure 10.

Another charge-sharing problem can appear if one constructs a circuit in which a clock is twice qualified and the intermediate qualified clock has a much larger fan-out than the final qualified clock. A race can occur between the first bleeder and the second pass gate during an overlap period when the first pass gate is turning off and the second pass gate is turning on. The final clock should instead be generated from the input clock with a single qualification gate.

4 Conclusion

We have used the NORA clocking methodology with qualified clocks to design the components of a combining switch for the NYU Ultracomputer network [4]. A combining switch contains two chip types: a forward path component that handles requests from the processor to the memory and a return path component for reponses from the memory to the processor. Four forward path components, one for each input/output pair, make a 2×2 forward path switch; likewise four return path components form the remainder of the switch.

The forward path component includes a systolic combining queue, an ALU and arbitration logic. The systolic combining queue is essentially the semi-systolic queue of Figure 1, with an additional row, which we call CHUTE; items in the IN row are transferred to the CHUTE when the address and operation code. match the adjacent item in the OUT row. The matched items in the OUT and CHUTE rows exit the queue simultaneously, the one from the CHUTE row going to the return path component and the one from the OUT row headed for memory, after passing through the ALU which combines the data from the two items. (See Gottlieb and Kruskal [5] on how fetch-and-ϕ operations, which include load and store, may be combined). The ALU is implemented using two-way carry look-ahead with multiple output Domino logic; the scheme is very similar to that in [7], but was developed independently.

The forward path component, containing over 30,000 transistors, has been fabricated in 2-micron CMOS using the MOSIS service with 132-pin PGA packages. We received 25 working parts out of 32. Its full functionality has been tested using an IMS tester at clock speeds up to 12 MHz. With the combining function turned off, the component can be used to construct a bidirectional non-combining switch. A wire-wrap board was built in January 1991 and functions reliably at clock speeds up to 10 MHz in a two-processor, two-memory-module Ultracomputer prototype running the operating system and applications.

The return path component includes an associative wait buffer [1] as well as two non-combining systolic queues and an ALU and contains over 50,000 transistors. It has been submitted to MOSIS for fabrication and should be used in a combining switch in our prototype this spring.

Our experience in designing with the NORA methodology has been positive overall. Care is required in using dynamic latches and gates, and design of the control logic that produces the clock qualifiers is occasionally tricky. Sections of control logic must often be designed double-rail in order to have the clock qualifying signals the correct number of inversions from a latch. However, the data paths, which take up most of the area of these designs, are compact and efficient.

Acknowledgements

This research was supported in part by U.S. Department of Energy grant number DE-FG02-88ER25052 and in part by National Science Foundation grant number MDS-8301768.

References

[1] Susan Dickey, Allan Gottlieb, and Richard Kenner. Using VLSI to reduce serialization and memory traffic in shared memory parallel com-

puters. In Charles E. Leiserson, editor, *Proceedings of the Fourth MIT Conference on Advanced Research in VLSI*, April 1986.

[2] Susan Dickey, Richard Kenner, Marc Snir, and Jon Solworth. A VLSI combining network for the NYU Ultracomputer. *International Conference on Computer Design*, pages 110–113, 1985.

[3] Nelson F. Goncalves and Hugo J. DeMan. NORA: A racefree dynamic CMOS technique for pipelined logic structures. *IEEE Journal of Solid-State Circuits*, SC-18(3):261–266, June 1983.

[4] Allan Gottlieb. An overview of the Ultracomputer project. In J. J. Dongarra, editor, *Experimental Parallel Computing Architectures*, pages 25–95, New York, 1987. Halstead Press.

[5] Allan Gottlieb and Clyde P. Kruskal. Coordinating parallel processors: A partial unification. *SIGARCH*, October 1981.

[6] Leo J. Guibas and Frank M. Liang. Systolic stacks, queues and counters. *MIT Conference on Advanced Research in VLSI*, 1982.

[7] Inseok S. Hwang and Aaron L. Fisher. Ultrafast compact 32-bit CMOS adders in multiple output Domino logic. *IEEE Journal of Solid-State Circuits*, 24(2):358–369, April 1989.

[8] R. H. Krambeck, C. M. Lee, and H. S. Law. High-speed compact circuits with CMOS. *IEEE Journal of Solid-State Circuits*, SC-17(3):614–619, June 1982.

[9] Charles E. Leiserson. *Area Efficient VLSI Computation.* MIT Press, Cambridge, Massachusetts, 1983.

[10] Carver A. Mead and Lynn A. Conway. *Introduction to VLSI Systems.* Addison-Wesley, Reading, Massachusetts, 1980.

[11] Masakazu Shoji. *CMOS Digital Circuit Technology.* Prentice Hall, Englewood Cliffs, New Jersey, 1988.

[12] Jiren Yuan and Christer Svensson. High-speed CMOS circuit technique. *IEEE Journal of Solid-State Circuits*, 24(1):62–70, February 1989.

A Fast Static CMOS NOR Gate

Thomas D. Simon
Department of Electrical Engineering
MIT
tsimon@sick-puppy.ai.mit.edu

Abstract

We have designed and simulated a new class of static CMOS NOR gates with speeds competitive with those of dynamic gates. We call this form of logic SFPL, Source Follower Pullup Logic. SFPL implements high fan-in gates in a single fast stage, as well as smaller gates with delays only fractionally larger than inverters. SFPL also has small input loading while maintaining reasonable output current. Complex combinational logic functions with small delays can be implemented without any of the disadvantages of dynamic logic.

1 Introduction

High fan-in logic gates have applications in PLAs, ROMs, address decoders, or as components in standard cell libraries. Static gate configurations such as fully complementary CMOS or N-channel NOR gates with grounded-P pullups are slow and do not scale well with fan-in. High fan-in gates can be implemented from smaller ones by tree cascading, but this incurs several gate delays.

Dynamic logic is faster and scales better with fan-in, but suffers from several drawbacks including tight constraints on the arrival time and polarity of input signals. It is awkward to use at asynchronous interfaces and in timing signal generation.

In addition, large numbers of dynamic logic cells heavily load the clock signals driving the precharge and evaluate devices. These heavily loaded clocks become unwieldy to drive at high frequencies with careful attention to edge placement and on-chip clock skew.

We have designed a fully static CMOS single-stage NOR logic form which we call SFPL, Source Follower Pullup Logic. It features delays comparable to the evaluation delays of dynamic logic families. SFPL does not require precharging or any clock inputs. SFPL gates of five inputs can be made with delays of less than twice an inverter delay. These gates scale up to as many as 30 inputs in a single stage without faster two stage implementations being possible.

We present this new gate design, analyze its performance, and discuss its noise tolerance and power dissipation. Simulations were run with HSpice in the Hewlett-Packard CMOS-26 $1\mu m$ drawn channel length process, the highest performance silicon process supported by the MOSIS service. A component is currently being fabricated to test this design.

2 Circuit Description

A 5-input SFPL gate is shown in figure 1. It resembles a grounded-P pullup configuration, except that (1) the P-channel pullup gate (node A) is controlled by the inputs, and (2) the N-channel pulldown devices can consequently be smaller while maintaining adequate noise margins and fall time. Node A is controlled by the inputs through N-channel pullup devices used as source followers.

When any input is high the voltage at A is raised, lowering the drive of the P-pullup and partially turning on the helper pulldown. On the other hand, since the voltage at node A does not reach the positive rail, the P-pullup is already partially turned on in preparation for a positive edge on the output. The sizing of the devices is chosen so that node A reaches an intermediate voltage (between 2 and 3 volts with a 5 volt supply) when only one input is high.

The advantage over grounded-P gates is that the slower turn on time of the P-pullup is traded for less fight against the N-pulldown devices on negative output transitions. This improves both the speed and the DC low output voltage. In turn, the N-channel pulldowns can be smaller than in a grounded-P configuration. The self-loading on the output node is therefore lower, which compensates for the slower turn on of the P-pullup, especially for large fan-in gates.

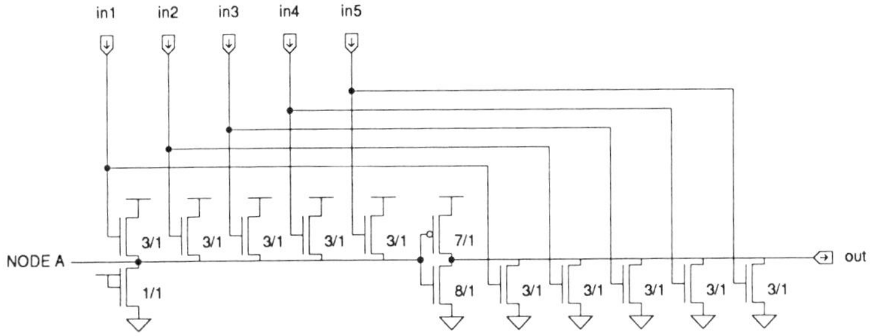

Figure 1: A 5-input gate.

When many inputs switch high, node A reaches a higher voltage (no higher than $V_{dd} - V_{tn}$). The capacitive coupling from the gates to the sources of the N-pullups helps bring node A down faster when the inputs fall. This compensates for the higher voltage change required to fully turn on the P-pullup.

The source follower configuration of the N-pullups also means that the loading on the inputs by these devices is relatively smaller per unit gate width than for N-channel devices used as pulldowns, since there is no Miller effect for the source followers. The total loading per input is quite small on the SFPL gate shown in figure 1.

3 Output Voltage

Figure 2 shows SPICE simulations of the five input SFPL gate under worst case processing, temperature, and supply voltage conditions. The simulations are with only one input switching on the left and all inputs switching on the right. The inputs are the square waveforms, the 2 to 3 volt amplitude signals correspond to node A, with the remaining signal being the output.

As with the grounded-P configuration, the output of SFPL gates are not rail to rail. The worst case is with only one input high. SFPL gates, however, have lower output low voltages for a given worst case delay than grounded-P gates. The device sizes on the SFPL gate shown were chosen so that the output low voltage is 250 mV with nominal processing and one input high. At the worst process corner, fast P-channel and slow N-channel devices, the output low voltage is 500 mV, still 200 mV below V_{tn}. The output levels and noise margins (discussed later) can be improved at the expense of slower rising edges on the output, just as with grounded-P gates.

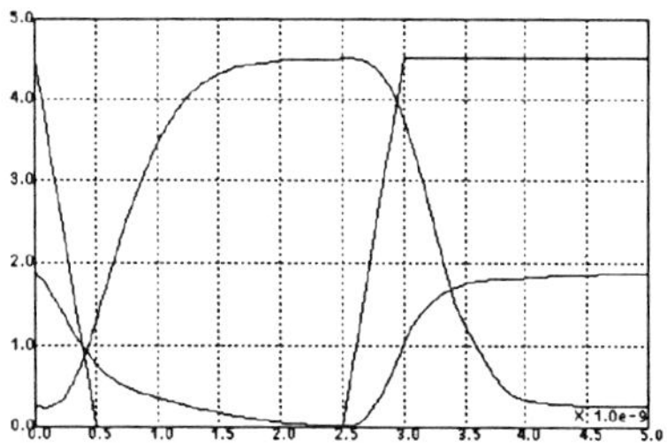

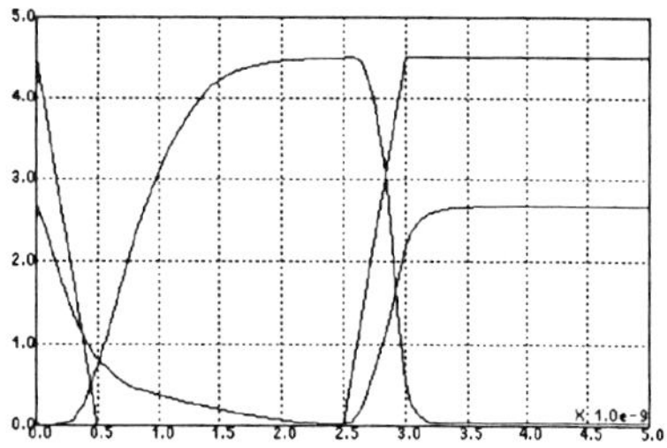

Figure 2: SPICE simulations of an SFPL gate under worst case processing, temperature, and supply voltage conditions. The figure on the left shows input, output, and node A with one input switching and the rest grounded. The figure on the right shows simulations with all five inputs switching together.

4 Speed

Figure 3 shows the inverter size used for comparison and as the test load in all simulations. The input or inputs are driven by voltage sources with 0.5 ns rise/fall times. Propagation delays are measured from inverter trip point to inverter trip point.

Figure 4 plots worst case delay normalized to the inverter delay with respect to fan-in, for gates with the same device sizing as the five input SFPL gate. The delay differences grow linearly since the self-loading grows linearly with fan-in, while the load and worst case drive remain constant.

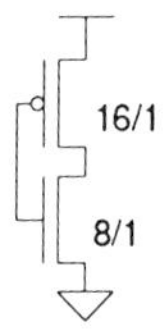

Figure 3: Inverter used as standard load and for speed normalization.

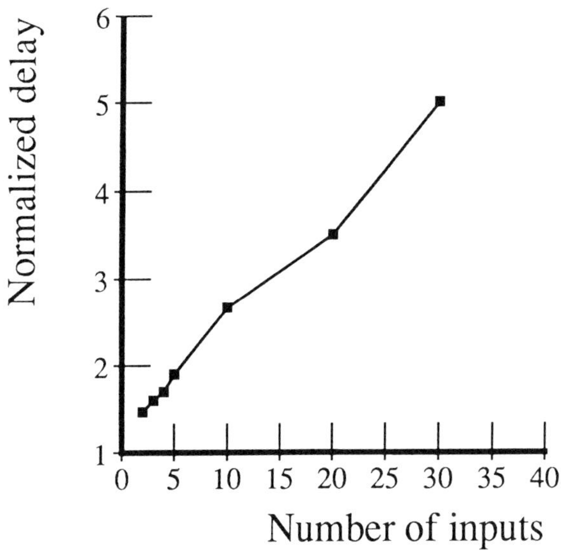

Figure 4: Normalized delay vs. fan-in for worst case conditions, including worst case input combinations.

By using $n+1$ n-input gates, an n^2-input NOR gate can be constructed in two stages. Inverters are required on the outputs of the gates in the first stage to restore polarity. The delay through such a n^2-input gate is twice the delay for the n-input gate plus some fraction of an inverter delay (a fraction because the loadings are smaller than for the test inverter). For the device sizings and load given, the cutoff above which two stage implementations are faster than single stage ones is about 30 inputs. There are also power dissipation tradeoffs involved which are discussed later.

We compare the speed of SFPL gates against grounded-P NOR gates, basic DOMINO OR gates [4], and fully complementary NAND gates. The five input versions of each are shown in figures 6 through 8. The NAND gates used for comparison have higher input loading than the others. The grounded-P has slightly worse output voltage than our gates (in the worst case V_{tn}) and about the same static power dissipation. These results are plotted in figure 5. SFPL is faster and scales better than the other static gates. The DOMINO gates scale better than SFPL, and are only slower for small numbers of inputs because of the inverter required for correct DOMINO pre-charged output.

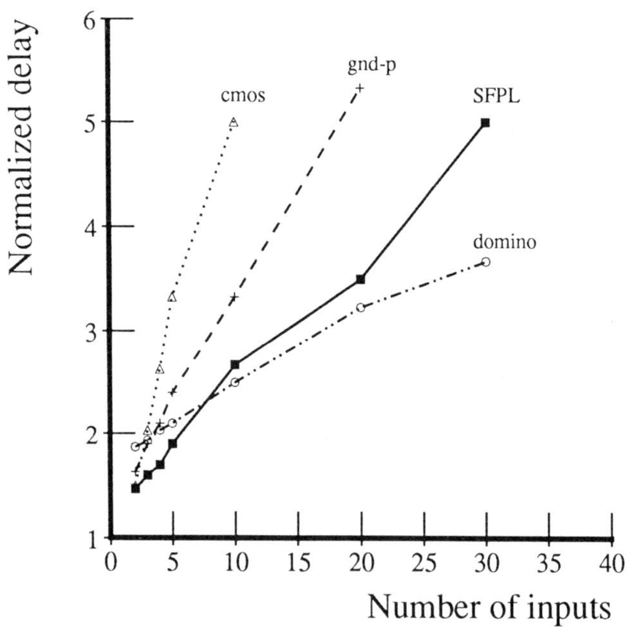

Figure 5: Comparison of worst case normalized delays vs. fan-in.

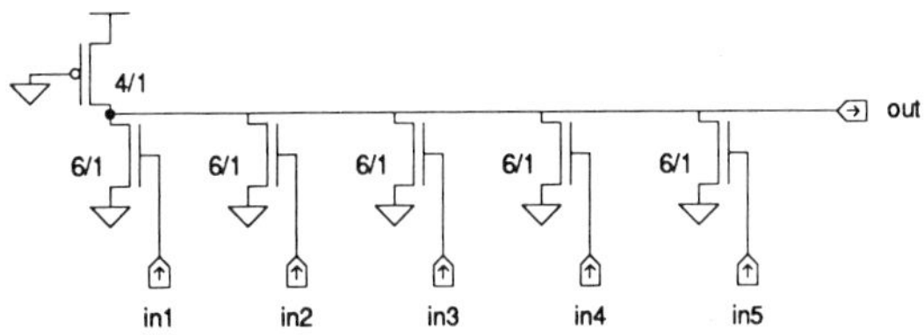

Figure 6: 5-input grounded-P NOR gate.

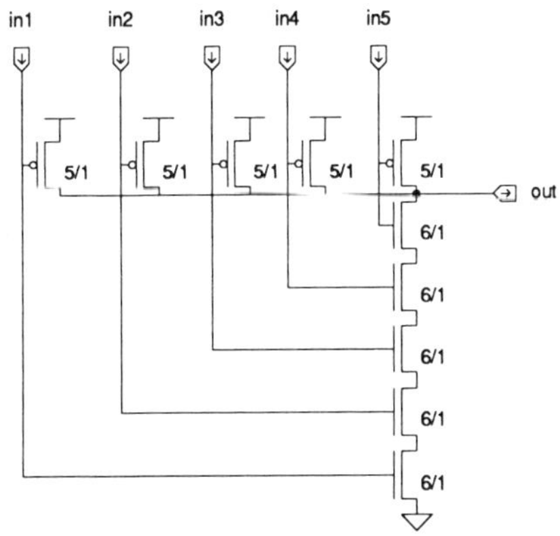

Figure 7: 5-input NAND gate.

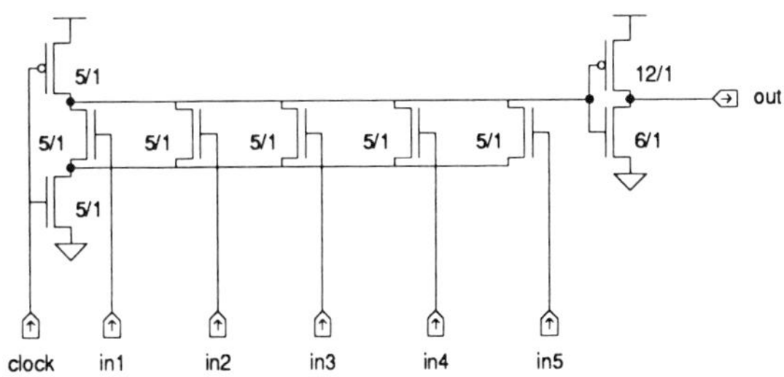

Figure 8: 5-input DOMINO OR gate.

5 Power Dissipation and Input Loading

The most severe problem in SFPL gate design is the static power dissipation, which is comparable to that of grounded-P gates. Static power is dissipated when the output is low by the N-pulldown at node A and by the P-pullup which is not fully turned off.

Worst case power dissipation occurs with one input high and the remainder grounded. With more inputs high, the higher current drawn at node A is compensated for by a higher node voltage, which lowers the drive of the P-pullup. For this reason, the static power dissipation does not go up with the number of inputs.

The power dissipated by SFPL gates is about 1.2 mW for the device sizings shown and the same process and temperature conditions used with the speed measurements. These same gates dissipate 4.8 mW at the worst power and lowest temperature process corner. This static power dissipation is five to ten times higher than the dynamic power used by SFPL or DOMINO gates with 20 inputs, at 100 MHz, with the given sizings and loads, and at the worst speed conditions.

Because static power dissipation does not increase with fan-in, the cost in power of using a two stage implementation of a large fan-in gate is high. A two stage n^2-input gate made of $n+1$ n-input gates consumes $n+1$ times the power of the single stage implementation, with worst case input combinations. Other power-speed tradeoffs are possible. For example, a $2n$-input gate can be made from 2 n-input gates and a 2-input fully complementary NAND gate to combine their outputs. This implementation will use twice the power of the single stage gate, under worst case input logic combinations.

Figure 9 plots delay versus static power for five input gates of varying device sizings and input loading. As expected, the delay-power product is not constant. The smaller, slower gates are more efficient. As devices are scaled up, static power increases linearly. Speed does not because the self-loading of the gates increases with the drive. Note that the fastest five input gate size shown is 10 percent slower than the test inverter. (In fact, the degenerate single input SFPL gate, which acts as an inverter, and has about the same input loading as the test inverter, is actually slightly faster than that inverter!)

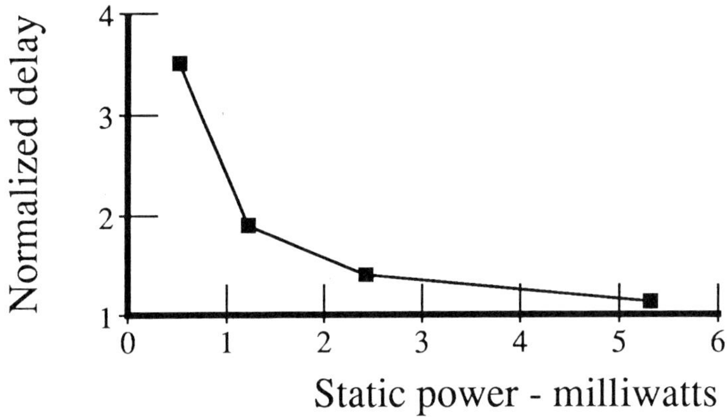

Figure 9: Power and delay of variously sized five input gates, at slowest process corner.

6 Noise Margins

SFPL, like other static gates, is not susceptible to the noise and timing problems of dynamic logic. Like grounded-P, however, they do have poorer noise margins than fully complementary logic. Figure 10 shows the DC transfer characteristics of the five input gate and those of the static fully complementary five input NAND gate, both with one input varying and with all inputs varying. These characteristics correspond to process variations with balanced N-channel and P-channel device thresholds. The output low noise margins are about 0.5 volt smaller for our gate. The worst case is with only one input varying. Figure 11 shows the DC characteristics at the fast P-channel and slow N-channel process corner.

These noise margins can be improved at the expense of slower rising edges on the output by changing the relative sizings of some of the devices. For example, the width of the N-channel source followers can be increased relative to the size of the N-pulldown on node A. With only one input switching, the voltage at node A will be higher, more fully turning off the P-pullup, and also more slowly turning it back on hard on the opposite transition. With many inputs switching, the voltage at A will remain about the same since the source followers will saturate at the same voltage independent of gate width.

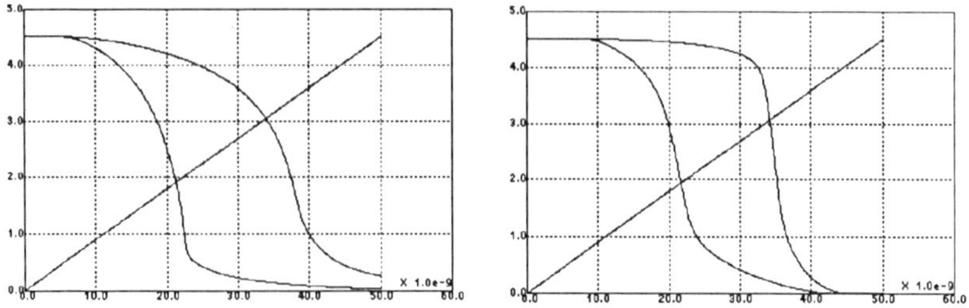

Figure 10: DC transfer characteristics for 5-input gates with balanced process variation and one or all inputs switching. Left figure shows SFPL gate. Right figure shows a conventional fully complementary CMOS NAND gate.

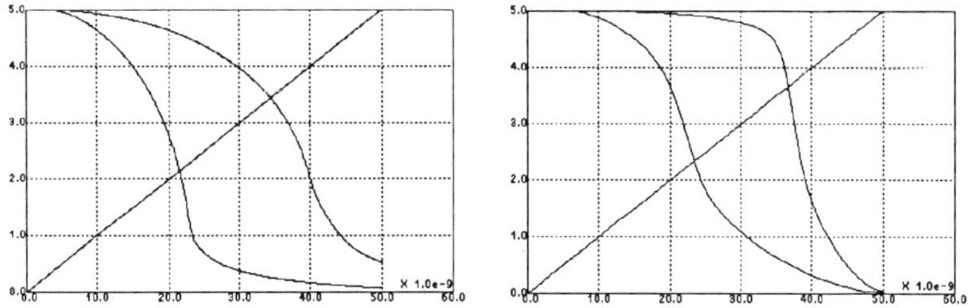

Figure 11: DC transfer characteristics for 5-input gates with faster P-channel devices and one or all inputs switching. Left figure shows SFPL gate. Right figure shows a conventional fully complementary CMOS NAND gate.

7 Sample Layout

Figure 12 shows a possible layout of the five input gate. The area is certainly less than twice that of alternative five input gates.

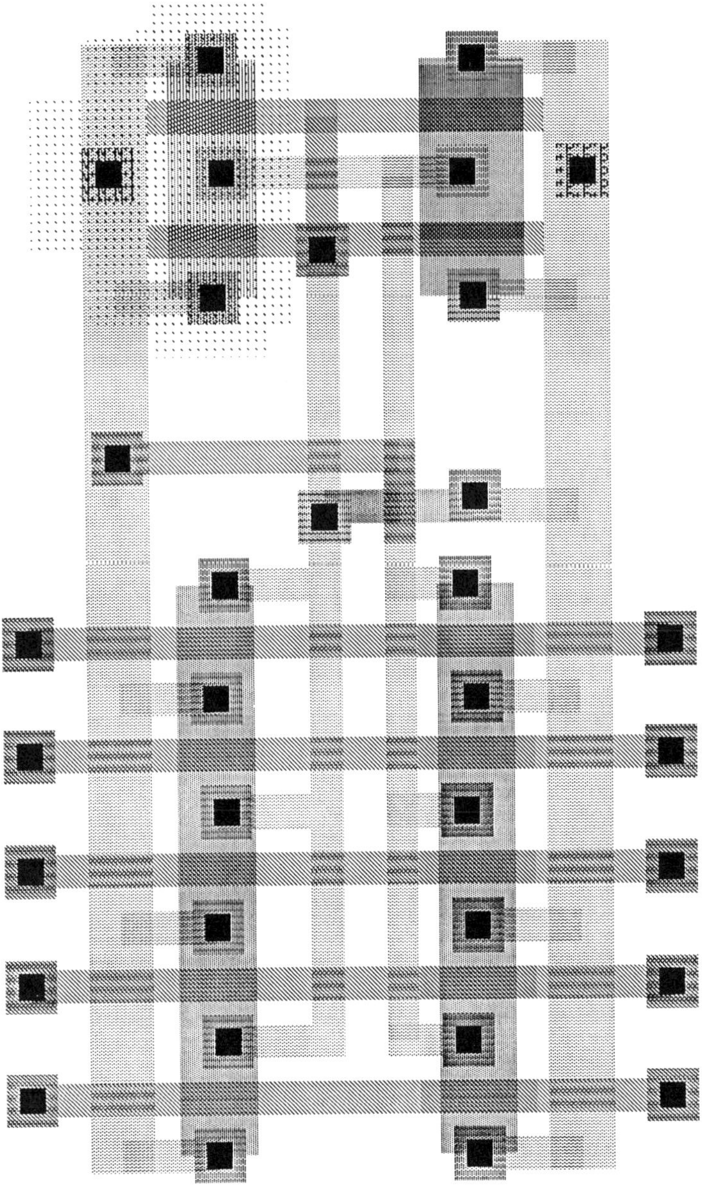

Figure 12: A sample standard cell layout of the five input SFPL NOR gate.

8 Applications

SFPL gates can be used to implement random combinational logic such as decoders. The propagation delays are the same for all inputs, as with grounded-Pgates but unlike complementary gates or any other logic family with series connected devices. This uniform delay may be useful in timing critical applications such as wave pipelined circuits. They are also very attractive options when both speed and static functionality is needed, such as for generating timing signals.

In addition, it is possible to make single stage and-or gates by stacking transistors in series, instead of using two stages of gates. There are tradeoffs of speed, area, and power. The one stage implementation uses less power and area, but is slower due to extra loading and the series connected resistances of both pullup and pulldown devices. One special case exception is a high fan-in selector with decoded select inputs. If constructed as shown in figure 13, the single stage implementation is faster than the two stage one, as well as being smaller and using less power. The select inputs are connected to the devices closer to the output nodes. When the select signals have settled, exactly one is guaranteed to be high and the capacitance of nodes on the other side of most of the select devices are isolated from the output nodes.

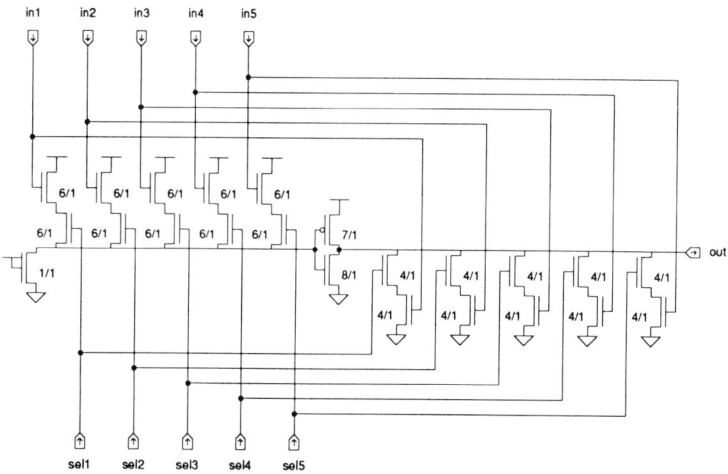

Figure 13: A 5-input selector with decoded select inputs.

9 Conclusion

SFPL has significant performance advantages over other static logic structures. While SFPL does not quite match the speed of dynamic logic, there can be important considerations which make SFPL and other static logic attractive, including design complexity, incompatibility of logic with the dominant clocking methodology, and excessive clock loading. In addition, future MOS processes will further scale transistor oxide thickness and channel lengths, requiring simultaneous reductions of circuit voltages. As we scale down both power supply and threshold voltages, the SFPL gate continues to function well. In contrast, the subthreshold leakage of MOS transistors becomes a significant problem for all forms of dynamic and precharged logic techniques as we scale to very low threshold voltages. In these highly scaled regimes, either reduction of leakage through low temperature operation, or a fully static methodology, perhaps including SFPL gates, will be required.

One disadvantage is the prohibitive power dissipation of SFPL, as with some alternative static logic families, which limits its use to a few thousand gates per die. In any case, SFPL gates are one more addition to the set of possible circuits to be considered by careful performance oriented logic designers.

Acknowledgments

Thanks go to Mark Johnson, Charles Sodini, Mark Santoro, Bob Sproull, André DeHon, Nick Carter, Mike Bolotski, and Henry Minsky for their generous suggestions and aid.

This research is supported in part by the Defense Advanced Research Projects Agency under contract N00014-91-J-1698.

References

[1] L. Heller, W. Griffin, J. Davis, and N. Thoma, "Cascode Voltage Switch Logic: A Differential CMOS Logic Family," International Solid State Circuits Conference, February, 1984, pp. 16-17.

[2] K. M. Chu and D. L. Pulfrey, "A Comparison of CMOS Circuit Techniques: Differential Cascode Voltage Switch Logic versus Conventional Logic," IEEE Journal of Solid State Circuits, Vol. 22, No. 4, August 1987, pp. 528-532.

[3] M. Johnson, "A Symmetric CMOS NOR Gate for High-Speed Applications," IEEE Journal of Solid State Circuits, Vol. 23, No. 5, October, 1988, pp. 1233-1236.

[4] B. Murphy, R. Edwards, L. Thomas, and J. Molinelli, "A CMOS 32b Single Chip Microprocessor," International Solid State Circuits Conference, February, 1981, pp. 230-231.

[5] R. H. Krambeck, C. M. Lee, and H. Law, "High-Speed Compact Circuits with CMOS," IEEE Journal of Solid State Circuits, Vol. 17, No. 3, June 1982, pp. 614-619.

[6] L. Pfennings, W. Mol, J. Bastiaens, and J. Van Dijk, "Differential Split-Level CMOS Logic for Subnanosecond Speeds," IEEE Journal of Solid State Circuits, Vol. 20, No. 5, October, 1985, pp. 1050-1055.

[7] J. Wang, C. Wu, and M. Tsai, "CMOS Nonthreshold Logic and Cascode Nonthreshold Logic for High-Speed Applications," IEEE Journal of Solid State Circuits, Vol. 24, No. 3, June, 1989 pp. 779-786.

Verification and Fault Tolerance

Compositional Verification of Systems with Synchronous Globally Timed Control

Filip Van Aelten
Jonathan Allen
Srinivas Devadas
Research Laboratory of Electronics
Department of Electrical Engineering and Computer Science
Massachusetts Institute of Technology
Cambridge, MA 02139, USA

Abstract

We address the problem of automatically verifying large digital designs at the logic level, against high-level specifications. We present a technique which allows for the verification of a specific class of systems, namely systems with synchronous globally timed control. In such systems, a single controller directs the data through the data path, and decides (globally) when to move the data. These systems are verified against a specification in an applicative language such as SILAGE, which can be viewed as describing a maximally parallel synchronous machine. Both the implementation and the specification have string functions (input/output mappings) associated with them, and correctness is taken to mean that a certain relation holds between these string functions. These relations extend beyond strict input/output equivalence, and provide room for various behavioral transformations.

In this paper, sufficient conditions for the validity of these relations are presented, which can be verified automatically and efficiently. They consist of the combinational correctness of the data path modules, and the validity of a number of past-tense Computation Tree Logic (CTL) formulae with respect to the controller. The latter can be verified using forward finite state machine (FSM) traversal algorithms. In this way, the verification of the composite machine is reduced to the separate verification of the data path modules and the controller.

1 Introduction

We address the problem of automatically verifying large digital designs at the logic level, against high-level specifications. Automatic verification procedures have recently been presented which apply to small to medium-sized circuits. First, procedures were presented which verify strict input/output equivalence between two finite state machines (FSM's) (e.g. [7]). This is done by exhaustively traversing the State Transition Graph of the product of the two machines, using implicit state enumeration techniques. Although these techniques are more efficient than explicit state enumeration, they are still impractical for circuits with many memory elements (latches), because of the exponential growth of the number of states with respect to the number of latches in the circuit. Recently, these procedures were extended to allow for differences in the input/output behavior of the specification and the implementation [1]. The correctness condition was cast as a relation between the string functions associated with the specification and the implementation, and these relations were verified by modifying the specification and the implementation, and then checking for input/output equivalence. This procedure was, again, practical for small to medium-sized circuits only.

The size of the circuit is less critical when theorem proving techniques are applied for verification, as opposed to state traversal methods. Several large circuits have been successfully verified with these techniques [12] [6] [13]. However, these techniques require extensive user interaction.

The work described in this paper aims both at automatic procedures and large circuits. This is realized by proving that, for a certain class of circuits, the correctness requirement can be reduced to certain sufficient conditions, which can be verified without building the State Transition Graph of the entire circuit. As a correctness requirement, we take two of the relations described in [1], the β (don't care times) and γ (parallelism) relations, and allow compositions of these relations. The β-relation relates a circuit, that processes inputs at certain relevant time points only, and produces outputs at certain relevant time points only, with a circuit of the same functionality that takes and produces relevant inputs and outputs at all time points. The γ-relation relates a serial implementation with its parallel counterpart, or vice versa.

The β and γ relations are verified between an implementation with synchronous globally timed control, and a fully parallel specification machine, as described by a program in an applicative language, or by a signal flow graph. Systems with synchronous globally timed control are synchronous systems in which a single controller directs the flow of data in the data path, and decides globally on the timing of data movement, as opposed to letting the data path modules take whatever time they need to finish a computation. These systems are verified against a specification program in an applicative language such as SILAGE [11]. Such a program can be viewed as representing a maximally parallel synchronous machine without a controller. The controller in the implementation emerges from the reuse of a limited set of data path modules to perform the same computation as the specification. Our verification strategy can be used for the independent verification of circuits produced by the CATHEDRAL-II synthesis system, described in [10]. This system produces, under user interaction, microcoded Digital Signal Processors with moderate throughput, starting from a specification in SILAGE.

A compositional technique for verification is obtained by requiring not only that the implementation has a correct input/output behavior, but that it realizes this behavior in a way consistent with the signal flow graph serving as a specification. The correctness requirement can then be reduced to the following. First, for every operator in the signal flow graph, there should be a data path module in the implementation which has the same combinational input/output behavior. Second, the sequence of inputs to, and outputs from, the controller has to satisfy certain conditions to obtain a correct flow of data between the data path modules. We express these conditions in past-tense Computation Tree Logic (CTL). We chose this formalism because past-tense CTL formulae can be verified by means of forward FSM traversal algorithms based on implicit enumeration techniques – the same algorithms as the ones that are used for FSM input/output equivalence checking. FSM traversal is now needed for the controller only, as opposed to the entire system, resulting in improved efficiency.

A verification method closely related to ours is the SFG-tracing method, described in [4]. The application targeted by this method is exactly the same as ours. Verification is done by symbolically simulating the implementation, thereby creating a parallel circuit, which can be matched, piece by piece, with the specification which is also maximally parallel. With our method, the data path modules available for synthesis can be verified once and for all, and attention is focused on the controller, which is completely different for every design.

The paper is organized as follows. Sections 2 and 3 contain preliminary information, on string function relations and past-tense CTL, respectively. Section 4 introduces the overall compositional verification strategy. In Section 5 the CTL requirements on the control sequence are presented. Section 6 applies our technique to a simple example, and Section 7 concludes the paper.

2 Preliminaries on String Function Relations

In this paper, both the specification and the implementation are synchronous machines. Synchronous machines realize a certain string function. They map a sequence (string) of input values into a sequence of output values. With Σ as alphabet, the set of strings, Σ^*, is defined as the set of all finite concatenations of elements of the alphabet. We use variables u, v, ... for characters, and x, y, ... for strings. The empty string is denoted by ε. String functions are functions from Σ^* to Σ^*. Useful operations on strings include:

- the concatenation operation, written as ., which concatenates two strings (or string and a character),
- an operation written as $\uparrow$, which takes a character and a number n, and returns a repetition of the character n times,

- and an operation written as $\downarrow$, which takes a string and a number n, and extracts the nth character from the string. We will use $x \downarrow i..j$ as an abbreviation for the string $(x \downarrow i). \ldots .(x \downarrow j)$.

An implementation is considered correct with respect to a specification, if a certain relation holds between the two machines. The two primitive relations verified in this paper are the "don't care times" relation, β, and the parallelism relation, γ. To define the β relation, a function *Relevant* has to be defined first.

Definition 2.1 *Relevant* : $\Sigma^* \underline{\times} B^* \rightarrow \Sigma^*$:

$$\begin{aligned} Relevant(\varepsilon, \varepsilon) &= \varepsilon \\ Relevant(x.u, y.v) &= Relevant(x, y) \quad &, v = 0 \\ &= Relevant(x, y).u \quad &, v = 1 \end{aligned}$$

The symbol $\underline{\times}$ represents the string cartesian product, which combines strings of equal length. The *Relevant*-function takes a string over an arbitrary alphabet and a Boolean valued string, and returns what remains of the first string after deleting all values for which there is a 0 in the corresponding place in the Boolean valued string.

Definition 2.2 *Let H be a string function from Σ^* to B^*, realizable by some synchronous machine.*
$F\ \beta_{H,n}\ G \Leftrightarrow \forall x \in \Sigma^*\ s.t. \mid x \mid \geq n, Relevant(F(x), R_{0\uparrow n} \circ H(x)) =$
$G(Relevant(x \downarrow 1..(\mid x \mid - n), H(x \downarrow 1..(\mid x \mid - n))))$

$\mid x \mid$ denotes the length of x. $R_{0\uparrow n}$ is the string function associated with a concatenation of n registers with initial value 0. For the case where $n = 0$, the equation reduces to:

$$Relevant(F(x), H(x)) = G(Relevant(x, H(x)))$$

In this case, F is in β relation with G if applying F to x, and then picking out the relevant outputs according to the function H, yields the same result as applying G to the relevant inputs. The complication in the definition arising from non-zero n-values is necessary to extend the β relation to implementation/specification pairs where the output stream of the implementation is delayed with respect to the one of the specification. An example of an implementation and a specification which are in β-relation is shown in Figure 1. The controller is not shown in the implementation. The specification, shown below, takes a new input every cycle and produces the corresponding output at the same cycle, whereas the implementation takes a new input only every 6th cycle, and produces the corresponding output 5 cycles later.

The parallelism relation, γ, which relates a serial implementation with a parallel specification, can be formally defined in a similar way. The β and γ relations may be combined. It can be shown that, without loss of expressive power, attention can be restricted to the $\gamma \circ \beta$ relation (leaving out $\beta \circ \gamma$) [1].

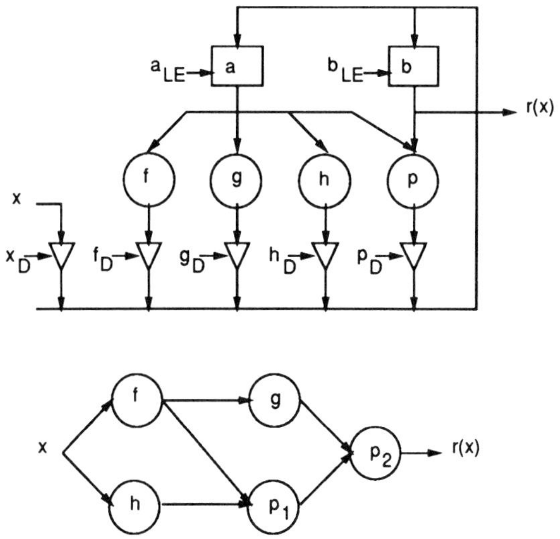

Figure 1: Example of an implementation and a specification pair which are in β relation

3 Preliminaries on Past-Tense CTL

The constraints on the sequence of inputs to, and outputs from, the controller are expressed in past-tense Computation Tree Logic (CTL). We first summarize the more common future-tense CTL, which is a subset of (future-tense) CTL* [5]. CTL* consists of state formulae, which are true in a specific state, and path formulae, which are true along a specific path. A state formula is either:

- an atomic proposition,
- $\neg f$, $f \wedge g$ or $f \vee g$, where f and g are state formulae, or,
- $A(f)$ or $E(f)$, where f is a path formula.

$A(f)$ is true when f is true along all paths leaving from the state under consideration. $E(f)$ is true when there exists a path from the state along which f is true.

A path formula is either:

- a state formula, or
- $\neg f$, $f \wedge g$, $f \vee g$, Xf, fUg, Ff, or Gf, where f and g are path formulae.

A path formula consisting of a state formula is true if the state formula is true in the first state. Xf, which is read as "next time f", is true if f is true in the second state on the path. fUg, read "f until g", is true if there is a state in the path in which g is true, and in all states prior to it f is true. Ff, read as "sometimes f", is true if there is a state in the path in which f is true. Gf, read as "always f", is true if f is true in all states on the path.

CTL is a subset of CTL* in which each path quantifier must be immediately followed by exactly one of the temporal operators G, F, X or U. This restriction reduces the complexity of verifying a formula against a State Transition Graph. A CTL state formula can be verified by evaluating it (i.e. computing the set of states in which it is true), and checking if the result is equivalent to the set of all reachable states in the State Transition Graph. A CTL state formula of the form $E(f)$, where f contains one temporal operator, can be evaluated by computing the fixed point of some state transformer. For instance, the fixed point characterization of $E(fUg)$ is given by ([3]):

$$E(fUg) = g \vee (f \wedge EX(E(fUg)))$$

In words, $E(fUg)$ is true in a state, if either g is true in it, or f is true in the current state, and $E(fUg)$ is true in a path starting from some next state. Evaluating the formula can be done by starting with the set of states in which g is true, and iteratively adding to this set the set of states from which a state in the current set can be reached, and in which f is true. Formulae containing the universal quantifier A can be reduced to ones containing only existential quantifiers E. For the evaluation of nested CTL formulae, several consecutive fixed point computations have to be performed.

CTL formulae with future-tense temporal operators require inverse image computations (i.e. given a set of states, a new set has to be computed from which a state in the original set can be reached in one step). Inverse image computations can be performed with the transition relation method [3], which requires manipulating Binary Decision Diagrams (BDD's) [2] which have both the inputs and the outputs of the transition function as variables. In [8] a method is presented which avoids the use of the transition relation method, but the authors conclude that the performance of the resulting algorithm is inferior to the one for forward traversal.

Backward FSM traversal can be avoided by using CTL formulae with past-tense operators. We use superscripts to distinguish between future and past-tense operators, as in [9] (p. 1006). F^+, G^+, X^+ and U^+ are the future tense operators; F^-, G^-, X^- and U^- are the past-tense operators. Future-tense operators are essential to express liveness properties (i.e. to require that something will happen sometime in the future). We will not need such a requirement. There is a clock which keeps ticking in the class of systems we consider, and there is an H-function which marks certain time points, and we will require that something happens at these specific time points. We will not have requirements for indefinite futures.

4 The Overall Compositional Strategy

The overall compositional strategy consists of verifying the following conditions:

1. The combinational blocks in the specification must all have corresponding combinational blocks in the implementation, which have to be functionally equivalent to it.
2. The flow of information between the combinational blocks in the implementation must be in accordance with the connections in the specification graph.

We make the following assumptions about the specification and the implementation.

1. The initial and final values of iteration indices in the specification are constants, and the iteration index is not modified inside the iteration (except for the addition of a fixed step at the beginning of every iteration).
2. All combinational blocks in the specification have an equivalent block in the implementation. Certain control inputs might have to be set in the implementation module for it to be equivalent to the specification module. Checking for functional equivalence is the first part of the verification process. In an automatic synthesis system, where the same modules are used for different design tasks, the combinational verification of these modules needs only to be done once.
3. Every pair of combinational operators (f, g) that is directly connected in the specification, is connected in the implementation via a multiplexer, a demultiplexer, a register file, and again a multiplexer and demultiplexer, as shown in Figure 2. Buses can be viewed as multiplexers, and the addressing portion of a register file as a demultiplexer. The verification system is provided with information as to how each pair of blocks is connected in the implementation. In Figure 2 the subscript D stands for "drive", and LE for "load enable". R_i is the particular register through which the two operators communicate. The control signals f and g only exist if the corresponding modules are ALU type modules which can perform different functions. One or both of the multiplexer/demultiplexer pairs may be absent if there is a direct connection between a register and a combinational block. There might even be a direct connection between f and g – in that case no work has to be done to verify correct flow of information. In case g takes more than one input from f, the symbols f_D, R_{iLE}, R_{iD} and g_{LE} will denote conjunctions of several control signals. For an interconnection in the specification between a combinational block and a register, only

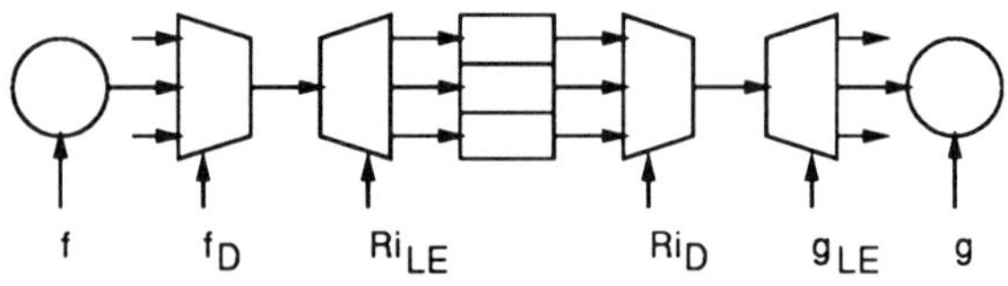

Figure 2: Connection between combinational blocks in the implementation

the portion between the f-block and the register file is relevant; for an interconnection between a register and a combinational block, only the portion between the register file and the g-block is relevant.

5 Correctness Conditions for the Controller

In this section sufficient conditions for the implementation to be in $\gamma_m \circ \beta_{H,n}$ relation with the specification are presented. These conditions are specified on the controller. The data path modules are assumed to be correct. H, n and m are specified by the user.

5.1 Iterative Loops, No If-then-elses

Consider first the case without if-then-elses in the specification. Assume that the sampling of inputs and the production of outputs are events just like the firings of operations in between. One can consider the firing of the first operation in a computation to be the sampling of an input, and the firing of the last operation as the production of an output. Let $s_{x[i]}$, $s_{y[i]}$, ... denote the sampling of inputs, and $p_{z[i]}$, $p_{w[i]}$, ... the production of outputs.

Let H_I be a signal in the implementation equivalent to the user specified H-function. The user might have specified H_I directly as H-function. If this is not the case, it has to be established first that H_I and H are equivalent. Similarly, let H_O be equivalent to H delayed over n cycles. Assume there is one iterative loop in the specification. It will become obvious how to modify the requirement for no iterative loop or more than one iterative loop. Assume the number of iterations in the loop is m. Consider a modulo m counter which increments its value every time H_I is one. If there is such a counter in the implementation, it has to be established that it works correctly. If not, it has to be constructed. Let $count_i$ be a proposition denoting that the value of this counter is i. We can now express the first requirement. For every value of i between 0 and $m-1$, the following conditions have to hold.

$$H_I \wedge count_i \Leftrightarrow s_{x[i]} \wedge s_{y[i]} \wedge \ldots \tag{1}$$

$$H_O \wedge count_{i+d} \Leftrightarrow p_{z[i]} \wedge p_{w[i]} \wedge \ldots \tag{2}$$

The increment d in $count_{i+d}$ accounts for the fact that the value of the counter might have increased by the time the outputs are produced. This will happen when pipelining is applied, i.e. when the processing of consecutive data samples overlaps.

Another set of requirements is as follows. Let f and g be adjacent combinational functions in the specification graph, with f feeding g. Let f' be $R_{jD} \wedge f_{LE} \wedge f \wedge f_D \wedge R_{iLE}$. R_j is the register feeding f, and R_i the one fed by f. This conjunction should be distinct for every operation. If it is not, more conditions should be ANDed with the above formula. g' is similar to f'. For every pair (f, g), the following condition has to be met:

$$g' \Rightarrow AX^- A(\neg R_{iLE} U^- f') \qquad (3)$$

This formula expresses that when g is fired, it should be true that R_i, from which g gets its inputs, has not been corrupted since f was fired and wrote a value in it.

These conditions are almost sufficient for correctness. Consider the vector $X = (x_j)_{0 \leq j \leq m-1}$ of inputs to the specification graph, and the vector $Y = (y_k)_{0 \leq k \leq p-1}$ of intermediate points in the specification. Let Y also contain the outputs of the specification graph. All the elements in this vector have corresponding nodes (mostly registers) in the implementation. Given a way of discriminating relevant time points, the values of these nodes at the relevant time points form a sequence $(y_k^l)_{l \in \omega}$ or $(x_j^l)_{l \in \omega}$.

The relevant time points for the outputs in Y are discriminated by $H_O \wedge count_{i+d}$. Let the relevant time points for the other nodes be determined by formula (3), as follows. Given a relevant time point for the result of g, a relevant time point for the input to g from f is the last firing of f prior to the firing of g, where it is understood that the contents of the register where the value is put does not change between the firing of f and the one of g. This results in the sequences $(X^l)_{l \in \omega}$ and $(Y^l)_{l \in \omega}$. It follows from formula (3) that for any $l \geq 1$, Y^l is a correct function of X^l and Y^{l-1}. Thus, given correct initial conditions, the output values at the relevant time points are a correct function of the input values at some time points. It remains to be shown that the relevant time points for the inputs, induced by formula (3), are identical to the ones marked by $H_I \wedge count_i$. It is still possible, for instance, that the first few operations in the specification graph fire once only, and that the subsequent operations work from the never-changing results of these first operations. It has to be required, therefore, that all operations fire equally many times, and in an appropriate order. More precisely, we require that for every adjacent pair (f, g), the following conditions be met.

$$f' \Rightarrow AX^- A(\neg f' U^- (g' \vee init)) \qquad (4)$$

$$g' \Rightarrow AX^- A(\neg g' U^- f') \qquad (5)$$

The first formula expresses that, prior to a firing of f, f should not have fired since g fired, unless it was the very first firing of f. This formula assumes that the initial state of the controller is never visited again. *init* is a proposition which is true at this initial state. If this is not the case, the controller can be viewed as having been running all the time. *init* should then be replaced with *true*. The second formula is similar, but without the initial state complication. Together, these formulae express that firings of f and g should alternate. This implies that the input times are strictly in the same order as the output times (i.e. that input events are in the same order as the corresponding output events, and that it is not the case that different input events map onto identical output events, or vice versa). Therefore, the first output event corresponds to the first input event, the second with the second, etc. (If this were not the case, different inputs events have to map to the same output event, or vice versa.) Conditions (1) and (2) say that if an input is taken, $H_I \wedge count_i$ is true for some i, and vice versa, and that when an output is produced, $H_O \wedge count_{i+d}$ is true for some i, and vice versa. Thus, the first $H_I \wedge count_i$ corresponds with the first $H_O \wedge count_{i+d}$, and so on, which is what remained to be shown.

The number of conditions to be verified is proportional to the size of the expanded specification graph. This could be problematic if long iterations occur in the specification, so that expansion leads to a large signal flow graph. To solve this problem, we amend the requirement that the conjuction of control signals f' (signifying the firing of f) should be distinct for every operator f. We now require that operations f at different iteration indices map to identical conjunctions f'. (The conjunctions $s_{x[i]}$, $s_{y[i]}$, $p_{z[i]}$ and $p_{w[i]}$, occurring in conditions (1) and (2), should still be distinct for different i values, however.) This requirement implies that all iterations should be implemented the same way in terms of data path operations. Note that this is an extension of the same requirement imposed on the implicit outer loop of the signal flow specification over consecutive input samples. The reason why identical operations at different iteration steps may map to identical conjunctions f' is as follows. The only thing that can go wrong when these, but not other operations, map to identical conjunctions f', is that a function g which takes inputs from a function f, takes inputs from an instance of f at the wrong iteration step. However, this would imply that a certain output of the system corresponds with inputs at the wrong times, but we have shown that this is impossible, due to the fact that adjacent operations alternate. Conditions (4) and (5) make up for the fact that in condition (3), the propositions denoting the firing of an operation are distinct within one iteration step only. If all operations ever to be performed had distinct marks f', conditions (4) and (5) would not have been necessary.

The new requirement leads to simultaneous verification of the m parallel segments in the specification graph. The number of conditions to be verified is now proportional to the length of the specification program.

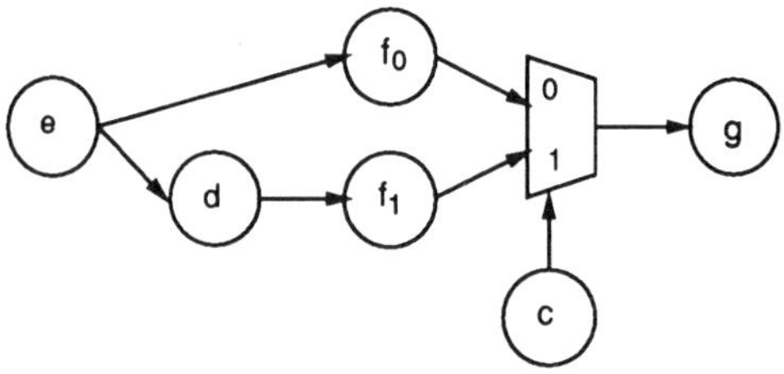

Figure 3: Decision making in the specification

5.2 If-then-elses

Consider now the case with if-then-elses in the specification. While selection is performed by the data path in the specification (Figure 3), we assume it is performed by the controller in the implementation. We assume that the controller not only performs the selection, but also computes the Boolean function c from the values of status registers in the data path at various time points. Let c'_{ij} be the operations that produce these values in the status registers. The meaning of the subscripts is as follows. For the selection of operation f_0, it has to be true that, for all values of i, there is a value of j such that c_{ij} fired with a result $cond_{0ij}$. Likewise, for the selection of f_1, it has to be true that, for all values of i, there is a value of j such that c_{ij} fired with a result $cond_{1ij}$. This corresponds with AND – OR computations on the inputs to c:

$$c(\ldots x_{ij} \ldots) = \bigwedge_i \bigvee_j (x_{ij} = cond_{1ij}) = \neg \bigwedge_i \bigvee_j (x_{ij} = cond_{0ij}) \tag{6}$$

For edge (d, f_1) the requirements are just as before. For edges (e, f_0) and (e, d), conditions (3) and (4) also remain as they are. It is still required that, for instance, when f_0 fires, the value of the register from which it takes its input should not have been corrupted since e fired. Also, going back in time from where f_0 fires, there should be a firing of e before another firing of f_0. Condition (5), however, needs to change because, going back in time from where e fires, there should be a firing of f_0 before another firing of e only when the values produced by the operations c_{ij} were such that f_0 has to fire. This is expressed by the following set of conditions (one for every i):

$$e' \Rightarrow AX^- A(\neg e' U^- (f'_0 \vee init \vee \bigvee_j (AX^-)^q (c'_{ij} \wedge cond_{1ij}))) \tag{7}$$

The integer q denotes the number of pipeline stages in the controller. $(AX^-)^q$ is short for a concatenation of (AX^-), q times.

Edges (f_0, g) and (f_1, g) can be combined by defining a global firing f' as $f'_0 \vee f'_1$. Conditions (3) through (5) are then the same as before, provided

f_0 fires only when the operations c_{ij} produced the appropriate values, and likewise for f_1. This can be expressed as follows:

$$f'_0 \Rightarrow AX^- A(\neg f' U^- \bigvee_j (AX^-)^q (c'_{ij} \wedge cond_{0ij})) \tag{8}$$

Likewise for f_1. For firings of f and c_{ij} to alternate, we also require, for all i, j:

$$(AX^-)^q c'_{ij} \Rightarrow AX^- A(\neg(AX^-)^q c'_{ij} U^- (f' \vee init)) \tag{9}$$

$$f' \Rightarrow AX^- A(\neg f' U^- (AX^-)^q c'_{ij}) \tag{10}$$

Condition (3) combined with (8) implies again that, given correct initial conditions, the output stream at times marked by $H_O \wedge count_i$, for values of i between 0 and $m-1$, is a correct function of input values at some time points. From there it again remains to be shown that these time points are exactly the ones marked by $H_I \wedge count_i$. Conditions (9) and (10) imply that firings of f (either f_0 or f_1) alternate with firings of c_{ij}. Firings of f also alternate with ones of g. Finally, both f_0 and d also alternate with e, except that there may be more than one firing of e between two firings of f_0, or between two firings of d. This will only be under conditions implying the firing of mirror operations (i.e. operations in the other path in the if-then-else construct). All this implies that the same argument can be made as before to show that the output stream is computed from the right input values.

6 A Simple Example

We illustrate our verification method with a simple example – the implementation/specification pair shown in Figure 1. The overall correctness requirement is that the implementation is in $\beta_{x_D,5}$ relation with the specification. The specification could be given in the form of a SILAGE program, which would look as follows:

```
f_result = f(x);
h_result = h(x);
g_result = g(f_result);
p1_result = p(f_result, h_result);
p2_result = p(g_result, p1_result);
```

The first step in the verification process consists of verifying that the combinational blocks f, g, h, and p are functionally equivalent the corresponding blocks in the specification. Block p in the specification should be equivalent to both $p1$ and $p2$ in the specification. The second step in the verification process consists of verifying a number of past-tense CTL formulae with respect to the controller. The controller sequences through the following control vectors:

a_{LE}	b_{LE}	x_D	f_D	g_D	h_D	p_D
1	0	1	0	0	0	0
0	1	0	0	0	1	0
1	0	0	1	0	0	0
0	1	0	0	0	0	1
1	0	0	0	1	0	0
0	1	0	0	0	0	1

After asserting the last control vector in the table, it starts again with the first one. We now consider the past-tense CTL formulae that have to be verified. Condition (1) in Section 5, saying that at every assertion of the H-function (which is x_d in this case), a new input should be taken, reduces to a tautology for the present circuit. Condition (2) is more complicated. H_O can be constructed as a function from the state variables to the effect that it is true in the last state. It first has to be established that H_O is equivalent to H, delayed over 5 cycles. This can be done by verifying the following CTL formula:

$$H_O \Leftrightarrow (AX^-)^5 H_I \tag{11}$$

where H_I equals x_D, and $(AX^-)^5$ is short for a concatenation of AX^- 5 times. Next, a number of conditions have to be verified for each edge in the specification graph. We only consider the edge between f and g here. Conditions (3) through (5) become:

$$(g_D \wedge a_{LE}) \Rightarrow AX^- A(\neg a_{LE} U^- (f_D \wedge a_{LE})) \tag{12}$$

$$(f_D \wedge a_{LE}) \Rightarrow AX^- A(\neg (f_D \wedge a_{LE}) U^- (g_D \wedge a_{LE})) \tag{13}$$

$$(g_D \wedge a_{LE}) \Rightarrow AX^- A(\neg (g_D \wedge a_{LE}) U^- (f_D \wedge a_{LE})) \tag{14}$$

It can be seen from the controller that all these conditions are satisfied. The same is true for the other edges in the specification graph. Therefore, the implementation is correct. Note that for large bit-widths, it will not be possible to use non-compositional techniques like those of [1, 7] to verify even the relatively simple example of Figure 1.

7 Conclusion

In this paper, a compositional strategy was presented for the automatic verification of digital designs with synchronous globally clocked control. The correctness condition is that the implementation be in a certain relation with the specification, where both the implementation and the specification are FSM's. The relations verified include the "don't care times" relation, the parallelism relation, and compositions thereof. The compositional strategy consisted of verifying the combinational blocks in the data path of the specification against the ones in the implementation, and verifying the sequence of control inputs and outputs against the connections in the specification. The latter is done by verifying a collection of past-tense CTL formulae with respect to the controller of the implementation. These formulae are proven to be sufficient conditions for correctness.

Acknowledgments

This work was supported in part by the Defense Advanced Research Projects Agency under contract N00014-91-J-1698, and in part by an IBM fellowship.

References

[1] F. Van Aelten, J. Allen, and S. Devadas. Verification of Relations between Synchronous Machines. In *Proceedings of the International Conference on Computer Aided Design*, pages 380–383, November 1991.

[2] R. Bryant. Graph-Based Algorithms for Boolean Function Manipulation. In *IEEE Transactions on Computers*, volume C-35, pages 677–691, August 1986.

[3] J. Burch, E. Clarke, K. McMillan, and D. Dill. Sequential Circuit Verification Using Symbolic Model Checking. In *Proceedings of the* 27th *Design Automation Conference*, pages 46–51, June 1990.

[4] L. Claesen, M. Genoe, F. Proesmans, E. Verlind, and H. De Man. SFG-Tracing: a Methodology for the Automatic Verification of MOS Transistor Level Implementations from High Level Behavioral Specifications. In P.A. Subrahmanyam, editor, *Lecture Notes in Computer Science*. Springer-Verlag, To appear.

[5] E. M. Clarke and O. Grumberg. Research on Automatic Verification of Finite-State Concurrent Systems. In *Annual Reviews of Computer Science*, volume 2, pages 269–290, 1987.

[6] A. Cohn. A Proof of Correctness of the VIPER Microprocessors: The First Level. In G. Birtwistle and P.A. Subrahmanyam, editors, *VLSI Specification, Verification and Synthesis*, pages 27–71. Kluwer Academic Publishers, 1988.

[7] O. Coudert, C. Berthet, and J. C. Madre. Verification of Sequential Machines Using Boolean Functional Vectors. In *IMEC-IFIP Int'l Workshop*

on Applied Formal Methods for Correct VLSI Design, pages 111–128, November 1989.

[8] O. Coudert, C. Berthet, and J. C. Madre. Verifying Temporal Properties of Sequential Machines Without Building their State Diagrams. In *Dimacs Workshop on Computer-Aided Verification*, June 1990.

[9] E. A. Emerson. Temporal and Modal Logic. In *Handbook of Theoretical Computer Science; Volume B: Formal Models and Semantics*, pages 995–1072. Elsevier, 1990.

[10] G. Goossens, J. Rabaey, J. Vandewalle, and H. De Man. An Efficient Microcode Compiler for Application Specific DSP Processors. In *IEEE Transactions of Computer-Aided Design*, pages 925–937, September 1990.

[11] P. Hilfinger. A High-Level Language and Silicon Compiler for Digital Signal Processing. In *Proceedings of the Custom Integrated Circuits Conference*, pages 213–216, May 1985.

[12] W. Hunt. *FM8501: A Verified Microprocessor.* University of Texas, Austin, Tech. Report 47, 1985.

[13] J. Joyce. Formal Verification and Implementation of a Microprocessor. In G. Birtwistle and P.A. Subrahmanyam, editors, *VLSI Specification, Verification and Synthesis*, pages 129–157. Kluwer Academic Publishers, 1988.

Functional Partitioning for Verification and Related Problems*

Jawahar Jain† James Bitner Jacob A. Abraham† Donald S. Fussell‡

Computer Engineering Research Center
The University of Texas at Austin
Austin TX, 78712

Abstract

We present *functional partitioning* methods to reduce the computational resources used in analyzing Boolean functions. One general partitioning strategy, *orthogonal partitioning*, divides a function's truth table into disjoint partitions; the Boolean **and** of any two subfunctions is always zero. Functional partitioning techniques can significantly augment current OBDD-based verification methods. The design and specifications are partitioned into smaller functions that can be independently verified in a pairwise manner; these functions can even be ordered separately. This allows verification in polynomial space and time of several functions whose OBDD representations have been proven to be exponential in the number of input variables such as Bryant's hidden weighted bit function, as well as much more efficient verification of other difficult functions such as multipliers. Functional partitioning improves the efficiency of probabilistic verification as well, where even further improvements can be realized for the above functions. In the presence of design errors, these partitioning techniques can provide even greater efficiency gains.

1 Introduction

Ensuring that a design correctly implements its specifications is now recognized as an issue of central importance in designing complex systems. There is an urgent and growing need for efficient methods that will guarantee equivalence between two different design descriptions. Boolean simulation proves increasingly inadequate in the error coverage that it provides as systems become larger and more complex. Recently, increasingly sophisticated deterministic and probabilistic verification methods have been proposed [3, 8]. The most popular current methods based on OBDDs [3] fail primarily because the space requirements are prohibitive, as in the case of ISCAS c6288, a 16-bit multiplier. In this paper, we discuss a functional partitioning strategy which can supplement current OBDD-based verification procedures by greatly reducing the space required for verification. It can also substantially reduce the time. Probabilistic methods [8] can utilize these partitioning techniques to achieve even more significant time and space advantages.

*This work was supported in part by the Texas Advanced Research Program and in part by the SDIO Innovative Science and Technology Office and managed by the Office of Naval Research (ONR) under Contract N00014-89-K-0089 at the University of Texas, Austin.

†also with the Dept. of Electrical and Computer Engg., University of Texas at Austin.

‡also with the Dept. of Computer Science, University of Texas at Austin.

Functional partitioning is a class of techniques which transform a Boolean function into a set of subfunctions with a well-defined logical relationship such as their Boolean **and** being always zero (*orthogonal partitions*), or into subfunctions whose Boolean **or** is always one (*disjunctive partitions*). Because functional partitioning can find equivalent partitions with known logical relationships in a pair of functions being analyzed, it allows verification to proceed on representations of the subfunctions without the need to build a single OBDD for the entire function. Thus we obtain a method which is efficient because it operates on these *partitioned OBDDs* rather than on the monolithic OBDDs currently in use.

Using partitioned OBDDs, we can verify a function given by Fortune, Hopcroft and Schmidt [6] in $\Theta(n^3)$ space and time which would require non-polynomial space without such partitioning. Similarly we can (probabilistically) verify the *hidden-weighted bit* function given by Bryant [4] in $\Theta(n^2 \log n)$ space and $\Theta(n^3 \log n)$ time. Partitioned OBDDs can use a set of $\Theta(n)$ OBDDs to represent the HWB function in $\Theta(n^3)$ space, while exponential space is required using a monolithic OBDD. They similarly allow polynomial time verification of the HWB function. Partitioned OBDDs allow verification of c6288, a 16-bit multiplier, by reducing the space required using monolithic OBDDs by three orders of magnitude. Through orthogonal partitioning, such a circuit was also (probabilistically) verified against its mathematical specification. In the presence of design errors, functional partitioning not only helps to locate the error but also substantially decreases the verification time. Finally, the method is well suited for parallel execution.

When both space *and* time are prohibitive, all verification methods fail. Our partitioning methods provide an opportunity in such cases to verify a part of a design - likely providing greater error coverage than possible through random Boolean simulation.

In Section 2 we will discuss a theory of functional partitions. We will demonstrate functional partitioning using theories of orthogonal and disjunctive partitions. There are three fundamental ways in which orthogonal partitions can be created. We will discuss each of them with their applications in Section 3, 4 and 5, demonstrating their utility on a multiplier, the FHS(n) and the HWB function, respectively. In Section 6 we contrast the relative restrictions on the use of orthogonal partitioning for deterministic and probabilistic methods. Section 7 outlines some heuristics for deciding which of these functional partitioning techniques, if any, will be applicable for a given circuit.

2 Fundamentals of Functional Partitions

Given the problem of checking for the satisfiability of a Boolean expression, we can partition the problem into a set of independent problems so that testing for satisfiability becomes checking whether *any* partition (subfunction) is satisfiable. Similarly to check the equivalence of two functions, F and G, if we can choose pairs of corresponding partitions of the given functions such that checking equivalence of F and G is transformed to only checking equivalence of these corresponding pairs, then (very likely) we will find that we have considerably simplified the computational complexity of this problem.

It should be intuitively clear that for any meaningful use of such partitions in verifying functional equivalence of two different functions, the partitions must have a well defined and useful relationship with respect to each other. *Mutually orthogonal* partitions form one such possibility. Two functions are *orthogonal* if and only if their Boolean **and** is always zero. Clearly two orthogonal functions do not have any minterm in common. Thus partitioning a function orthogonally is equivalent to dividing its truth table into disjoint parts. We can divide the truth tables into k parts, where the ith part is specified by a *window function*, w_i. To compare the equivalence of F and G when both functions are partitioned in the same way, we need only compare the equivalence of the corresponding partitions. Formally, we have

Orthogonal Correspondence Theorem: Given k mutually orthogonal Boolean functions $w_1, \ldots, w_k$ such that $w_1 \vee \ldots \vee w_k = 1$, then $F \equiv G$ if and only if $F \wedge w_i \equiv G \wedge w_i$ for $1 \leq i \leq k$.

Then $(F \wedge w_1, \ldots, F \wedge w_k)$ are the k orthogonal partitions of F.

Such a partitioning approach enjoys the typical advantages of a divide and conquer strategy. In addition, by breaking the original problem into several independent cases, each subfunction or *pair* of subfunctions can be usefully subjected to different analytical techniques and representations. By creating a "balanced" number of partitions, we will demonstrate that one can gain not only in space but often in time as well. As we will see, these time and space gains can often be dramatic.

To derive *corresponding* orthogonal partitions between two functions, one would look for a characteristic common to the two functions. One characteristic that is guaranteed to be common between the two functions is their variable set. Thus one convenient way to derive corresponding orthogonal partitions is through the primary inputs. Shannon's expansion theorem provides one easy and practical way by which a function F that has a variable x appearing in it can be written as $F = \overline{x}F_{x=0} \vee xF_{x=1}$. Now two functions F and G are equivalent if and only if $F_{x=0} \equiv G_{x=0}$, and $F_{x=1} \equiv G_{x=1}$. Notice that the Boolean **and** of $\overline{x}F_{x=0}$ and $xF_{x=1}$ is always zero as required of orthogonal partitions.

The pairwise correspondence of partitions presented above allows particularly convenient and efficient comparison of two functions. Orthogonal partitioning can also be used for testing for functional equivalence when the functions are not partitioned into corresponding orthogonal partitions, as shown in the following theorem. We let $\mathbf{m}(f)$ denote the number of satisfying assignments to function f (i.e., the number of minterms).

Minterm Matching Theorem: Let Boolean functions F and G be written in terms of mutually orthogonal Boolean functions as $F = f_1 \vee \ldots \vee f_m$ and $G = g_1 \vee \ldots \vee g_n$. Then $F \equiv G$ if and only if

$$\text{for all } 1 \leq i \leq m,\ \mathbf{m}(f_i) = \sum_{j=1}^{n} \mathbf{m}(f_i \wedge g_j) \quad \text{and}$$

$$\text{for all } 1 \leq j \leq n,\ \mathbf{m}(g_j) = \sum_{i=1}^{m} \mathbf{m}(f_i \wedge g_j).$$

Note that $F \equiv G$ if and only if $F \rightarrow G$ and $G \rightarrow F$. The first condition above tests $F \rightarrow G$, which is true if and only if every minterm in F is also present in G. To determine this, we examine each function f_i and see if all its minterms occur in G. $\mathbf{m}(f_i \wedge g_j)$ gives the number of minterms common between f_i and g_j. Since $g_1, \ldots, g_n$ are orthogonal, they have no minterms in common, and hence a minterm in f_i cannot be present in two different g_j's. Therefore if $\mathbf{m}(f_i) = \sum_{j=1}^{n} \mathbf{m}(f_i \wedge g_j)$, every minterm in f_i must have a corresponding minterm match in G. Otherwise there is a minterm that does not match, and the functions are not equivalent. Similarly, the second condition tests if $G \rightarrow F$.

The Minterm Matching Theorem suggests that partitioned OBDDs can be very generally useful for efficient verification even in cases where it might not be feasible to guarantee that two functions have corresponding partitions, or where the correspondence is not known. It can be seen that in testing for $F \rightarrow G$, the theorem can be also applied by matching only those minterms of f_i in g_j which have not been matched with any partition of G examined so far. For example, $f_i \wedge \overline{g_1}$ gives minterms of f_i which did not match with g_1; $\mathbf{m}(f_i) - \mathbf{m}(f_i \wedge \overline{g_1})$ gives us the number of matching minterms,[1] and now it suffices to match $f_i \wedge \overline{g_1}$ with g_2.

Orthogonal partitioning is particularly useful in conjunction with *probabilistic verification.*[2] For each function an integer *hash code* is generated, based on a weighted sum of the 2^n truth values of the function. The hash code is formally defined in terms of an integer-valued *algebraic transform,* $\mathbf{A}[F]$ of the original Boolean function F. Due to this definition of the hash code, if F is written in terms of k mutually orthogonal Boolean functions as $\beta_1 \vee \ldots \vee \beta_k$, the hash codes for the β_i's can simply be summed to yield the hash code for F. In fact, if a function is described using any combination of a number of orthogonal subfunctions, its hash code can be computed using only the hash codes of the individual subfunctions using the Orthogonal Partitioning Theorem [8]. Using probabilistic methods, the two functions can be very conveniently partitioned and processed independently.

Another technique analogous to orthogonal partitioning is *disjunctive partitioning.* We can write $F = (\overline{x} \vee F_{x=1}) \wedge (x \vee F_{x=0})$ where the Boolean **or** of the above two partitions ($\overline{x} \vee F_{x=1}$ and $x \vee F_{x=0}$) is always one. By applying the principle of duality, the theories and procedures for orthogonal partitions can be used for disjunctive partitions as well. For example, in probabilistic techniques, if a function can be broken into a set of disjunctive partitions, we can compute hash codes for each partition separately to yield an integer-valued representation for the original function. Due to lack of space, we will not elaborate on the uses of disjunctive partitioning.

In the following sections, we discuss three practical techniques for creating orthogonal partitions, namely, *Partial Assignments*, *Circuit Division*, and *Graph Division.* We describe each technique, illustrate its application on a particular circuit, and present the time and space advantages achieved.

3 Partial Assignments

As observed earlier, Shannon's expansion theorem orthogonally partitions a Boolean function f where $f = \overline{x} f_{x=0} \vee x f_{x=1}$. Thus, if C is the circuit for f, we can represent (evaluate) it by representing (evaluating) the two *reduced circuits,* $C_{x=0}$ and $C_{x=1}$, where input x is set to 0 and 1, respectively. Deterministic methods take the corresponding partitions in the two design descriptions (circuits) and verify the equivalence of such a pair.

[1] The number of minterms in an OBDD graph can be computed in time linear in the size of the graph. Also see section 5.

[2] For the theory and terminology of probabilistic verification see [7, 8]. Also see [2] for probabilistically checking equivalence of two free BDDs.

Algebraic transforms of Boolean functions can be expanded in a field in a manner analogous to Shannon's expansion of Boolean functions. Specifically, we can write $\mathbf{A}[f] = (1 - R_x) \cdot \mathbf{A}[f_{x=0}] + R_x \cdot \mathbf{A}[f_{x=1}]$, where R_x is the integer assignment given to the variable x [8]. The results of the computations performed for $C_{x=0}$ and $C_{x=1}$ can then be combined to yield the result for C simply by computing a weighted sum of the results.

One method for creating such partitions is to generate the reduced circuits $C_{x=0}$ and $C_{x=1}$ by setting x to the appropriate Boolean value and simplifying. Another approach is simply to substitute the given Boolean value for x and evaluate the original circuit C. If we choose to evaluate C by either of these methods, the space required is clearly reduced, since both $C_{x=0}$ and $C_{x=1}$ are smaller than C, and they can be constructed and evaluated sequentially rather than simultaneously as with *apply* [3] or the *collapse-with-compose* algorithm [8]. The time required to evaluate either $C_{x=0}$ or $C_{x=1}$ will be less than the time for C as well. The *sum* of these times might not be less, however, in which case we have gained space and lost time. Thus, we often have a time-space trade-off consideration when using this technique.

The latter approach can be extended to partitioning on more than one variable using a further Shannon's expansion of f. The result is a binary *assignment tree*, whose interior nodes represent the $\vee$ operations introduced in the expansion and each of whose leaves is an OBDD representing the reduced function resulting from the variable assignments corresponding to the given path.[3] We call the functions corresponding to various paths from the root to the leaves *partial assignments.* A simple and reasonable strategy is to choose a value for k and use all 2^k partial assignments for the first k variables in the ordering used for OBDD construction. Note however, that all paths in the tree need not be of the same length and that the variables used along different paths need not be the same. Thus, we have considerable freedom in choosing the assignment tree.

The efficiency of this method relies on the assumption that the representations for the reduced circuits will be much smaller than the representation for the entire circuit. For functions that require an exponentially large OBDD, such as multipliers, the structure of the first several levels of the OBDD can be expected to resemble a binary tree. In this case, the two subgraphs of the root node can be expected to be mostly disjoint, under any reasonable ordering, Thus, it is intuitively reasonable that making partial assignments to the first few variables in the OBDD ordering will divide the OBDD for the entire function into smaller and nearly disjoint pieces.

We will now show that this assertion does in fact hold for the multiplier function. A program was written to perform orthogonal partitioning by partial assignments and run on the first 16 outputs of the multiplier circuit c6288. We chose to use a assignment tree on the first v variables in the most reasonable-seeming ordering, $(a_0, b_0, a_1, b_1, \ldots)$. However, paths in the tree were cut off once they contained z zeros, provided they were of length at least l. The choice of these three parameters determines the number of partial assignments, m. We chose the strategy of cutting off the tree in this way because it was known *a priori* for a multiplier function that using the value 0 for a variable reduces the circuit much more than the value 1. Thus, it seemed worthwhile to cut off the assignment tree when a path had many zeros in order to balance the space requirements of the partitions.

[3] In the extreme case, we can make a complete decomposition into 2^n partitions, resulting, effectively in Boolean simulation. This illustrates the extent of the time-space trade-off possible in this procedure.

output	Monolithic OBDD	Partitioned OBDDs			redundancy ratio, r	ratio of space improvement, $2D/P_{\max}$
	D	m	$P_{\max}$	$\sum_{i=1}^{m} P_i$		
10	5,248	7	1137	6,483	1.235	9
12	29,400	32	1582	40,126	1.364	37
14	162,768	256	1075	226,474	1.391	303
16	930,000*	512	2757	1,199,697	1.289	675
16	930,000*	1024	1587	1,348,134	1.449	1172
16	930,000*	3942	487	1,520,146	1.634	3819

*estimated by extrapolation from the D's of earlier outputs.

Table 1: Graph Size Comparison for Orthogonal Partitioning

The program collected the following statistics for a particular output and set of m partial assignments: D, the size of the OBDD for the entire function, and P_i, the size of the OBDD computed for the ith partial assignment. In addition to computing $\sum_{i=1}^{m} P_i$, it also computed the size of the largest OBDD, $P_{\max} = \max_{i=1}^{m} P_i$. The results are given in Table 1.

We can calculate several interesting measures from this data. Unless the OBDD for the function is a tree, an overlap will occur between the OBDD subgraphs resulting from different partial assignments. We quantify the redundant computation by $r = (\sum_{i=1}^{m} P_i)/D$, which we will call the *redundancy ratio.* Table 1 shows that increasing m appropriately makes the graph sizes for these partitions quite small, and that r grows only slowly with the number of partitions.

We also calculate the *ratio of space improvement* between a monolithic OBDD method and a probabilistic or partitioned OBDD method. In the probabilistic approach, each partition is computed separately and then hashed to an integer. Thus the maximum memory requirement is the size of the largest OBDD, i.e., $P_{\max}$. If partitioned OBDDs are used for deterministic verification, each partition of F must be compared with one or more partitions of G, depending on whether partition correspondences are known. In either case, at most one partition from each of F and G must exist at once, so the maximum memory requirement will equal $2P_{\max}$. Monolithic OBDD methods compare the complete OBDDs for the two functions being verified. Thus their space cost is $2D$. Hence the ratio of improvement in space is $2D/P_{\max}$ for probabilistic verification and $2D/2P_{\max}$ for partitioned OBDDs. Table 1 shows that the partial assignment approach can show a 3 order of magnitude improvement in the space required to verify a 16-bit multiplier.

4 Circuit Division

One straightforward manner to exploit the principle of orthogonality is to physically locate subcircuits which represent orthogonal functions, or create them by transforming the circuit. Such subcircuits can be independently represented and either a pairwise verification can be performed or the Minterm Matching Theorem can be applied. Similarly their hash codes can be combined to yield the final hash code according to the Orthogonal Partitioning Theorem [8]. Though, in general, it is difficult to determine whether two subcircuits represent orthogonal functions, we can detect this occurrence in numerous cases.

4.1 A Simple Circuit Division Algorithm

If we determine that gates g_1 and g_2 represent functions F_1 and F_2 which can be written as $\overline{x} \wedge f_1$ and $x \wedge f_2$, for some variable x, we can conclude F_1 and F_2 are orthogonal. If g_1 and g_2 are determined to represent orthogonal functions, we must also consider whether a useful partitioning can be derived. If the set of gates which reach g_1 is *disjoint* from the set reaching g_2, the circuit can be partitioned into disjoint subcircuits. This will be the case if g_1 and g_2 are *logic dominators,* i.e., each dominates (in the graph-theoretic sense) all the gates which reach it. Logic dominators can be found in almost linear time using the algorithm in [10]. Even if there is some overlap, it is worthwhile partitioning the circuit if only a small amount of redundant computation is required.

The power of this approach is that it divides the problem into several simpler independent problems. The subcircuits might have a simpler structure which makes them more amenable to further analysis, like (recursive) partitioning or some other hashing technique. Since the subcircuits are evaluated independently, different methods of representation and different hashing techniques can be possibly used on each one.

To illustrate the partitioning process and demonstrate the power of this technique, let us consider the function defined in [6, p. 234]. For any n, a function of $2n - 1$ inputs is defined which we will call FHS(n). It is proved that exponential space is required to represent it as an OBDD, for any possible ordering of the input variables. Hence exponential time is required to compute it. However, a free BDD is given to represent it in $\Theta(n^2)$ space.

From the description of the function, it is simple to construct a circuit representation. One possibility is shown in Figure 1. The upper part of the circuit consists of an $(n - 1) \times n$ grid of positions. A position either does nothing (represented by an empty circle) or computes $c_i \equiv d_j$ (represented by $\circledast$). The results of these equivalence tests are then AND'ed along the diagonal lines, with a logic 1 fed in at the top. The selection logic simply takes the c_i's and produces the product term shown on each of the 8 outgoing selection lines, using 8 AND gates. The results from the selection logic are used to choose which of the diagonals is to be the output of the circuit by AND'ing each with the appropriate selection line and OR'ing the results. Note that the intermixing of these control and data inputs and the pairing of each c_i with nearly every d_j is what leads to an exponentially large OBDD representation.

Our circuit division algorithm begins by identifying orthogonal subcircuits, using the technique mentioned above, as follows: For every variable x, each gate is labeled with a pair (f_0, f_1) where $f_0, f_1 \in \{0, 1, \mathbf{f}\}$, such that the function of the gate can be written as $\overline{x} f_0 \vee x f_1$, where $\mathbf{f}$ is used to denote an arbitrary function. For variable c_1, gates $A_1, \ldots, A_4$ are labeled with $(\mathbf{f}, 0)$ and gates $A_5, \ldots, A_8$ with $(0, \mathbf{f})$. The label of $(\mathbf{f}, 0)$ for variable c_1 signifies that the function at those gates can be written in the form $\overline{c_1} \wedge \mathbf{f}$; the $(0, \mathbf{f})$ label for c_1 identifies a gate to represent a function of the form $c_1 \wedge \mathbf{f}$. The variable c_1 in above may be regarded as an *ideal splitting variable.*

Since each A_i is a logic dominator, the logic of the circuit can be partitioned cleanly into two subcircuits. We can form these two subcircuits by

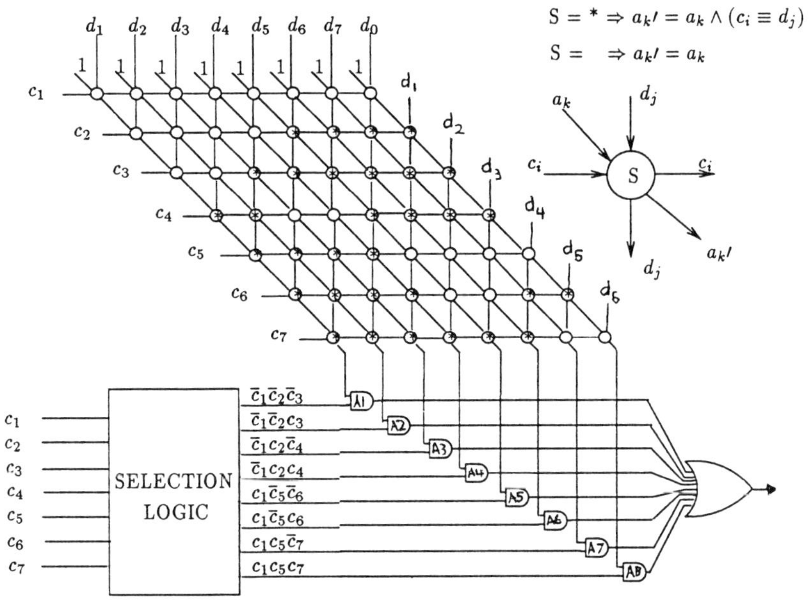

Figure 1: A Circuit Realization of FHS(8).

deleting the output gate and introducing two OR gates which combine the results of $A_1, \ldots, A_4$ and $A_5, \ldots, A_8$, respectively.

We continue recursively by partitioning each subcircuit. Thus we identify c_2 and c_3 as the respective ideal splitting variables for each. After another level of recursion, the partitioning terminates, and each resulting subcircuit consists of a single diagonal AND'ed with a single selection line. It is therefore a *tree*. It can be thus easily represented deterministically.

The running time can be easily computed. Note that the circuit has $2n-1$ inputs and $\Theta(n^2)$ gates. Hence labeling it requires time $\Theta(n^3)$, and finding the dominators requires time $\Theta(n^2\alpha(n^2))$. Given this information, the circuit is quickly decomposed, and the information need not be recomputed to partition the subcircuits. Finally, discovering that each of the n reduced circuits is a tree requires time $\Theta(n)$. Computing its deterministic representation will now require only $\Theta(n^2)$ time. Therefore, the entire process takes $\Theta(n^3)$ space and time.

The simple algorithm outlined above could label a gate with 9 different values. This can be now augmented so that more information about a function can be stored. By adding another symbol (say, $\mathbf{g}$) to $\{0, 1, \mathbf{f}\}$, we can now improve the algorithm to also capture cases where the function represented at a gate can be written as $(\overline{x} \vee f_1) \wedge (x \vee f_2)$. An algorithm can be thus easily written which by using only the fundamental rules of Boolean algebra can create any of the 16 such labels for all gates, with respect to every variable, to capture (most) such cases where two inputs to a gate represent orthogonal or disjunctive functions. For circuit of size C with n inputs, such an algorithm can determine labels for all gates in $O(n \cdot C)$ time.

4.2 Deterministic Verification Using Circuit Division

In general, for deterministic verification using partitioned OBDDs, employing the Correspondence Theorem is more efficient than employing the Minterm Matching Theorem. The Correspondence Theorem requires that the functional divisions within the given specification correspond to the partitions computed from the circuit design. A reasonable verification strategy can then be to first compute orthogonal partitions from one of the design descriptions, such as the FHS(8) circuit above. The sequence of Boolean values provided to the splitting variables in computing such partitions can now be treated as a *partial assignment.* The Boolean expression specification of the FHS(8) function can then be orthogonally partitioned by applying the same set of partial assignments. For example, we identify $\overline{c_1}c_2c_4$ as one *pattern* on the splitting variables in the particular circuit implementation, which created an orthogonal partition. If the specification is simulated using the same pattern, then we will have created in it a corresponding orthogonal partition. This pair of partitions can now be verified for equivalence. For the FHS(n) function such a process can leave us with verifying equivalence between n different pairs of tree circuits with n such partitions in the specification. Since each of the n reduced circuits is a tree, a FHS(n) circuit can be easily verified probabilistically [8] or deterministically.

Similarly, the Achilles-heel function and two other related functions given in [1] can be verified by augmented deterministic methods requiring only linear space representations and $\Theta(n^2)$ time using circuit division techniques. An 8-bit Achilles-heel function can be written as

$$f = \overline{mux_0}\cdot\overline{mux_1}\cdot\overline{mux_2}\cdot f_0 \vee mux_0\cdot\overline{mux_1}\cdot\overline{mux_2}\cdot f_2 \vee \ldots \vee mux_0\cdot mux_1\cdot mux_2\cdot f_7$$

In the above, f_i is given by $\Pi(x_j \vee y_{(j+i)mod 7})$ for $0 < j \leq 7$. Thus $f_0 = x_0y_0 + \ldots + x_7y_7$; mux_i provide $\log(n)$ control inputs. An n-bit function of this nature has a $\Theta(n^2)$ circuit implementation. Constructing an OBDD representation for the complete function requires exponential resources. Note the circuit (function) has n orthogonal partitions. Each partition can be ordered independently and thus represented in linear space and hence verified deterministically in quadratic time.[4] Thus the complete circuit can be verified in $\Theta(n^3)$ time. The probabilistic method will be similarly successful. These classes of circuits provide examples which requires exponential resources using deterministic techniques which compute a monolithic OBDD for a given output function but polynomial time verification algorithms can be written using orthogonal partitioning. Notice if the two design descriptions break into noncorresponding orthogonal functions, circuit division can still be employed for verification by the application of the Minterm Matching Theorem.

Nevertheless, the extension of such techniques to OBDD based deterministic methods has obvious limitations. For example, consider verifying two FHS(n) circuits in which the $\vec{c}$ and $\vec{d}$ inputs have been interchanged,

[4] Representing all partitions together as a partitioned BDD, with each partition ordered independently, leads to a $\Theta(n^2)$ free BDD representation.

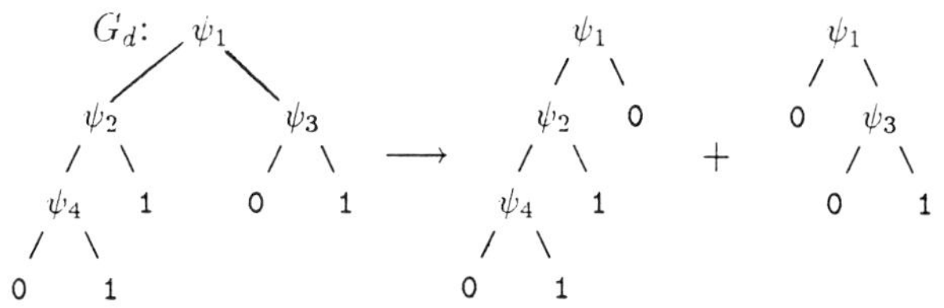

Figure 2: Orthogonal Partitioning of a Decomposed Representation.

due to a design error. A probabilistic method can hash each circuit independently through circuit division. Our partitioned OBDD methods will not be successful here since the two circuits do not have identical splitting variables. Constructing the OBDD graphs for even one partition will now require exponential resources in one of the circuits. Ordering both design descriptions differently and applying the Minterm Matching Theorem similarly has corresponding limitations in computing the common minterms between the two partitions. Thus even the partitioned OBDD techniques fail. Note the failure in efficiently identifying congruent splitting variable pairs is not sufficient to declare the two circuits inequivalent.

Some verification methods rely on comparing points of internal equivalences [5, 9] in two circuits. Such methods can use circuit division even when only a subset of the internal points can be partitioned through such circuit division.

5 Graph Division

As all paths in a BDD that originate from a common node represent mutually orthogonal functions, orthogonal partitions can also be derived from a given BDD representation of a function. While there is no benefit if all the variables in the BDD are input variables, consider the case when the nodes in the BDD represent *subfunctions* (Figure 2). Here graph G_d represents overall function $F = \overline{\psi_1}(\psi_2 \vee \psi_4) \vee \psi_1\psi_3$ in terms of four subfunctions $\psi_1, \ldots, \psi_4$. We will refer to such a graph as a *decomposed representation* of F. Orthogonal partition of F can be now computed by simply partitioning its decomposed representation (G_d) along paths originating from a common node; Figure 2 shows the result of partitioning G_d into two graphs representing functions $\overline{\psi_1} \wedge (\psi_2 \vee \psi_4)$ and $\psi_1 \wedge \psi_3$, which are then the two partitions of the decomposed graph. Since these two functions are orthogonal, each can be represented and evaluated separately. The probabilistic method will simply directly add the hash codes to yield the final hash code. Augmenting deterministic methods using the Correspondence Theorem is more difficult. However, the Minterm Matching Theorem is directly applicable.

We will illustrate the utility of partitioning the decomposed graph by presenting an efficient procedure to verify the hidden weighted bit (HWB) function, described in [4]. To illustrate the generality of the partitioning principles we will also show how we can not only efficiently partition the circuit representation but an abstract specification for the HWB function as well. The behavioral specifications of many functions is rather easily broken into orthogonal partitions because different aspects of a function's behavior usually correspond to different parts of its truth table, and hence naturally form orthogonal partitions.

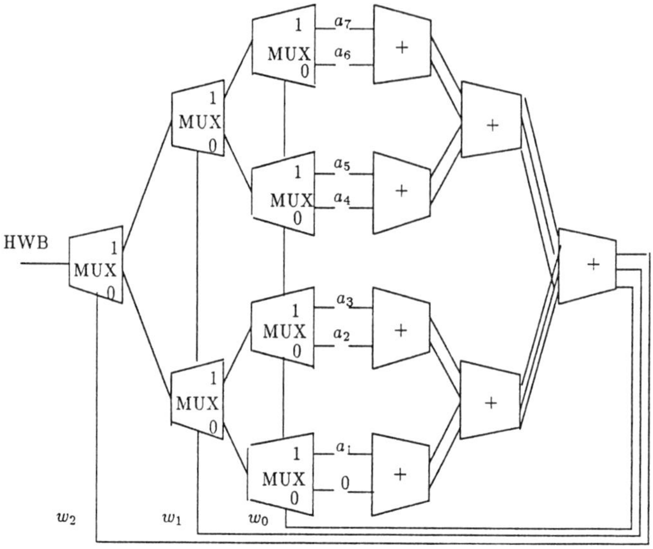

Figure 3: The circuit for a hidden weighted bit function

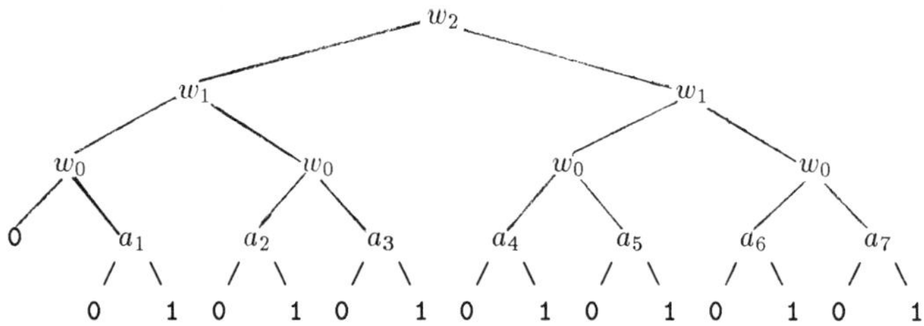

Figure 4: A decomposed graph for the hidden weighted bit function

5.1 Partitioning an HWB Circuit

The HWB function can be considered as a selector function. Given n inputs, $a_1, \ldots, a_n$, the function selects input a_i as its output if exactly i inputs are 1. If all inputs are 0, the output of the function is also 0. A circuit for a 7-input HWB function [4] is shown in Figure 3. The circuit has a rather natural decomposition which leads to the tree structure of the decomposed graph shown in Figure 4. Since we have $\Theta(\log n)$ w_i's, the decomposed graph has a size only linear in number of inputs, and we can easily partition it separately along each possible path. For example, we can partition the HWB function shown in Figure 4 in seven orthogonal partitions: $\overline{w}_2 \wedge \overline{w}_1 \wedge w_0 \wedge a_1, \ldots, w_2 \wedge w_1 \wedge w_0 \wedge a_7$. The OBDD for each w_i requires $\Theta(n^2)$ nodes.[5] Thus each partition can be computed in $\Theta(n^2 \log n)$ space. However, representing OBDDs for all partitions in only one graph leads to an expo-

[5]Note that each w_i is simply a bit in the sum $a_1 + \cdots + a_n$. Due to the symmetry between the a_i's, all variable orderings result in an OBDD of the same size for any w_i.

n		Probabilistic Method		Partitioned OBDDs		Monolithic OBDDs			
						apply/compose		apply only	
		Time	Space	Time	Space	Time	Space	Time	Space
15	Ckt	2.3	883	3.7	732	4.4	3,586	5.9	2,861
	Spec	0.3	(1)	(3)	(3)	0.7	806	0.7	806
31	Ckt	14.1	4,640	29.1	3892	29.5	32,882	121.4	111,601
	Spec	0.9	(1)	(3)	(3)	5.5	19,782	5.5	19,782
63	Ckt	131.8	28,015	300.2	24,005	unable (2)		unable (2)	
	Spec	6.0	(1)	(3)	(3)	unable (2)		unable (2)	
127	Ckt	1483.4	188,422	3938.0	164,191	unable (2)		unable (2)	
	Spec	47.8	(1)	(3)	(3)	unable (2)		unable (2)	

Table 2: Verification of an n-bit HWB function using Graph Division

Time is in seconds. Space is the peak memory requirement for all dynamic data structures, and is measured in units of BDD nodes.

(1) The space required is an integer array of size n.

(2) We estimate by extrapolation that for HWB(63) and HWB(127), OBDDs of 2.5 million and 239 billion nodes would be required. The 'space' required to compute these OBDDs is expected to be 10 to 15 times as great.

(3) Time and space for circuit and specification combined for this method.

nential blow-up in the OBDD size. Bryant showed that the function has no efficient OBDD representation [4]. By hashing each partition separately and directly adding their respective hash codes, we can compute the final hash code in $\Theta(n^3 \log n)$ time. We will show by applying the Minterm Matching Theorem and generating a polynomial number of orthogonal partitions from the specification, efficient deterministic verification is also achieved.

5.2 Partitioning the Mathematical Specification

A formal mathematical definition of the HWB function is easily made in terms of $\mathbf{O}_k(S)$, a Boolean function which is true if and only if set S contains k ones: $\text{HWB}_n = \bigvee_{k=1}^{n} a_k \wedge \mathbf{O}_k(\{a_1, \ldots, a_n\})$. Note that the n clauses are mutually *orthogonal:* Since each vector of a_i's has a specific number of 1-bits, no vector of a_i's can satisfy two clauses, and hence $C_1 \wedge C_2 = 0$ for any two clauses C_1 and C_2.

To generate a hash code from this specification, each clause can be rewritten using the relation $\mathbf{O}_k(S) = [\overline{a}_k \wedge \mathbf{O}_k(S - \{a_k\})] \vee [a_k \wedge \mathbf{O}_{k-1}(S - \{a_k\})]$ to yield $\text{HWB}_n = \bigvee_{k=1}^{n} a_k \wedge \mathbf{O}_{k-1}(\{a_1, \ldots, a_n\} - \{a_k\})$. Using properties of the $\mathbf{A}$-transform [8], we have

$$\mathbf{A}[\text{HWB}_n] = \sum_{k=1}^{n} a_k \cdot \mathbf{A}[\mathbf{O}_{k-1}(\{a_1, \ldots, a_n\} - \{a_k\})] \tag{1}$$

The hash code for $\mathbf{O}_k(S)$ can be quickly computed using the relation:

$$\mathbf{O}_k(S \cup \{a_i\}) = [\overline{a}_i \wedge \mathbf{O}_k(S)] \vee [a_i \wedge \mathbf{O}_{k-1}(S)]$$

Applying the **A**-transform yields:

$$\mathbf{A}[\mathbf{O}_k(S \cup \{a_i\})] = (1 - a_i) \cdot \mathbf{A}[\mathbf{O}_k(S)] + a_i \cdot \mathbf{A}[\mathbf{O}_{k-1}(S)] \qquad (2)$$

Equation (2) can be used to compute the hash code for $\mathbf{O}_k(S)$ using an integer array, *code*, indexed from 0 to k. Initially, position 0 is assigned 1, and all others 0. As we sequence through the elements of S, *code*$[i]$ stores the value of $\mathbf{O}_i(S')$, where S' is the subset of S that has been examined so far. Thus the kth term in equation (1) can be evaluated in time $\Theta(nk)$, for a total time of $\Theta(n^3)$ with a space requirement of $\Theta(n)$.

Table 2 shows experimental evidence of the space-time advantage obtained in verifying the HWB(n) function using the graph division partitioning technique. For this function it turns out that the natural partitions of the mathematical specification represent the same functions as obtained from the graph division on the circuit. To take advantage of such possibilities, one can generate hash codes for each partition in both design descriptions (a linear time operation in the size of any free graph). If a pair of partitions map to the same hash code, they are (probabilistically) likely to be equivalent, and we apply the Minterm Matching Theorem to match the minterms of that pair of partitions. In general, however, it cannot be assured that both functions are likely to have a large number of partitions which are functionally equivalent. Hence to give a more meaningful idea of the computational requirements of deterministic techniques under noncorresponding orthogonal partitions, we report the verification time using partitioned OBDDs by matching a partition of one design with the partitions of the other arranged in the order of increasing size.

Corresponding partitions are likely to be generated when two implementations of a circuit with much structural similarity are being compared. For example, given a circuit C, we often optimize some aspect (such as the area overhead) to produce an another circuit C_m. Checking for equivalence is an essential part of the above synthesis step. For each circuit, we can compute the decomposed graph G_d. The functional equivalence check between C and C_m can now be carried out in pairs of the corresponding partitions; the extent of the graph broken into corresponding partitions is given by the decomposition points whose functionalities have remained unchanged.

6 Deterministic vs. Probabilistic Verification

In using functional partitioning, the probabilistic method appears to have significantly more flexibility and computational advantages. For each function, it combines the hash codes resulting from all the partial assignments into a single integer, and *then* compares the two resulting integers. Since the final comparison in the probabilistic method is between *integers,* and not OBDDs, it has complete freedom in the method it uses to evaluate each function. It is not required, for example, to use the same ordering in computing (intermediate) representations for each. For example, using partial assignments to compare two functions which require different orderings for optimal representation can be very inefficient. Such restrictions are clearly illustrated by our experiments to verify a 16-bit unsigned multiplier (c6288) against its abstract mathematical specification (see Table 3).

The circuit c6288 was hashed using partial assignments. Following the approach described in Section 3, the program was given an assignment tree on the first seven variables $(a_0, b_0, \ldots, a_3)$, and the paths in the tree were cut

	Probabilistic Method		Partitioned OBDDs			
			apply & compose		apply only	
	Time	Space	Time	Space	Time	Space
c6288	2:58 hr.	288,260	unable		5:48 hr.	307,146
Spec	35 min.	384	unable		unable	

Table 3: Verification of a Multiplier against an Abstract Specification using Partial Assignments

off once they contained four 0's. Thus, 99 partial assignments were given to the program.[6] Partitions were constructed sequentially and hashed to an integer value. For the multiplier specification, a program was written which used the mathematical properties of the unsigned integer multiplier function to generate the hash code from a given set of random integer inputs. The program sequenced through all 2^{16} assignments to the a_i's. A hash code was computed for each such "partial assignment" using a strategy somewhat similar to that for generating the HWB hash code.

Deterministic verification using OBDDs require either the use of the Correspondence Theorem or the Minterm Matching Theorem. In using the Correspondence Theorem it is required that identical partitions be generated from both descriptions of the function (see Section 2). However, the partial assignments effective for partitioning the circuit are quite ineffective in partitioning the specification. Similarly using the Minterm Matching Theorem proves very inefficient in *either* space or time. In matching any partition from the first functional description against partitions from the second description, storing all partitions of the second description in memory at any given time is very space intensive. Generating all partitions from the second design for every partition of the first design is very time intensive. This precluded an efficient deterministic verification of c6288 against the mathematical specification. However as we show this circuit can be deterministically compared against itself using partial assignments. This deterministic result has considerable significance as it allows, for the first time, an explicit yet concise OBDD representation of a 16-bit multiplier.[7]

The probabilistic approach can also use functional partitioning in the evaluation of subfunctions. Consider two functions F and G which have been partitioned into two disjoint subfunctions, say $F = f_1 \vee f_2$ and $G = g_1 \oplus g_2$. The probabilistic approach computes the hash code for each subfunction and combines the results. Functional partitioning can now easily be used in computing any of these hash codes. Since F and G are not partitioned into corresponding disjoint pieces, a piece-wise comparison of their subfunctions using functional partitioning cannot be made.

[6]The choice of these numbers was based on the measurement of the space and time required for nine "benchmark" partial assignments to determine the space-time balance for various depths of the decomposition tree.

[7]The programs were written in $C{+}{+}$ and run on a Sun-4/280 with 32 MBytes of memory. Deterministic runs were made using *apply* only as well as using both *apply* and *compose*. Probabilistic method was run using *collapse-with-compose*. The details of implementational issues is same as in [7, 8].

7 Heuristic Selection of Partitioning Techniques

These functional partitioning methods may be often seen to exchange space for time. A very useful characteristic of such partitioning is that for functions whose space requirements is prohibitive, the space can be reduced almost linearly by the number of partitions and the time performance degrades only very slowly. For functions which require space bounded by only a small polynomial, verification can be done easily without partitioning, and it is only for such functions that creating functional partitioning charges an excessive time penalty. Thus a natural question is whether given an arbitrary circuit (function), can the suitability of functional partitioning (and identification of the particular technique) be determined *a priori?* Quantifying an exact measure of the feasibility of such partitioning is likely to be difficult but their possible suitability in various scenarios can be determined heuristically.

Circuit division techniques are applicable when ideal splitting variables (see Section 4) can be located for large subcircuits. As noted earlier, using our 16-valued simulation and the dominator algorithm of [10], subcircuits representing orthogonal functions can be located easily. Also note that even when a variable does not split a (sub)circuit into two disjoint pieces but the two circuits share logic only near the inputs, the redundant computation (time penalty) will be small. Orthogonal subcircuits are also good candidates for being represented with different global orderings in a partitioned OBDD.

Failing to identify orthogonal functions in any (sub)circuit we can use the graph division techniques for further partitioning. Such techniques are especially applicable when the function's decomposed representation has a small tree structure, for example, the HWB function. When paths are easily enumerable and they share relatively small subgraphs, significant redundant computation will be avoided. As graphs on decomposition points can usually be constructed quickly, determining the suitability of graph division is quite easy. On the other hand, if the decomposed graph has a small diameter (such as a BDD for a multiple input exclusive-or function), most orthogonal paths will require redundant computation which will reduce the effectiveness of such partitioning. For example, though the multiplier c6288 is easily orthogonally partitioned using partial assignments, its decomposed graphs have a small diameter under almost all meaningful decompositions, and are not suitable for the graph division technique.

If special structures cannot be identified in either the function's description (graph division) or in the function's implementation (circuit division), we may wish to resort to using the general properties of Boolean functions (Shannon's expansion) and create partitions using partial assignments.

For circuits that fail the above two tests, the good candidates for partitioning through partial assignments (such as multiplier circuits) are easily distinguished from the unsuitable circuits (such as a tree of 2-input exclusive-or gates). First we identify good decomposition points and a global ordering, just as for monolithic OBDD methods [8]. The best candidates for partial assignments are the first few variables in this order. Note the structure of the decomposed graph and the nature of each decomposed Boolean function usually "outlines" the computational details of how the function will be composed. Thus, to find out how much computation is shared between two different partial assignments, one computes any meaningful measure which determines how much the functionality of each decomposition point has changed between the two partial assignments. For example, one can select a good cutset for each decomposition point and enumerate the number of edges in each cutset where the structure of a possible OBDD [11] is significantly changed. If most of the edges remain essentially unchanged between two partial assignments, then these two instances (partitions) of the problems will likely have a large common computation.

The above heuristics are not only intuitively meaningful, but can also distinguish between various hard and easy circuits considered in this paper.

The use of partial assignments is clearly not a panacea. It is effective only when the redundancy ratio r is low. Consider again the function $f_1 = x_1 \oplus \cdots \oplus x_n$. Partial assignments divide the OBDD for f_1 into pieces which have a substantial overlap, and r is prohibitively large. Clearly, when two subgraphs of the root node have a large overlap, making a partial assignment on the root variable will not create a very effective partition. For difficult functions, or functions being represented with an inefficient ordering, the space reduction can be orders of magnitude larger than the increase in the time required for verification. We corroborated our analytical arguments on ISCAS circuits, where we studied the partitioning of the function graph for the last output function of each circuit. Under efficient orderings, where these functions can be easily represented by a single OBDD, the partial assignment approach was not very effective in partitioning the graph. Nevertheless, when using inefficient orderings, such was not the case. Using an ineffective ordering for c3540, the OBDD manipulation programs exceeded the available memory of 32 Mbytes while still processing the graphs at intermediate gates. The same function could be represented in about 4Mbytes when the circuit was partitioned using 64 partial assignments.

Finally, consider verifying the circuit implementing $F = f_1 \wedge f_2$; f_1 is given above and $f_2 = Mult^n(\vec{y}, \vec{z})$. Though an OBDD for F will require exponential resources (since f_2 is a difficult function), it is insensitive to whether f_1 is ordered before f_2 or vice-versa. Note f_1 is not at all amenable to partitioning using partial assignments. Hence, for such partitioning one would prefer the multiplier function f_2 to be ordered earlier. In general, for the most effective use of partial assignments in the partitioned OBDD approach, one must first consider whether the circuit has any (simple) structurally disjoint partitions and order earlier the partitions representing relatively difficult functions. Note the relatively difficult functions can be rather easily distinguished using methods such as OPBDDs [11].

8 Conclusions and Future Work

We have shown how functionally partitioning a Boolean circuit can be used to enhance the efficiency of probabilistic and deterministic verification. For functions such as *Fortune-Hopcroft-Schmidt* function, or Bryant's *hidden-weighted-bit* function, verification in polynomial time and space is possible. Such functions were intractable using monolithic OBDD based methods. Similarly we show how such partitioning can easily reduce the space required for verifying a 16-bit multiplier by as much as three orders of magnitude.

Since the techniques allow corresponding pairs of partitions to be verified independently, functional partitioning is even more efficient in the presence of design errors, as a discrepancy between design and specification can be expected to appear in any given partition with reasonably high probability. For example, design errors were randomly introduced in c6288. The errors were introduced by simply inverting the functionality of a gate after every 100 lines in the circuit file; in each run one gate was erroneously set. Partitioned into 1024 orthogonal circuits, the inequivalence was verified by both probabilistic and deterministic methods in less than 20 seconds. Similar results were obtained for checking inequivalence in c3540 and the other difficult ISCAS circuits as well.

Functional partitioning leads to extremely efficient parallel algorithms. Partitioned into 1024 subfunctions, it can also help verify c6288, where the largest graph requires no more than 1587 nodes, with the total intermediate space of any partition less than 18,000 nodes.[8] On a Sun 4/280, no partition required more than one CPU minute to be verified. For difficult functions, parallel verification can offer a speedup linear in the number of partitions.

Functional partitioning is a useful approach in managing the computational resources for functions difficult to verify. It also allows the possibility of verifying only a part of design, when the resources required to verify the complete design are prohibitive. Similarly, time-space trade-offs exist. For example, in using partial assignments, one can partition the decomposed functions rather than the circuit and reduce time in exchange for space.

We are currently implementing algorithms which can dynamically determine the suitability of orthogonal partitioning and also construct different orderings for different orthogonal partitions. Other classes of functional partitions are also being investigated.

References

[1] P. Ashar, A. Ghosh, and S. Devadas. Boolean satisfiability and equivalence checking using general binary decision diagrams. *ICCD*, pages 259–264, October 1991.

[2] M. Blum, A. K. Chandra, and M. N. Wegman. Equivalence of free Boolean graphs can be decided probabilistically in polynomial time. *Information Processing Letters*, 10:80–82, March 1980.

[3] R. E. Bryant. Graph based algorithms for Boolean function representation. *IEEE Transactions on Computers*, C-35:677–690, August 1986.

[4] R. E. Bryant. On the complexity of VLSI implementations and graph representations of Boolean functions with application to integer multiplication. *IEEE Transactions on Computers*, C-40:206–213, February 1991.

[5] E. Cerny and C. Mauras. Tautology checking using cross-controllability and cross-observability relations. *ICCAD*, pages 34–37, 1990.

[6] L. Fortune, J. Hopcroft, and E. M. Schmidt. The complexity of equivalence and containment for free single variable program schemes. *Goos, Hartmanis, Ausiello and Bohm, Eds., Lecture Notes in Computer Science 62, Springer-Verlag*, pages 227–240, 1978.

[7] J. Jain, J. Bitner, D. Fussell, and J. Abraham. Probabilistic design verification. *ICCAD*, 1991.

[8] J. Jain, J. Bitner, D. Fussell, and J. Abraham. Probabilistic verification of Boolean functions. *Formal Methods in System Design*, 1992. to appear.

[9] S.-W. Jeong, B. Plessier, G. Hachtel, and F. Somenzi. Stuctural BDDs: Trading canonicity for structure in verification algorithms. *ICCAD*, 1991.

[10] T. Lengauer and R. Tarjan. A fast algorithm for finding dominators in a flowgraph. *ACM Transactions on Programming Languages*, 1:121–141, July 1979.

[11] D. Ross, K. M. Butler, R. Kapur, and M. R. Mercer. Fast functional evaluation of candidate OBDD variable orderings. *The Proceedings of The European Conference of Design Automation*, pages 4–10, 1991.

[8]The above discussion is with respect to the 16th output of c6288 which Bryant shows to require as much resources for verification as all the other outputs together [4].

Fault Tolerance and Performance of Multipath Multistage Interconnection Networks

Frederic Chong, Eran Egozy, and André DeHon
Artificial Intelligence Laboratory
Massachusetts Institute of Technology
545 Technology Square
Cambridge, MA 02139
ftchong@ai.mit.edu, eran@ai.mit.edu, andre@ai.mit.edu

Abstract

In building a multiprocessor system, we can minimize the system's mean time to failure by providing an architecture resilient to component faults. We compare the fault tolerance and performance characteristics of various fault-tolerant multistage interconnection networks. We primarily focus on networks composed of *dilated* routing components. A dilated router features redundant outputs in each logical direction, and can thus be used to construct *multipath* networks. Multipath networks have multiple paths from any input to any output. An *interwired* multipath network disperses its routers' redundant outputs to input ports of physically distinct components. We introduce a deterministic wiring scheme for routing interwired networks that maximizes the number of routing paths available for each endpoint. We compare the deterministically-wired network to both randomly-interwired networks and non-interwired multipath networks. We characterize fault tolerance by measuring the probability of a single endpoint disconnection as faults accumulate in the network. Our performance simulations are based on traffic from shared memory applications and include barrier synchronization to expose the effects of localized performance degradation. We find that at a minor performance cost, deterministically-interwired networks are more fault tolerant than randomly-interwired networks and non-interwired networks.

1 Introduction

Sufficient fault tolerance is an important part of any practical, large-scale computer system. In this paper, we focus on selecting wiring schemes for Multistage Interconnection Networks (MINs) which offer the most fault tolerance and performance for a fixed amount of hardware.

We examine *interwired* and *non-interwired* multipath networks. An interwired network connects redundant outputs of each routing component to physically distinct components in the next stage of the network. Such interwiring was first used by Bassalygo and Pinsker [1]. Upfal [18] later used interwiring to form what he called *multibutterflies.* We introduce a deterministic algorithm for interwiring networks and compare it to the random interwiring methods presented by Leighton and Maggs [15]. We also examine replicated networks where multiple single-path networks are attached together at the endpoints to form a multipath network.

To measure fault tolerance, we use the simple fault metric of *completeness* coupled with random fault generation to probabilistically quantify the number of faults a given network can sustain. A network is said to be complete as long as there is some non-faulty path between every pair of network endpoints. This metric is conservative if the system can actually tolerate endpoint isolation or network partitioning. Nonetheless, it provides a simple decision test for comparing the effects of faults in various networks.

For our performance comparisons, we simulate message traffic over faulty and non-faulty networks to quantify the effective aggregate bandwidth of such networks in the presence of faults. We model several shared memory applications by simulating message distributions and processor synchronization.

Section 2 describes the basic networks which we study, as well as the routing models assumed. Section 3 details the various network wirings considered in this paper. Section 4 describes our fault tolerance experiments and Section 5 presents the data resulting from our experiments. Section 6 explains our performance modeling and Section 7 summarizes our performance results. We discuss related issues to this paper in Section 8 and draw final conclusions in Section 9.

2 Network and Routing Models

2.1 Networks

This paper focuses on MINs built from several stages of crossbar style routing elements. For simplicity, we will assume crossbar routing switches with an equal number of inputs and outputs. Each crossbar switch can connect any of its inputs to any of its outputs with the restriction that only one input can be connected to each output at any given moment. A particular crossbar router is characterized by both its *radix* and its *dilation*. The *radix* of a router is the number of distinct logical directions distinguished by the routing component. The *dilation* is the number of logically equivalent outputs in each distinct direction. Thus, with a dilation greater than one, redundant routing is provided in each direction. Figure 1 shows an example of such a MIN.

We further limit ourselves to multipath networks with redundant connections to each endpoint. The redundant endpoint connections make it possible to guarantee that no single fault will isolate the endpoint from the network. Figures 1 through 5 show MINs each of which has two connections from and to each endpoint.

Much previous work has been done on this class of networks. Kruskal and Snir [10] suggest using multiple copies of a MIN to provide fault tolerance. They also describe a network with dilated routing components and Grondalski [6] implements such a network. The Kappa network [9] also uses dilated routers to achieve fault-tolerant interconnect. Leighton and Maggs [15] suggest effective ways of wiring and using randomly-wired multibutterfly networks.

An alternative approach to achieving fault-tolerant MINs is to construct *extra-stage* networks with more switching stages than are actually required to uniquely specify a destination ([13], [3] *et. al.*). The set of routing specifications that reach the same physical destination defines a class of equivalent paths. So long as one path of each such class remains intact in a faulty extra-stage network, any endpoint will be able to successfully route to its destination. However, such schemes require the endpoint to specify a path through the network. The extra stages in these schemes also result in larger latencies than the corresponding multipath network, even in the absence of faults. We do not include this class of networks in our analysis of multipath networks.

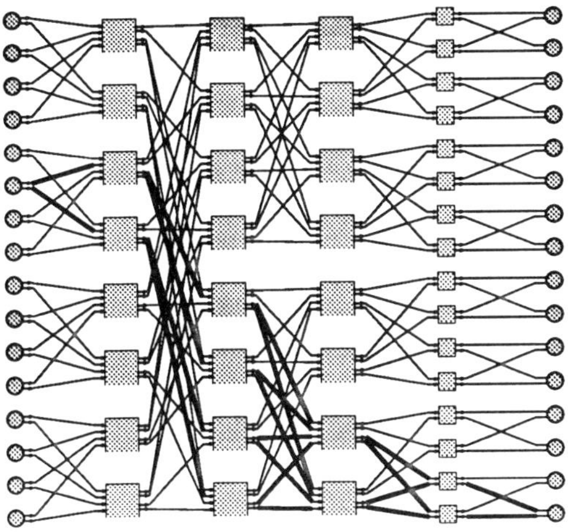

A multipath MIN constructed from 4 × 2 (inputs×radix) dilation-2 crossbars and 2 × 2 dilation-1 crossbars. Each of the 16 endpoints has two inputs and outputs for fault tolerance. Similarly, the routers each have two outputs in each of their two logical output directions. As a result, there are many paths between each pair of network endpoints. Paths between endpoint 6 and endpoint 16 are shown in bold. This network was wired using the deterministic interwiring algorithm described in Section 3.4.

Figure 1: 16 × 16 Multipath Network

2.2 Routing

Our fault tolerance results described below are independent of the details of routing and fault-identification. That is, routing can be circuit-switched or packet-switched using any number of routing strategies. The fault tolerance results only depend on network topology and the routers' ability to use their redundant connection in each logical direction to avoid faults.

Our performance modeling assumes additional implementation details. The routers perform source-responsible, pipelined, circuit-switched routing. Data are pipelined between routers in adjacent stages of the network to improve throughput and available bandwidth. The circuit-switched, source-responsible protocol is simple and can be implemented very efficiently in VLSI hardware. The routing protocol assumed is described in detail in [5]. We have used this routing protocol in the VLSI routing component RN1 [17]. An RN1 routing component can be configured to act as a single 8-input, radix-4, dilation-2 routing component or as a pair of independent 4-input, radix-4, dilation-1 routing components.

3 Network Wiring

We consider four basic wiring strategies for building multipath multistage routing networks. In this section, we describe these strategies and show an example of each network.

As with all bidelta [11] networks, each stage in these multipath networks routes by successively subdividing the set of possible destinations into a number of *equiv-*

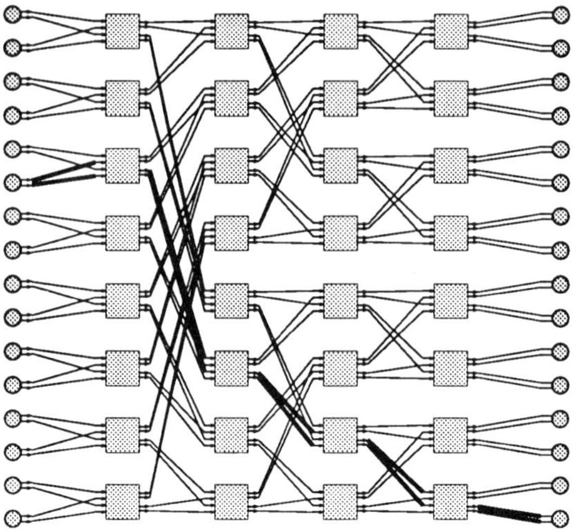

A dilated, non-interwired four-stage network connecting 16 endpoints. Each component is a 4×2, dilation-2 crossbar.

Figure 2: Dilated Non-Interwired Network

alence classes equal to the radix of the routing components. For example, consider a radix-4 network. When connections enter the network, any input can reach any destination. The first stage of routing components divides this class into 4 different equivalence classes based on desired destination. Each succeeding network stage further subdivides a previous stage's equivalence classes into 4 more equivalence classes. When there is a single destination in each equivalence class, the network has uniquely determined the desired destination and can connect to the destination endpoints. This successive subdivision can be easily seen in the networks shown in Figures 1 through 5.

3.1 Dilated Non-Interwired Network

The dilated networks described in [10] [6] connect all outputs in a given logical direction to the same physical routing component in the subsequent stage of the network. The topology is thus identical to the corresponding non-dilated bidelta network [11]. Such networks gain performance by having multiple paths. These networks can also tolerate some faults in the wiring between components. Unfortunately, these networks are just as susceptible to component failures as non-multipath networks. An example of a dilated non-interwired network composed of 4×2, dilation-2 routers is given in Figure 2.

3.2 Replicated Network

In a replicated network, such as the ones proposed by Kruskal and Snir [10], multiple single-path networks are used to achieve fault tolerance. Each single-path network

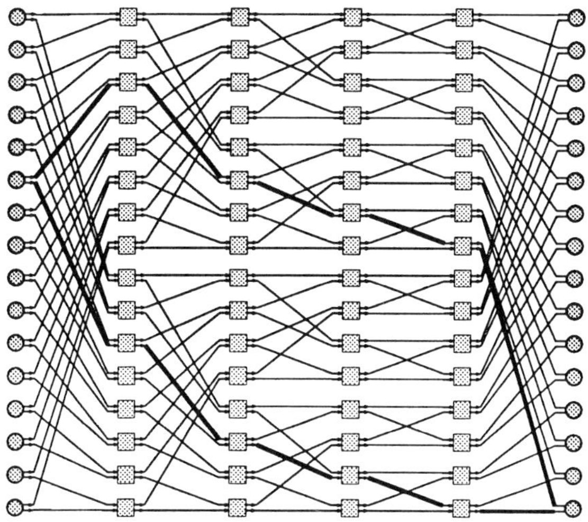

Two non-dilated four-stage networks connecting 16 endpoints are attached together at the endpoints. Each component is a 2×2, dilation-1 crossbar.

Figure 3: Replicated Network

is a bidelta network using dilation-1 crossbar switches. Each endpoint gets one connection to and from each of the single-path networks. An endpoint chooses which network to use when it needs to make a connection. As long as each pair of endpoints is connected by a path in at least one of the networks, the replicated network will remain complete. Replicated networks are used in the telecommunications field to minimize contention and increase available bandwidth over single-path networks [7]. A replicated network, where each subnetwork is constructed from 2×2 routing components, is shown in Figure 3.

3.3 Randomly-Interwired Network

An interwired network takes advantage of the redundant outputs of dilated routing components and connects them to physically distinct components in the next stage of the network. Bassalygo and Pinsker [1] used such interwiring to construct the first non-blocking networks of size $O(N \log N)$ and depth $O(\log N)$. Upfal [18] later used interwiring to form networks he called *multibutterflies*. Leighton and Maggs [15] further analyzed the high *group expansion* (or *α-β expansion*) of these randomly-interwired networks. Group expansion is a description of the degree to which any subset of components in one stage will fan out into the next stage. More formally, any subset of α components from one stage must connect to at least $\alpha \times \beta$ components in the next stage, where β is the expansion factor and $\alpha \times \beta$ is less than the number of components in each stage. Good fault tolerance is an inherent property of high group expansion [15].

Deterministic construction of graphs with high group expansion is a difficult problem. Recent work [8] describes a deterministic algorithm based upon Ramanu-

```
▷ in_set contains all the input ports of a single equivalence class in the next stage.
▷ connections is an array matching in_ports and out_ports, initially empty.
▷ out_ports_list lists the output ports of a single equivalence class in the currect stage.
wire_eq_class(in_set, connections, out_ports_list)
foreach out_port
    in_port ← choose and remove a random input port from in_set
    while(connected(router#(in_port), router#(out_port), connections))
        put in_port back in in_set
        in_port ← choose and remove a random input port from in_set
    connect(in_port, out_port, connections)
return(connections)

connected(in_router, out_router, connections_array)
if in_router is in already connected to out_router
    return(true)
else return(false)
```

This algorithm randomly interwires an equivalence class. To interwire a whole stage, the algorithm is repeated for each class (boundary cases omitted for clarity).

Figure 4: Pseudo-code for Random Interwiring

jan graphs which, unfortunately, does not provide as much expansion as random multibutterfly generation. A randomly generated multibutterfly has a high probability of possessing high group expansion. This high probability results from the high fraction of networks in the class of multibutterflies that have high group expansion. Consequently, it is common practice to randomly generate multibutterfly networks until one is found with the desired group expansion.

We use a heuristic algorithm based on the random interwiring ideas presented in [15]. Specifically, we constrain every router to send its redundant output paths to physically distinct components in the next stage, thus improving group expansion over a purely random interwiring scheme. Figure 4 shows the algorithm used to generate a randomly interwired network.

We use dilation-1 routers in the final stage of the network to achieve a high degree of fault tolerance. The two ends of the network are the weakest links to achieving a high degree of fault tolerance since fewer alternative components are available in these stages to make any given connection. To maximize fault tolerance, we must maximize the number of distinct components connected to each endpoint. The heuristic described above guarantees that the input connections from any endpoint connect to distinct components. Similarly, using dilation-1 routers in the final stage maximizes the number of distinct components that provide outputs from the network to each endpoint. Unfortunately, the cost of using dilation-1 routers in the last stage is a small performance decrease (see Section 7). However, if dilation-1 routers were not used, a single component failure in the network could render the network incomplete.

An example of a randomly-interwired four-stage network using radix-2 routing components is shown in Figure 5.

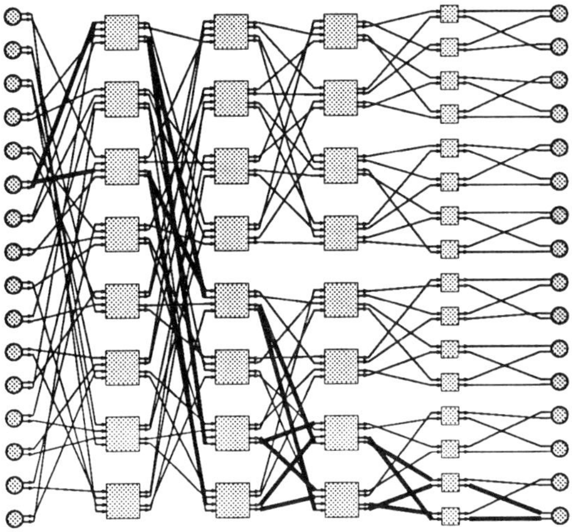

A randomly-interwired four-stage network connecting 16 endpoints. Each component in the first three stages is a 4×2, dilation-2 crossbar. To prevent any single component from being in an endpoint's critical path, the last stage is composed of 2×2, dilation-1 crossbars.

Figure 5: Randomly-Interwired Network

3.4 Deterministically-Interwired Network

Although random interwiring provides for a reasonable degree of group expansion, it does not guarantee each endpoint the maximum number of redundant paths through the network. We have developed an algorithm that provides the maximum amount of *path expansion* through the network at the expense of some group expansion. Path expansion refers to the number of distinct components in the network that any one endpoint can reach.

To achieve maximum path expansion, we connect the network with the algorithm listed in Figure 6. The paths from any input to any output may fanout by no more than a factor of d, the dilation of the routers, at each stage. This fanout may also become no larger than the size of the routing equivalence classes at that stage. The routine *groupsz* returns the maximum fanout size allowed by both of these factors. Each stage is partitioned into *fanout classes* of this size, which are then used to calculate network wiring.

As with the randomly-interwired network, the last stage is composed of dilation-1 routers to increase fault tolerance. Figure 1 (Section 2) shows a deterministically-interwired network composed of radix-2 routers.

4 Fault Modeling

We concern ourselves only with complete component failures rather than to individual wire failures. We characterize network completeness through Monte Carlo simulation. For each trial, we randomly add component faults until the network is incomplete. Results are tabulated to approximate the probability of network com-

```
▷ Returns the next-stage router to which to wire for maximum path expansion
wire_to_port(n,d,s) ▷ n=router number, d=dilated port number, s=router stage
  outgrpsz ← groupsz(s)
  ingrpsz ← groupsz(s + 1)
  eq_start ← ingrpsz × ⌊n/(radix × dilation × outgrpsz)⌋▷ offset to beginning of
     fanout class
  eq_router ← ((n × dilation + d) mod ingrpsz)▷ offset to specific chip within
     fanout class
  return(eq_start + eq_router)

▷ Calculates size of fan-out class
groupsz(s)
  expansion ← dilation^s ▷ maximum fanout due to dilation
  eq_class ← radix^(stages−s) ▷ equivalence class size
  return(min(expansion, eq_class))
```

An algorithm designed to maximize path expansion. Each endpoint will have the maximum number of redundant paths possible through this type of network.

Figure 6: Pseudo-code for Deterministic Interwiring

pleteness for each fault level. We also derive the expected number of faults each network can tolerate.

Because the routing components in the final stage of our interwired networks are half the size of routers in the previous stages, we assign two such routers to one physical component package and label both routers faulty if the physical component is chosen to be faulty. Furthermore, the two routers are assigned so that removing any such pair will not cut off an endpoint. We make this assignment so that fault increments will be of constant hardware size. This assignment also simulates how the pair of 4×4, dilation-1 routers in an RN1 routing component may be assigned.

We generated three-stage and four-stage networks for each of the four types of networks described above, each connecting 64 and 256 endpoint nodes respectively. Each endpoint has two connections to and from the network to provide for the minimal amount of redundancy necessary to achieve fault tolerance. Every network uses radix-4 routers of either dilation-1 or dilation-2 and hence could be implemented using the RN1 component. Thus, all the networks with a given number of stages contain the same amount of hardware. Network organization and component configuration are solely accountable for the fault tolerance and performance differences of these networks.

5 Fault Tolerance of Networks

For each network, we plot the probability of the network remaining complete for a given number of uniformly distributed random faults. Since the non-dilated routers of the replicated network are half the size of dilated routers in the rest of the networks, we normalize the component count on the graph so that the same amount of hardware is faulty at each fault level. Results for the three-stage and four-stage networks are shown in Figure 7. The expected number of faults that each network can tolerate is summarized in Table 1.

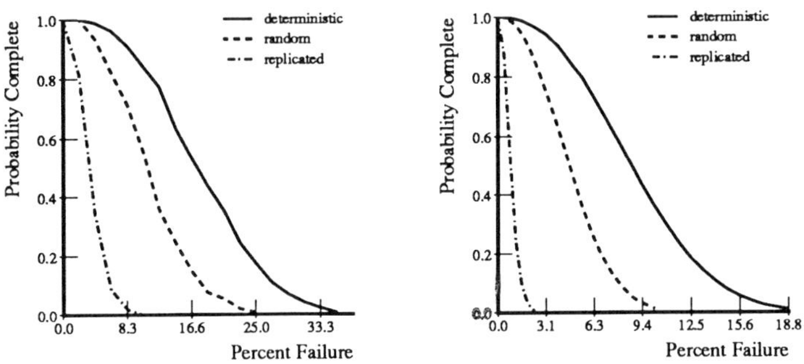

(A) 3-Stage Network Completeness

(B) 4-Stage Network Completeness

The probability that a network with a given number of faults is complete for the replicated, the randomly-interwired, and the deterministically-interwired 3-stage and 4-stage networks. (A) Each 3-stage network uses 48 radix-4 components that interconnect 64 endpoints. (B) Each 4-stage network uses 256 radix-4 components that interconnect 256 endpoints.

Figure 7: Completeness of (A) 3-stage and (B) 4-stage Multipath Networks

Network Stages	Wiring	Total # Comp.	Test Trials	Expected Failure Tolerated # Faults	% Network	Error Bound # Faults
3	Non-Interwired	48	—	0	0.0%	0
3	Replicated	96	2500	3.1	3.1%	0.025
3	Random	48	1000	5.0	10%	0.063
3	Deterministic	48	1000	8.1	16%	0.079
4	Non-interwired	256	—	0	0.0%	0
4	Replicated	512	5000	4.1	0.8%	0.024
4	Random	256	5000	11.8	4.6%	0.075
4	Deterministic	256	5000	22.6	8.8%	0.130

The above table shows the expected number of faults each network can tolerate while remaining complete. Each network was fault tested by the method described in section 4 for the indicated number of trials.

Table 1: Fault Tolerance of Various Multipath Networks

Dilated Non-Interwired Network Because every component in this network is in some critical path between two endpoints, it cannot sustain a single component fault and is tolerant only to wire failures. It is easy to see that this network will tolerate exactly as many wire faults as an equivalently-sized replicated network. Both networks have exactly as many redundant paths between each pair of successive stages as there are redundant connections to each endpoint (see Figures 2 and 3). The dilated non-interwired network always has fewer connections between internal routing stages than the interwired networks and is thus more susceptible to wire faults (see Figures 1, 2, and 5).

Replicated Network For each pair of endpoints, the replicated network provides exactly as many paths through the network as there are connections to each endpoint. This provides a degree of fault tolerance as shown in Figure 7 and Table 1.

Randomly-Interwired Network Since these networks are non-unique, ten random networks were generated, and the one with the highest fault-tolerant characteristics was selected for comparison. The differences in expected number of tolerable faults between the best and worst of the ten three-stage and four-stage networks is 0.3 faults and 0.2 faults, respectively. Because the difference in expected number of faults between the random networks is small, the choice of one particular network over another makes no difference when comparing randomly-interwired networks against other wiring schemes. When actually implementing a network, one is certainly free to generate a large number of random networks and select the best one.

Randomly-interwired networks provide significantly more tolerance to component faults than replicated networks, as shown in Figure 7 and Table 1. The expansive properties allow the number of paths between each pair of endpoints to increase towards the middle of the network. In the first and final routing stages, the number of components available to make a connection between any pair of endpoints is still limited by the number of redundant connections to each endpoint. However, within the network, more components are available in each routing stage and more wires are available between every pair of routing stages.

Deterministically-Interwired Network This network achieves the highest fault tolerance of all the networks presented here for two reasons. First, the grouping of endpoint connections makes it less likely that losing a small number of routers from the first stage will result in an incomplete network. Second, the deterministic wiring guarantees maximal path expansion. The deterministic wiring thus guarantees that a maximum number of components in each stage are available for routing between each pair of endpoints. Similarly, the maximum number of wires between each pair of stages is available for connecting any pair of endpoints.

6 Performance Modeling

Given the completeness characteristics of the networks we have examined, we are interested in how their performance degrades in the presence of faults. For randomly-wired multibutterflies, Leighton, Lisinski and Maggs [14] found promising results from simulating 1024 node networks under some simple models of network architecture and utilization.

We investigate the empirical performance of several multipath networks under a specific architectural model. We attempt to derive efficient, yet realistic, simulations of network utilization by modeling several 64-processor, shared memory applications. As our performance metrics, we measure the I/O bandwidth utilization and average message latencies of these applications. We also simulate processor synchronization to expose the effects of localized network performance degradation.

6.1 Performance Metrics

We calculate average I/O bandwidth utilization by observing the percentage of router cycles for which processor nodes are successfully receiving or replying to messages. The performance of a network is measured by the amount of useful communication it provides. For our analysis, useful communication occurs when a processor node is receiving or replying to a message. An I/O bandwidth utilization of 100 percent indicates that the processor nodes are receiving maximum usable performance from the network. Any more traffic the network can handle will not result in any increase in useful work. Since we model an architecture in which memory is contained within each processing node, our I/O bandwidth metric is analogous to memory bandwidth metrics used by [14] and [12].

We also calculate average message latencies. Latencies are measured from time of message injection into the network until time of message completion, which includes acknowledgment from the receiver. Note that our latencies vary with the message lengths of each application. Also realize that a processor node does not necessarily suffer the full latencies given. For certain types of messages, some useful work could begin before the entire message is received.

6.2 Modeling Shared Memory Applications

Our goal is to provide a realistic model of network utilization that can be used to compare many different networks and parameters. To keep simulation time tractable, we use a variant of uniform traffic. Our simulation sends messages to random destinations in a uniform distribution. However, message lengths are randomly generated according to distributions derived from specific parallel applications. These application were taken from caching studies done in [2]. These studies simulate a shared memory architecture with coherent caches at each processing node.

Cache-coherent shared memory systems utilize a small set of message types, each of a fixed length. The messages sent through the network read cache lines, write cache lines, and maintain cache coherency. A cache-line read, for example, requires an 8-byte read request followed by a reply containing the data. The reply consists of an 8-byte header followed by 16 bytes representing the desired cache line.

6.3 Synchronization

Our performance simulation includes barrier synchronization to eliminate *skew*. Our simulation models applications which assume some degree of synchronization between the multiple processor nodes of the system. When a simulation violates this assumption, results are skewed. We prevent this skew by performing periodic barrier synchronization according to grain sizes estimated for each application.

It is important to eliminate simulation skew because it can mask the effects of localized network degradation. Analytic models suggest that faults and congestion may severely affect the performance observed by specific destination nodes while leaving others largely unaffected [12]. Without synchronization, such localized degradation would be lost in the *average* I/O bandwidth utilization. Modeling synchronization, however, forces all processors to wait for those falling behind, resulting in a more realistic decrease in I/O bandwidth utilization.

6.4 Application Descriptions

Message distributions were taken from trace simulations of four applications: SIMPLE, SPEECH, FFT, and WEATHER. The trace simulations involved the following system configuration: 64 processors, full-map directories, and 16-byte cache line. This configuration corresponds to 3-stage, radix-4 networks. SIMPLE models the hydrodynamic behavior of fluids using finite difference methods to solve the equations in two dimensions. SPEECH is the lexical decoding stage of a phonetically-based spoken language understanding system. It uses a variant of the Viterbi search algorithm. FFT is a radix-2 Fast Fourier Transform. WEATHER uses finite-difference methods to solve partial differential equations which model the atmosphere around the globe.

We also simulate a uniform message distribution, FLAT24. FLAT24 uses 24-byte messages and other simulation parameters which are similar to the messages and parameters of SIMPLE and SPEECH. FLAT24 shall serve as a basis for network comparison.

7 Performance Results

In this section we summarize our performance results for each network in the presence of faults. Performance was measured for complete networks only. For each fault level shown, multiple trials were run in which faults were randomly chosen. After fault insertion, applications were simulated on those networks which remained complete. Data shown represent the average I/O bandwidth utilization and latencies over those trials involving complete networks. These results do not represent the actual performance of these applications on hardware. Rather, our data provides a basis for network comparison by illustrating performance trends in the presence of faults.

First, we compare the performance of interwired and non-interwired dilated networks. Since the non-interwired network can tolerate no faults, we only compare the fault free case. To isolate the effects of interwiring, the deterministic and random networks presented use dilation-2 components in the last stage. We studied 3-stage non-interwired, randomly-interwired, and deterministically-interwired networks. I/O bandwidth utilization varied by less than 2 percent, a variation not significant for our simulation accuracy.

However, to avoid single-point disconnections in the last stage, the interwired networks we shall analyze for fault performance are constructed from dilation-1 components in the last stage. Although dilated components provide better performance, the use of these dilation-1 components substantially increases fault tolerance (see Section 3.3). For applications studied on 3-stage networks, the decrease in I/O bandwidth utilization was less than 6 percent.

Figure 8 details the fault performance of our applications on 3-stage, radix-4, networks which can withstand faults. The application FLAT24, shown as a solid line, is representative of graph trends and will be used in network comparisons.

Figure 9 compares the performance of those networks which can tolerate faults. The number of faults on the replicated network has been normalized to reflect an equivalent hardware loss. Each component in the replicated network is half the size of the components used in the other two networks. Our results show that deterministically and randomly-interwired networks deliver superior performance to the replicated network of equivalent hardware. The performance of the random network is slightly better than that of the deterministic network. However, recall that our figures are for complete networks only. For each fault level shown, the random networks have a lower probability of remaining complete than the deterministic networks.

8 Discussion

Although we have concentrated upon the fault tolerance of multipath networks, these networks also perform well under unbalanced loading. Intuitively, the effects hot-spots are very similar to component failures. Patterns such as matrix transpose have been shown perform well on randomly-wired multibutterflies by [14]. Further investigation of non-uniform loading is merited.

Additionally, our completeness metric assumes that a network is unusable when any input is disconnected from any output. This may not be true in the case of reconfigurable systems. In such systems, other metrics such as the number of surviving processors after network disconnection are important as well.

Fault tolerance becomes increasingly important as networks become large. The wiring techniques we have presented are applicable to a wider range of networks than traditional MINs. Fat-trees [16] [4] are an example of a highly scalable network which would benefit from the interwiring described here. As system size increases, a number of additional issues arise. We can use hybrid interwiring schemes which combine the high fault tolerance of deterministic wiring with the robust performance of random wiring. For instance, we can wire the the first portion of a network deterministically and the remainder randomly. The disparity between non-interwired networks and interwired networks also increases with system size. While non-interwired networks have a constant number of paths between endpoints, interwiring exploits dilation to create an exponentially increasing number of paths with each stage within the first half of the network.

9 Conclusion

We have compared the fault tolerance and performance of several multipath networks and have found a number of tradeoffs. The use of non-dilated components in the final stage of the network increases fault tolerance with some loss in performance. With a smaller decrease in performance, deterministic interwiring can tolerate more faults than random interwiring under our completeness metric. In general, interwired networks provide significantly more fault-tolerance than comparably-sized, non-interwired networks.

10 Acknowledgments

Much of the work from this paper grew from the ideas of our advisor, Tom Knight. Additionally, we would like to thank David Chaiken for providing the data describing the shared memory applications. We would also like to thank Mike Bolotski,

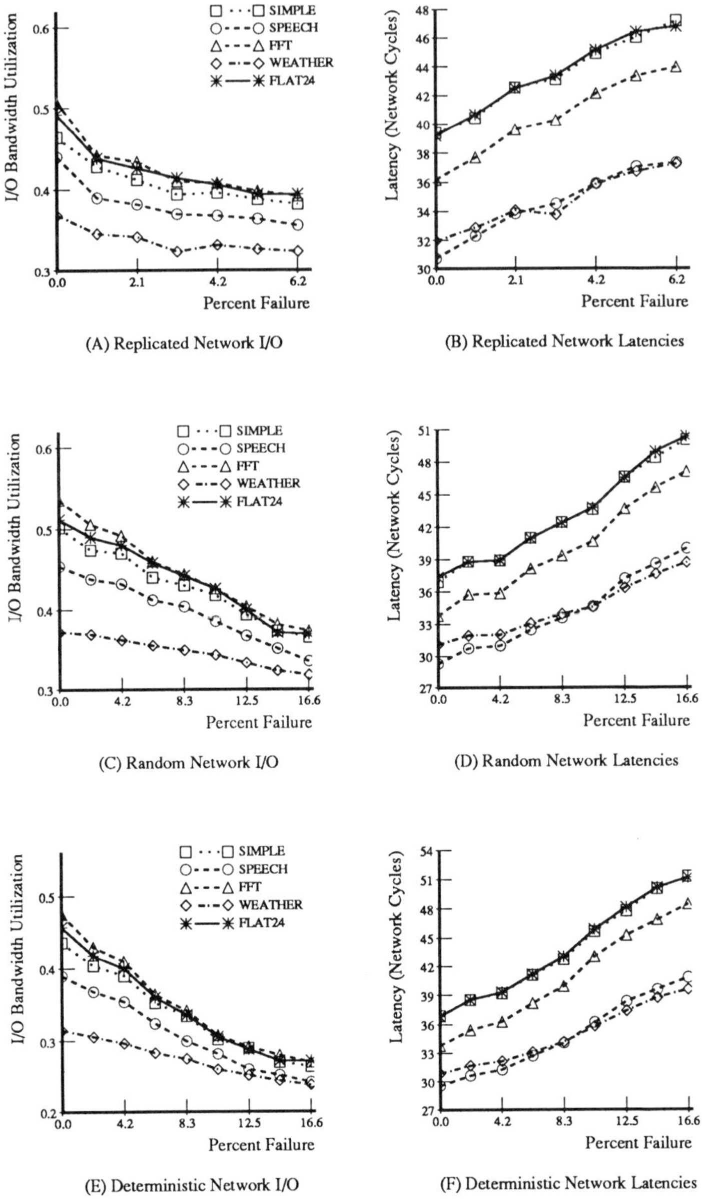

I/O bandwidth utilization and latencies for applications on 3-stage replicated, random, and deterministic networks. The application FLAT24, shown as a solid line, is representative of graph trends and will be used for network comparison.

Figure 8: Applications on 3-stage Replicated, Random, and Deterministic Networks

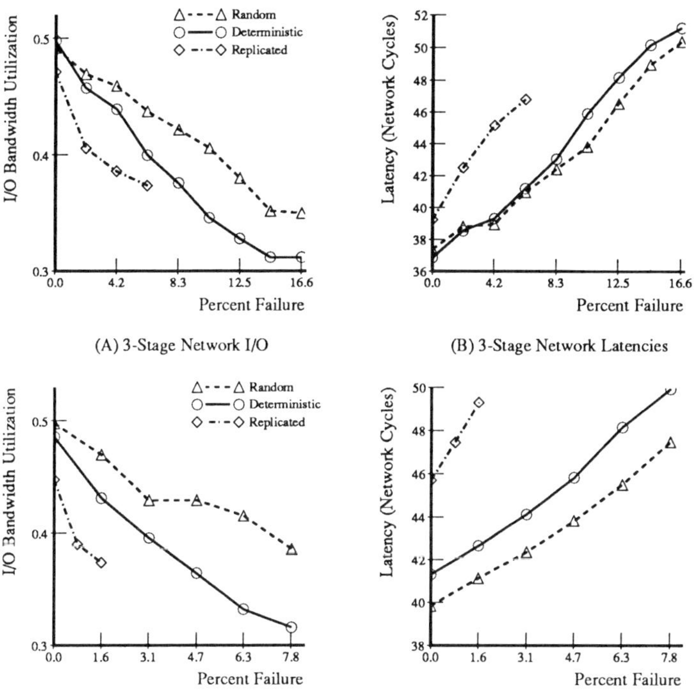

(A) 3-Stage Network I/O

(B) 3-Stage Network Latencies

(C) 4-Stage Network I/O

(D) 4-Stage Network Latencies

Comparative I/O bandwidth utilization and latencies for 3-stage and 4-stage random, deterministic, and replicated networks on FLAT24. Recall from Table 1 that expected percentages of failure tolerated by replicated, random, and deterministic networks are, respectively: 3.1%, 10%, and 16% for 3-stages; and 0.8%, 4.6%, and 8.8% for 4-stages. Note that the performance degradation appears to level off because only complete networks are measured. Although the surviving networks suffer less degradation as percentage of failure increases, the number of surviving networks is becoming substantially smaller.

Figure 9: Comparative Performance of 3-Stage and 4-Stage Networks

David Chaiken, Steve Keckler, Bruce Maggs, Patrick Sobalvarro, Thomas Simon, and Deborah Wallach for their helpful comments.

This research is supported in part by the Defense Advanced Research Projects Agency under contract N00014-91-J-1698. This material is based upon work also supported under an Office of Naval Research Graduate Fellowship, a National Science Foundation Fellowship, and the MIT Undergraduate Research Opportunities Program.

References

[1] L. A. Bassalygo and M. S. Pinsker. Complexity of optimum nonblocking switching networks without reconnections. *Problems of Information Transmission*, 9:64–66, 1974.

[2] David Chaiken, Craig Fields, Kiyoshi Kurihara, and Anant Agarwal. Directory-Based Cache-Coherence in Large-Scale Multiprocessors. *IEEE Computer*, 23(6):41–58, June 1990.

[3] Chin Chi-Yuan and Kai Hwang. Connection principles for multipath packet switching networks. In *10th Annual Symposium on Computer Architecture*, pages 99–108, 1984.

[4] André DeHon, Thomas F. Knight Jr., and Henry Minsky. Fault-tolerant design for multistage routing networks. In *International Symposium on Shared Memory Multiprocessing*. Information Processing Society of Japan, April 1991.

[5] André DeHon, Thomas F. Knight Jr., and Henry Minsky. RNP: Fault tolerant routing protocol. Transit Note 41, MIT Artificial Intelligence Laboratory, March 1991.

[6] Robert Grondalski. A VLSI chip set for massively parallel architecture. In *IEEE International Solid-State Circuits Conference*, pages 198–199, 1987.

[7] Joseph Y. Hui. *Switching and Traffic Theory for Integrated Broadband Networks*. Kluwer Academic Publishers, Boston, 1990.

[8] Nabil Kahale. Better expansion for Ramanujan graphs. In *32nd Annual Symposium on Foundations of Computer Science*, pages 398–404. IEEE, October 1991.

[9] S.C. Kothari, G.M. Prabhu, and R. Roberts. The kappa network with fault-tolerant destination tag algorithm. *IEEE Transactions on Computers*, 37(5):612–617, May 1988.

[10] Clyde P. Kruskal and Marc Snir. The performance of multistage interconnection networks for multiprocessors. *IEEE transactions on Computers*, C-32(12):1091–1098, December 1983.

[11] Clyde P. Kruskal and Marc Snir. A unified theory of interconnection network structure. In *Theoretical Computer Science*, pages 75–94, 1986.

[12] Vijay P. Kumar and Andrew L. Reibman. Failure dependent performance analysis of a fault-tolerant multistage interconnection network. *IEEE Transactions on Computers*, 38(12):1703–1713, December 1989.

[13] Duncan Lawrie and Krishnan Padmanabhan. A class of redundant path multistage interconnection networks. *IEEE Tr. on Computers*, C-32(12):1099–1108, December 1983.

[14] Tom Leighton, Derek Lisinski, and Bruce Maggs. Empirical evaluation of randomly-wired multistage networks. In *International Conference on Computer Design: VLSI in Computers and Processors*. IEEE, 1990.

[15] Tom Leighton and Bruce Maggs. Expanders might be practical: fast algorithms for routing around faults on multibutterflies. In *IEEE 30th Annual Symposium on Foundations of Computer Science*, 1989.

[16] Charles E. Leiserson. Fat-trees: Universal networks for hardware efficient supercomputing. *IEEE transactions on Computers*, C-34(10):892–901, October 1985.

[17] Henry Minsky, André DeHon, and Thomas F. Knight Jr. RN1: Low-latency, dilated, crossbar router. In *Hot Chips Symposium III*, 1991.

[18] E. Upfal. An $O(\log N)$ deterministic packet routing scheme. In *21st Annual ACM Symposium on Theory of Computing*, pages 241–250. ACM, May 1989.

Timing

Optimizing Two-Phase, Level-Clocked Circuitry

(Extended Abstract)

Alexander T. Ishii
NEC C&C Research Laboratories
Princeton, New Jersey 08540

Charles E. Leiserson
Marios C. Papaefthymiou
MIT Laboratory for Computer Science
Cambridge, Massachusetts 02139

Abstract

We investigate two strategies for reducing the clock period of a two-phase, level-clocked circuit: *clock tuning*, which adjusts the waveforms that clock the circuit, and *retiming*, which relocates circuit latches. These methods can be used to convert a circuit with edge-triggered latches into a faster level-clocked one.

We model a two-phase circuit as a graph whose vertex set V is a collection of combinational logic blocks, and whose edge set E is a set of interconnections. Each interconnection passes through 0 or more latches, where each latch is clocked by one of two periodic, nonoverlapping waveforms, or *phases*.

We give efficient polynomial-time algorithms for problems involving the timing verification and optimization of two-phase circuitry. Included are algorithms for

- verifying proper timing: $O(VE)$ time.
- minimizing the clock period using clock tuning: $O(VE)$ time.
- retiming to achieve a given clock period when the phases are symmetric: $O(VE + V^2 \lg V)$ time.
- retiming to achieve a given clock period when either the duty cycle (high time) of one phase or the ratio of the phases' duty cycles is fixed: $O(V^3)$ time.

By characterizing the set of possible clock periods under any retiming of the circuit, we are able to obtain polynomial-time algorithms for clock period minimization by

- retiming when the phases are symmetric: $O(V^2E)$ time.
- retiming when either the duty cycle of one phase is fixed or the ratio of the phases' duty cycles is fixed: $O(V^2E + V^3 \lg V)$ time.
- simultaneous retiming and clock tuning: $O(V^{11})$ time.

Obviously, this last algorithm is not practical. For this problem, however, we give a fully polynomial-time approximation scheme that achieves the optimal period to within any given relative error bound $e > 0$ and runs in $O(V^3(1/e)\lg(1/e) + (VE + V^2 \lg V)\lg(V/e))$ time. We also give fully polynomial-time approximation schemes for the first two of these three period-minimization retiming problems. These two algorithms run in $O((VE + V^2 \lg V)\lg(V/e))$ and $O(V^3 \lg(V/e))$ time, respectively, and can be used to obtain the optimum clock period in the special case where all propagation delays are integers.

We generalize most of the results for two-phase clocking schemes to simple multiphase clocking disciplines, including ones with overlapping phases. Typically, the algorithms to verify and optimize the timing of k-phase circuits are at most a factor of k slower than the corresponding algorithms for two-phase circuits.

1 Introduction

A VLSI designer often has the choice of whether to use level-clocked latches or the more conventional edge-triggered latches to implement clocked storage elements in his circuit. An edge-triggered latch directly supports the abstraction of a storage element that is synchronized by the ticking of a clock. When the clocking waveform rises (*i.e.*, the clock ticks), an edge-triggered latch instantaneously samples its input and asserts that value on its output.

A level-clocked latch operates somewhat differently. While the clock input to a level-clocked latch is low, the latch output maintains its value from the most recent time that the clock was high. While the clock is high, however, the input flows unimpeded to the output, unsynchronized with either edge of the clock. In order to avoid problems with race conditions, it is common for level-clocked circuits to adopt clocking disciplines which involve multiple clock waveforms, or *phases*.

In a *two-phase clocking scheme* [15], two clocking waveforms, or *phases*, denoted ϕ_0 and ϕ_1, are employed, as is shown in Figure 1. Formally, we denote a two-phase clocking scheme by a 4-tuple $\pi = \langle \phi_0, \epsilon_0, \phi_1, \epsilon_1 \rangle$ of strictly positive real numbers. In this context, ϕ_0 denotes the *duty cycle* of the first phase, *i.e.*, the length of time during which the phase is high, while ϵ_0 denotes the *gap* of the first phase, *i.e.*, the amount of time between each falling edge of the first phase and the rising edge of the "next" pulse of the second phase, where clock pulses are ordered according to their falling edges. The duty cycle and gap of the second phase are, similarly, denoted by ϕ_1 and ϵ_1, respectively. The ratio $\rho = \phi_1/\phi_0$ is the *duty ratio* of the clocking scheme. The symbol π is overloaded to also denote the sum

$$\pi = \phi_0 + \epsilon_0 + \phi_1 + \epsilon_1, \tag{1}$$

or *period* of the clocking scheme $\pi = \langle \phi_0, \epsilon_0, \phi_1, \epsilon_1 \rangle$. Observe, that since ϵ_0 and ϵ_1 are required to be strictly greater than zero, phases ϕ_0 and ϕ_1 must be nonoverlapping, *i.e.*, not simultaneously high.

In this paper we investigate methods for verifying the proper timing of circuits that utilize two-phase clocking schemes, and we present algorithms for optimizing their clock periods. The application domain of these methods, however, is not restricted only to the domain of such two-phase, level-clocked circuits. For example, since an edge-triggered latch can be implemented by two back-to-back level-clocked latches, our algorithms also provide an automatic way for speeding-up

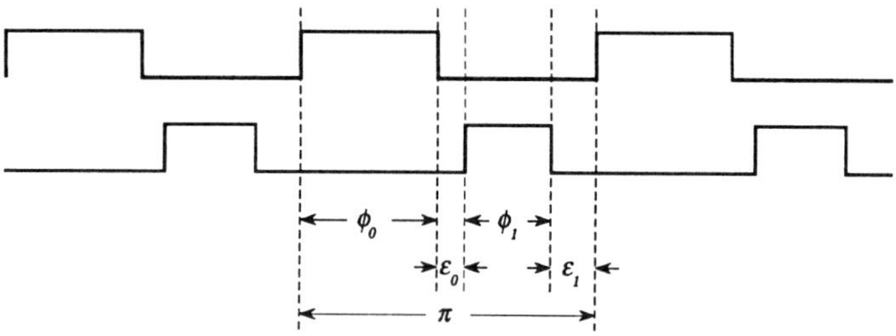

Figure 1: A two-phase, nonoverlapping clocking scheme.

edge-triggered circuits. In addition, most of our algorithms can be generalized to level-clocked circuits utilizing k clock phases.

We model a circuit as a connected graph $G = (V, E)$ in which the vertex set V is a collection of combinational logic elements with associated propagation delays, and the edge set E is a set of interconnections each of which passes through 0 or more latches. Each latch is clocked by either ϕ_0 or ϕ_1. A general framework for the timing verification of level-clocked circuits has been given in [7]. Extensive details of [7], and some details of this work, appear in [6].

An example of a two-phase, level-clocked circuit is given in Figure 2(a). The integers in the vertices signify propagation delays. For simplicity, assume that $\epsilon_0 = \epsilon_1 = 0$. (In practice and in our later analysis for two-phase circuits, we require that ϵ_0 and ϵ_1 be strictly positive; handling the case when either or both are zero simply causes various inequalities to be strict instead of nonstrict.) If the two phases ϕ_0 and ϕ_1 have equal duty cycles, then the circuit cannot be clocked with a clock period shorter than 36 units of time, since the path DEA has propagation delay 54, and intuitively, a datum has at most 3/2 clock periods to propagate from the latch preceding D to the latch succeeding A.

Our first new result is an $O(VE)$-time algorithm that verifies whether a level-clocked circuit is properly timed by a given two-phase clocking scheme. This result improves the $O(E^2)$ bound obtained when the general algorithm from [7] is applied to the special case of two-phase, level-clocked circuits. (The bound in [7] is also $O(VE)$, but the circuit model used in that paper represented both functional elements and latches as vertices, and interconnections between them as edges. Translating their bound to the model presented here yields the $O(E^2)$ bound. The algorithm in [7] applies to more general circuits than two-phase ones, however.)

Our next result deals with modifying, or *tuning*, the clocking scheme of a circuit—that is, providing the circuit with new clocking waveforms. The two clocking schemes in Figure 2(e) and Figure 2(f), for example, illustrate how the clocking scheme of a circuit can be tuned. Observe, that the circuit shown in Figure 2(a) is properly timed by either clocking scheme, but the first clocking scheme has a period of 36, while the second has a clock period of 33. Notice that the notion of clock tuning encompasses more than a simple increase in the frequency of the clock. In particular, clock tuning allows the duty ratio of the clocking scheme to be adjusted as well. Observe also that the circuit cannot be properly timed by any clocking scheme whose period is less than 33, since the delay of the path CDEA is 66 and must be distributed among at most two full clock periods. The clocking scheme shown in Figure 2(f) is thus an *optimal* clock tuning for the circuit in Figure 2(a). We give an $O(VE)$-time algorithm that computes an optimal clock tuning for any given two-phase circuit. Previous algorithms for clock tuning have either addressed alternate types of clocking methodologies [23, 20, 4], or been uncharacterized with respect to worst-case running time [2, 21].

Another way to optimize a circuit is by *retiming*: a method for relocating latches within the circuit without affecting the functionality of the circuit. Retiming has been well studied in the context of edge-triggered circuits [9, 12, 19], and has been the subject of study in the context of single-phase, level-clocked circuits [14, 22]. We extend this technique to encompass the optimization of two-phase, level-clocked circuits. We consider three retiming problems.

The first problem asks whether, for a given two-phase circuit G and clocking scheme π, the circuit G can be retimed to be properly timed by π. As an example, consider the circuit in Figure 2(a). If we retime the circuit to be properly timed by the clocking scheme in Figure 2(g), we obtain the circuit in Figure 2(b). For

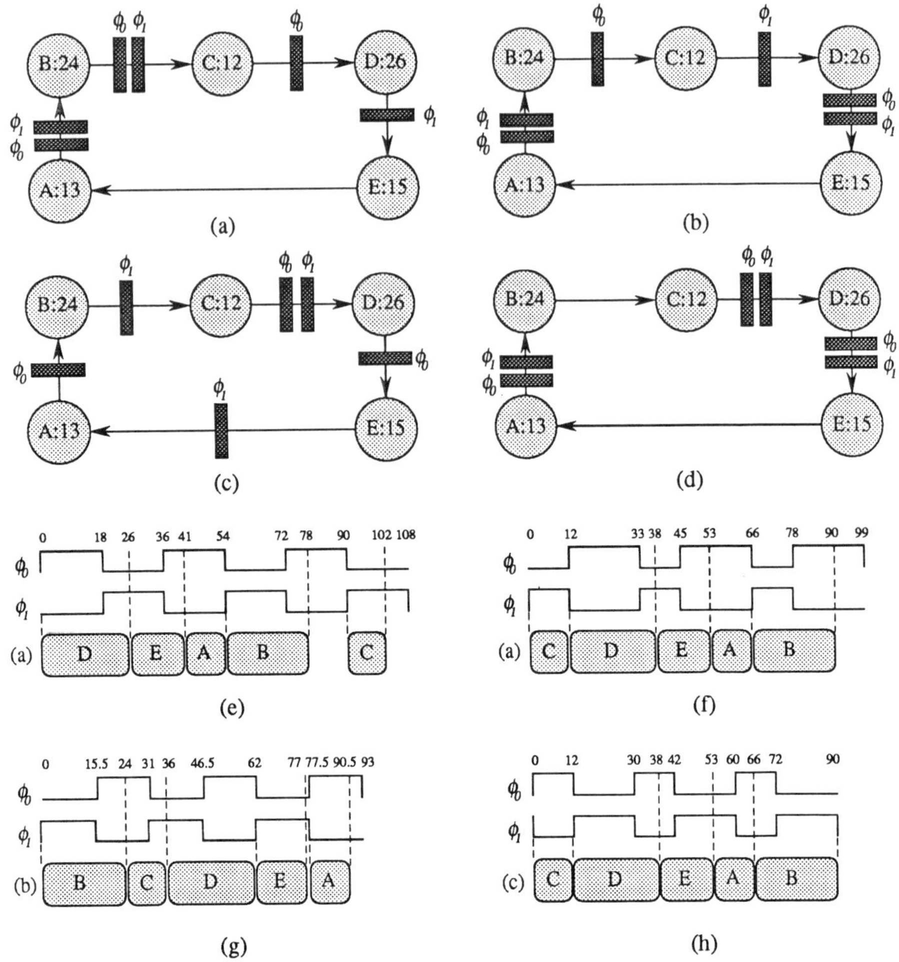

Figure 2: An illustration of the various techniques for optimizing two-phase, level-clocked circuits. Part (a) of the figure shows a level-clocked circuit. When the duty cycles of the phases ϕ_0 and ϕ_1 are equal, as is shown in part (e) of the figure, the clock period of the circuit cannot be made smaller than 36. By tuning the clocks of the circuit to have a duty ratio of 4:7, as shown in part (f), a clock period of 33 can be achieved. Part (b) of the figure shows a retimed version of the original circuit. When clocked with the symmetric clocking scheme shown in part (g), it achieves a clock period of 31, which is optimal for symmetric clocking schemes under any retiming. The optimal combination of retiming and clock tuning for the circuit is shown in part (c). When clocked with the waveforms in part (h), which have a duty ratio of 3:2, the circuit achieves a clock period of 30, which is the optimal for any combination of clocking scheme and retiming. Part (d) shows a typical level-clocked implementation of an edge-triggered circuit which is equivalent under retiming to the other circuits. Each pair of level-clocked latches in the circuit implements an edge-triggered latch. This edge-triggered circuit has a clock period of 36, which is the best that can be obtained by any retiming using edge-triggered latches.

symmetric clocking schemes, such as that in Figure 2(g), in which the two phases are identical and 180 degrees out of phase with each other—and thus, the clocking scheme has the form $\pi = \langle \phi, \epsilon, \phi, \epsilon \rangle$—we provide an algorithm for retiming that runs in $O(VE + V^2 \lg V)$ time. When the clocking scheme π has general form, we provide an algorithm for retiming that runs in $O(V^3)$ time.

The second problem is the *fixed duty-ratio retiming* problem, which asks how a circuit can be retimed to achieve the minimum clock period possible over all clocking schemes with a given duty ratio. For example, with a duty ratio of 1, the optimal retiming of the circuit in Figure 2(a) and tuning of its clock yields the circuit in Figure 2(b) with the clocking scheme in Figure 2(g). We give an algorithm for the fixed duty-ratio retiming problem which runs in $O(V^2 E + V^3 \lg V)$ time and produces the optimal retimed circuit. In addition, we give an algorithm that, for any given $e > 0$, runs in $O(V^3 \lg(V/e))$ time and produces a retimed circuit that is properly timed by a clocking scheme whose period is at most $(1 + e)$ times the optimal. If the clocking scheme is required to be symmetric, the algorithms can be improved to run in $O(V^2 E)$ time and $O((VE + V^2 \lg V) \lg(V/e))$ time, respectively. All these algorithms can be adapted to optimally retime circuits when one of the two duty cycles of the clock is fixed and the other can be tuned.

The third problem asks how a circuit can be retimed to achieve the minimum clock period possible when no duty ratio is specified. For example, the optimal retiming of the circuit in Figure 2(a) yields the circuit in Figure 2(c) with the clocking scheme in Figure 2(h). We provide an efficient approximation algorithm, but unfortunately, only an impractical polynomial-time algorithm for this *variable duty-ratio retiming* problem. For any given $e > 0$, the approximation algorithm runs in $O(V^3 (1/e) \lg(1/e) + (VE + V^2 \lg V) \lg(V/e))$ time and produces a retiming that is properly timed by a clocking scheme whose period is at most $(1 + e)$ times the optimal.

Our optimization techniques cannot only be used to facilitate the design of level-clocked circuits, but also to convert edge-triggered circuits into faster level-clocked circuits. The basis for the conversion is that an edge-triggered latch can be implemented by a pair of level-clocked latches. In Figure 2(d) we illustrate the typical level-clocked implementation of an edge-triggered circuit. This circuit has a clock-period of 36, but the algorithms presented in this paper can automatically produce the optimal level-clocked circuit in Figure 2(c) which, with the clocking scheme in Figure 2(h), is timed properly with a clock period of 30. As a testament to the additional power gained by level-clocking, it should be noted that no edge-triggered retiming of the circuit from Figure 2(d) improves upon its period of 36.

The remainder of this paper is organized as follows. Section 2 describes the necessary and sufficient conditions for a two-phase, level-clocked circuit to be properly timed. Section 3 describes the $O(VE)$ algorithm for verifying the proper timing of a two-phase, level-clocked circuit. Section 4 presents the $O(VE)$ algorithm for tuning clocking schemes. This algorithm adjusts the duty cycles of the two phases, so that the clock period is minimized. In Section 5, we show how to retime a circuit to achieve a specified clocking scheme π. We describe the $O(VE + V^2 \lg V)$-time algorithm for retiming when the duty cycles are symmetric. This algorithm is based on the reduction of the retiming problem to an efficiently solvable mixed-integer linear program. We also describe the $O(V^3)$-time algorithm for retiming to achieve an arbitrary clocking scheme π. Both algorithms also determine when the given clocking scheme cannot be achieved by any retiming. In Section 6 we present algorithms for retiming that achieve the optimal clock period to within a relative error of $e > 0$. Specifically, we give an $O((VE + V^2 \lg V) \lg(V/e))$-time algorithm for the case when duty cycles are symmetric, and an $O(V^3 \lg(V/e))$-time algorithm for the case when either the duty cycle of one of the phases is fixed, or the duty ratio is fixed. When the duty ratio is not fixed, we present an $O(V^3(1/e)\lg(1/e) + (VE + V^2 \lg V)\lg(V/e))$-time algorithm that retimes the circuit to achieve the optimal period to within a relative error of e. In Section 7 we describe fully-polynomial algorithms for the problems of optimal retiming. Specifically, we give an $O(V^2E)$-time algorithm for optimal retiming when duty cycles are symmetric, an $O(V^2E + V^3 \lg V)$-time algorithm for optimal retiming when either the duty cycle for one of the phases is fixed or the duty ratio of the clocking scheme is fixed, and an $O(V^{11})$-time algorithm for optimal retiming combined with clock tuning. Section 8 extends many of our techniques to circuits that utilize more than two phases. The algorithms for k-phase circuits are generally at most a factor of k slower than the corresponding algorithms for two-phase circuits. Finally, Section 9 presents some concluding remarks. Due to space limitations, most proofs in this extended abstract have been omitted. These will appear in the final version of the paper.

In independent work, Lockyear and Ebeling [13] have also obtained algorithms for retiming multiphase, level-clocked circuits. Their results include an $O(V^3)$-time algorithm to determine whether a given clock period can be achieved by retiming when all clock phases are symmetric. They use this algorithm as a subroutine for determining a retiming that achieves the minimum clock period possible when phases are symmetric. They also determine a set of constraints for retiming with asymmetric clock phases, and they sketch a Bellman-Ford-like algorithm for solving the constraints.

2 Preliminaries

In this section we give necessary and sufficient conditions for a two-phase, level-clocked circuit to be properly timed by a given clocking scheme. The set of conditions, or *timing constraints*, is based on the general formulation from [7], but is substantially simplified by the additional structure inherent in two-phase, level-clocked circuits.

A level-clocked circuit is a set of interconnected functional elements, and level-clocked latches. A functional element (*i.e.*, combinational logic block) is characterized by its maximum propagation delay, while a level-clocked latch is characterized by the phase that controls it. A level-clocked circuit is *two-phase* level-clocked if it

1. utilizes a two-phase clocking scheme,
2. contains no latch-free cycles, and
3. contains no combinational paths between latches controlled by the same phase,

where a path is *combinational* if the first and last components in the path are the only path components that are not functional elements. Observe that every cycle in a two-phase circuit must contain at least two latches—one for each of the two different clock waveforms.

A fundamental property of two-phase, level-clocked circuits is that it is typically possible to associate clock waveforms with each circuit component. In particular, if a circuit has no sources or sinks, *i.e.*, if the output of every component is an input to some component and each input to a component is the output of some component, then for any functional element B, the latches from which B can be reached via combinational paths must all be controlled by the same clock phase ϕ. The clock ϕ is the *input phase*, or simply, the *phase* of the functional element B. One can also associate with B the unique clock waveform ϕ' which controls all latches that can be reached from B via a combinational path. The clock ϕ' is the *output phase* of B.

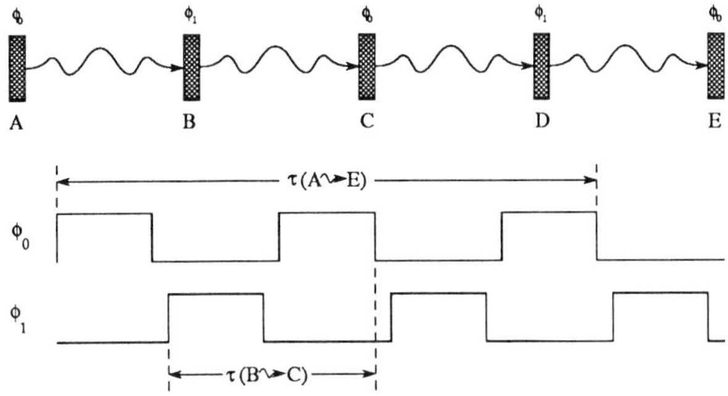

Figure 3: The rise-to-fall time for the path $A \rightsquigarrow E$ is $\phi_0 + 2(\epsilon_0 + \phi_1 + \epsilon_1 + \phi_0) = \phi_0 + 2\pi$. The rise-to-fall time for the path $B \rightsquigarrow C$ is $\phi_1 + \epsilon_1 + \phi_0$. Each of the squiggly lines in the top part of the figure represents a path of zero or greater combinational delay.

It is easy to confirm that the input phase of a functional element must be different from the output phase. For succinctness, we use the function χ to indicate the (input) phase of a component B. If $\chi(B) = 0$, then the phase of B is ϕ_0. Similarly, if $\chi(B) = 1$, then the phase of B is ϕ_1. We generally denote the phase of B as $\phi_{\chi(B)}$. Using this notation, the output phase of B can be denoted as $\phi_{1-\chi(B)}$. If a circuit contains sources or sinks, then the existence of well-defined input phases and output phases must be guaranteed by adding an explicit condition to the definition of a two-phase, level-clocked circuit.

Intuitively, a level-clocked circuit is properly timed if whenever a latch holds a value (*i.e.*, whenever its clock input is low), it holds the same value it would in an identical circuit in which all functional elements have zero propagation delay. Ishii and Leiserson [7] give a set of "Δ-constraints" that serve as a necessary and sufficient set of conditions for the proper timing of a general class of level-clocked circuits. Although this formulation is somewhat complicated by the generality of the circuits which it handles, for two-phase circuits, it leads to a set of necessary and sufficient conditions that can be expressed in a simple manner.

Let p be any path, not necessarily simple, from a latch A to a latch B in the circuit. The *rise-to-fall* time $\tau(p)$ of p is the time it takes in a zero-delay circuit for the effect of a rising edge of A's phase to propagate along p and be stored in B by the falling edge of B's phase. For example, the rise-to-fall time of path $A \rightsquigarrow E$ in Figure 3 is $\phi_0 + 2(\epsilon_0 + \phi_1 + \epsilon_1 + \phi_0) = \phi_0 + 2\pi$. The rise-to-fall time of path $B \rightsquigarrow C$ is $\phi_1 + \epsilon_1 + \phi_0$. The following proposition states the conditions for the proper timing of a circuit. Its proof will be given in the full version of the paper.

Proposition 1 *A two-phase circuit is properly timed if and only if for all latches A and B in the circuit, the propagation delay $d(p)$ along any path p from A to B is no greater than the rise-to-fall time $\tau(p)$ of the path.*

It should be noted that, as in [7, 6], the notion of proper timing assumed in this paper is *structural*, in the sense that we require that a circuit be properly timed regardless of the functions computed by the functional elements. This requirement avoids potential difficulties with computational intractability.

A two-phase, level-clocked circuit is formally represented by a connected, directed multigraph $G = \langle V, E, d, w, \chi \rangle$, where the *delay* function d is a mapping from V to the nonnegative real numbers, the *weighting* function w is a mapping from E to the nonnegative integers, and the *phase* function χ is a mapping from V to the set $\{0,1\}$. For convenience, we use the terms "graph" and "circuit" interchangeably. A vertex $v \in V$ corresponds to a functional element whose maximum propagation delay is equal to $d(v)$. Level-clocked latches are not represented as vertices, but rather, are represent implicitly by the weights $w(e)$ of edges $u \xrightarrow{e} v \in E$. A value $w(e) = 0$ indicates that a direct connection exists between the output of the functional element represented by u and an input of the functional element represented by v. A value $w(e) > 0$ indicates that the connection between the two functional elements consists of $w(e)$ level-clocked latches connected in series. One can verify that a given graph $G = \langle V, E, d, w, \chi \rangle$ corresponds to a two-phase, level-clocked circuit if and only if, for all paths $u \overset{p}{\rightsquigarrow} v$ in G, the value $w(p) - \chi(u) + \chi(v)$ is an even number.

Given Proposition 1, we can formulate a set of conditions for the proper timing of a two-phase, level-clocked circuit. Specifically, we have the following lemma.

Lemma 2 *Let $G = \langle V, E, d, w, \chi \rangle$ be a circuit that utilizes a clocking scheme $\pi = \langle \phi_0, \epsilon_0, \phi_1, \epsilon_1 \rangle$. Then G is properly timed if and only if, for every path $u \overset{p}{\rightsquigarrow} v$ in G, we have*

$$d(p) \leq \pi \left(\frac{1 + w(p)}{2} \right) + \phi_{1-\chi(v)} \tag{2}$$

if $\chi(u) \neq \chi(v)$, and

$$d(p) \leq \pi \left(\frac{2 + w(p)}{2} \right) - \epsilon_{1-\chi(v)} \tag{3}$$

if $\chi(u) = \chi(v)$. □

The reader should note that the lemma requires *all* paths in the circuit to be considered, not just simple ones.

3 Verifying proper circuit timing

In this section we describe an $O(VE)$-time algorithm to verify whether a level-clocked circuit is properly timed by a given two-phase clocking scheme. This bound is better than the $O(E^2)$ bound that one obtains by applying the general verification algorithm in [7] to two-phase circuits.

The verification algorithm VERIFY is shown in Figure 4. The algorithm determines whether a circuit is properly timed by checking certain constraints which can be shown to be equivalent to the path constraints in Lemma 2.

In order to prove the correctness of Algorithm VERIFY, we first show that Step 1 computes a new circuit G' with at most 3 latches per wire and such that G is properly timed if and only if G' is properly timed. The reason for performing this transformation is that the running time of Steps 2 and 3 depends on the number of latches on each wire of the circuit. Thus, reducing this number to at most 3 results in a more efficient verification algorithm. Note that the new circuit G' does not in general compute the same function as G.

Lemma 3 *Let $G = \langle V, E, d, w, \chi \rangle$ be a circuit, and let $G' = \langle V, E, d, w', \chi \rangle$ be the circuit obtained by replacing the $w(e)$ latches on each edge $e \in E$ with $w'(e)$ latches, where*

$$w'(e) = \begin{cases} w(e) & \text{if } w(e) \leq 3 \,, \\ (w(e) \bmod 2) + 2 & \text{if } w(e) \geq 4 \,. \end{cases}$$

VERIFY(G, π) $\quad \triangleright\ G = (V, E, d, w, \chi)$; $\pi = \langle \phi_0, \epsilon_0, \phi_1, \epsilon_1 \rangle$

1. Generate $G' = (V, E, w', d, \chi)$ by replacing $w(e)$ on each edge $e \in E$ with $w'(e)$, where

$$w'(e) = \begin{cases} w(e) & \text{if } w(e) \leq 3 \ , \\ (w(e) \bmod 2) + 2 & \text{if } w(e) \geq 4 \ . \end{cases}$$

2. Compute $D(v, i)$ for all $v \in V$ and $i = 0, 1, 2, \ldots, 3|V|$, from the recurrence

$$D(v, i) = d(v) + \max\left\{ D(u, i - w'(e)) : u \xrightarrow{e} v \text{ and } i \geq w'(e) \right\} .$$

3. Check the following constraints for each vertex $v \in V$ and $i = 0, 1, \ldots, 3\,|V|$:

$$D(v, i) \leq \pi \left(\frac{1+i}{2} \right) + \phi_{1-\chi(v)} \quad \text{if } i \text{ is odd;}$$

$$D(v, i) \leq \pi \left(\frac{2+i}{2} \right) - \epsilon_{1-\chi(v)} \quad \text{if } i \text{ is even.}$$

   If any constraint is violated, then return **no**.

4. Determine whether the graph G with weight $w(e) - 2d(u)/\pi$ on each edge $u \xrightarrow{e} v$ has a negative-weight cycle. If a negative-weight cycle exists, then return **no**; otherwise, return **yes**.

Figure 4: Pseudocode for the $O(VE)$ verification algorithm.

Then for any clocking scheme $\pi = \langle \phi_0, \epsilon_0, \phi_1, \epsilon_1 \rangle$, G is properly timed by π if and only if G' is properly timed by π. □

The following lemma, whose proof relies on Lemma 2, justifies Steps 2, 3, and 4 of the algorithm. The lemma shows that we can substitute $O(V^2)$ inequalities for the infinitely many path-constraints (2) and (3) that describe the conditions for proper circuit timing. This simpler set of inequalities motivates the computation of the various $D(v, i)$ values.

Lemma 4 *Let $G = \langle V, E, d, w, \chi \rangle$ be a circuit that uses a clocking scheme $\pi = \langle \phi_0, \epsilon_0, \phi_1, \epsilon_1 \rangle$. Then G is properly timed if and only if for every vertex $v \in V$ and $i = 0, 1, \ldots, 3\,|V|$, we have*

$$D(v, i) \leq \pi \left(\frac{1+i}{2} \right) + \phi_{1-\chi(v)} \tag{4}$$

if i is odd, and

$$D(v, i) \leq \pi \left(\frac{2+i}{2} \right) - \epsilon_{1-\chi(v)} \tag{5}$$

if i is even; and for every simple cycle p, we have

$$d(p) \leq \pi \left(\frac{w(p)}{2} \right) . \tag{6}$$

□

Inequalities (4) and (5) directly justify Steps 2 and 3 of the algorithm. The values $D(v,i)$, which are computed in Step 2 and used in Inequalities (4) and (5), are defined for $v \in V$ and $i = 1, 2, \ldots, 3|V|$ by

$$D(v,i) = \max\{d(p) : u \overset{p}{\rightsquigarrow} v, \text{ and } w(p) = i\} ,$$

where, by convention, $D(v,i) = 0$ if there are no paths with exactly i latches to v from any vertex. Intuitively, for a given vertex v and integer i, a value $D(v,i)$ embodies the largest left-hand side of any constraint described by Inequalities (2) and (3). This formulation has the advantage of reducing the number of constraints to be checked, since every path between two vertices need not be considered.

Inequality (6) justifies Step 4 as follows. Sum the edge weights around a cycle. If the cycle has a negative weight, then Inequality (6) is violated. Conversely, if there is no negative-weight cycle, then every cycle in the graph satisfies the inequality. It follows that Algorithm VERIFY correctly determines whether a circuit is properly timed by a given two-phase clocking scheme.

Algorithm VERIFY terminates in $O(VE)$ steps. The circuit transformation of Step 1 can be computed in $O(E)$ time. In Step 2, the algorithm performs $3|V|+1$ passes over the edge set E. In each pass, it first considers the edges with $w'(e) > 0$, and then the edges with $w'(e) = 0$ in topological sort order. Since a topological sort can be computed in $O(E)$ time [1], Step 2 terminates in $O(VE)$ time. Step 3 checks $O(V^2)$ constraints in $O(V^2)$ time. Finally, Step 4 can be performed in $O(VE)$ time using the Bellman-Ford algorithm [1]. Thus, the total is $O(E) + O(VE) + O(V^2) + O(VE) = O(VE)$.

4 Period minimization by clock tuning

In this section we describe an $O(VE)$-time algorithm which, given a circuit G and a clocking scheme $\pi = \langle \phi_0, \epsilon_0, \phi_1, \epsilon_1 \rangle$, computes a clocking scheme $\pi^* = \langle \phi_0^*, \epsilon_0, \phi_1^*, \epsilon_1 \rangle$ that properly times G and whose period π^* cannot be further reduced without reducing either ϵ_0 or ϵ_1.

The basic idea behind the period minimization algorithm is to view π, ϕ_0, and ϕ_1 as variables. By Equation (1), $\phi_1 = \pi - \epsilon_0 - \epsilon_1 - \phi_0$, and thus, the constraints defined by (4), (5), and (6) can be viewed as inequalities in the two variables π and ϕ_0. In fact, they form a linear program in which the objective is to minimize π. This program can be described by a set of lines in the (ϕ_0, π) plane. We distinguish three kinds of lines corresponding to Inequalities (4), (5), and (6). Let us look more closely at each of these three sets of constraints that form the linear program.

The constraints defined by Inequality (4) depend on both π and ϕ_0. We can separate these constraints into two sets depending on the value of $\chi(v)$. If $\chi(v) = 1$, then the inequality becomes

$$\pi \geq \frac{D(v,i) - \phi_0}{(i+1)/2} , \tag{7}$$

which defines a half-plane of feasible (ϕ_0, π) points for each fixed i. If $\chi(v) = 0$, the inequality becomes

$$\pi \geq \frac{D(v,i) + \epsilon_0 + \epsilon_1 + \phi_0}{(3+i)/2} , \tag{8}$$

which once again defines a half-plane of feasible (ϕ_0, π) points for each fixed i, since ϵ_0 and ϵ_1 are constants. Each inequality holds for each $v \in V$, and for each odd i in $1 \leq i \leq 3|V|$. The $O(V^2)$ constraints defined by Inequality (4) can be determined in $O(VE)$ time by computing the values $D(v, i)$, as in Step 2 of Algorithm VERIFY. These constraints can be reduced in $O(V^2)$ time to an equivalent set of $O(V)$ constraints by selecting, for each odd i in the range $1 \leq i \leq 3|V|$, the particular constraint for which $D(v, i)$ is maximized.

Constraints defined by Inequality (5) depend only on π, and not on ϕ_0. Consequently, these constraints determine a lower bound on π^* which is independent of the duty cycle of either phase. A single lower bound on π^* can thus be determined from these $O(V^2)$ inequalities in $O(V^2)$ time, after having computed the values of $D(v, i)$ in $O(VE)$ time.

Constraints defined by Inequality (6) depend only on π, and not on ϕ_0, and they also determine a lower bound on π^* which is independent of the duty cycles. This lower bound can be found in $O(VE)$ time by solving a "tramp steamer" problem. The *tramp steamer problem* (also known as the *minimum cost-to-time ratio cycle problem*) was formulated in [3] as follows. Let $G = \langle V, E, c, t \rangle$ be a directed graph in which each edge (u, v) has an integer cost $c(u, v)$ and an integer transit time $t(u, v)$, such that for any cycle p in G, we have $\sum_{(u,v)\in p} t(u, v) > 0$. For any cycle p in G, define the cost-to-time ratio of the cycle by

$$R(p) = \frac{\sum_{(u,v)\in p} c(u, v)}{\sum_{(u,v)\in p} t(u, v)} .$$

The problem is to find the minimum value $R(G)$ of the cost-to-time ratio over all cycles in the graph.

Lemma 5 *Let $G = (V, E, d, w, \chi)$ be a circuit, and let $\tilde{G} = \langle V, E, c, t \rangle$ be the graph obtained by assigning to each edge $(u, v) \in E$ a cost $c(u, v) = -d(u)$ and a transit time $t(u, v) = w(u, v)$. Let $R(G)$ be the minimum cost-to-time ratio of any cycle in G. Then Inequality (6) is satisfied if and only if*

$$\pi \geq 2 \left| R(\tilde{G}) \right| . \tag{9}$$

□

Using the algorithm for the tramp steamer problem given in [5], the cycle constraints can be checked in $O(VE)$ time. In order to obtain this running time, we must guarantee that the transit time of any path with $|V|$ edges is $O(V)$. Indeed, this requirement is met, since from Lemma 3 the number of latches on any edge can be reduced to at most 3.

Besides these three kinds of constraints, we additionally require the following:

$$\pi \geq \epsilon_0 + \epsilon_1 + \phi_0 . \tag{10}$$

This inequality guarantees that $\phi_1 > 0$, and thus that only (ϕ_0, π) combinations representing valid clocking schemes are considered.

By linear programming theory [17], an optimal (ϕ_0^*, π^*) combination can be obtained at a point of intersection of the constraints defined by Inequalities (5), (7), (8), (9), and (10). As we have mentioned, Inequalities (7) and (8) can be reduced to a set of $O(V)$ constraints, and the remaining three inequalities can each be reduced to a single constraint. Megiddo's algorithm [16] can solve such a "two-dimensional" linear program in $O(V)$ time. Alternatively, one can simply compute the $O(V^2)$ intersections between the lines defined by Inequality (7) and the lines defined by Inequalities (8) and (10). If (ϕ_0', π') is the intersection point of maximum π value, and if π'' is the lower bound derived from Inequalities (5) and (9), then an optimal point (ϕ_0^*, π^*) is given by $(\phi_0', \max\{\pi', \pi''\})$. In either case, the overall running time is $O(VE)$.

5 Retiming

In this section, we show how to determine whether a circuit can be retimed so that it is properly timed by a given clocking scheme. We present two algorithms that take a circuit $G = \langle V, E, d, w, \chi \rangle$ and a clocking scheme $\pi = \langle \phi_0, \epsilon_0, \phi_1, \epsilon_1 \rangle$, and either compute a retiming of G that is properly timed by π, or determine that no such retiming exists. The first algorithm is optimized for the special case of a symmetric clocking scheme, where $\phi_0 = \phi_1$ and $\epsilon_0 = \epsilon_1$. Like other previous retiming results [10, 22] the algorithm casts retiming for a symmetric clocking scheme as a mixed-integer linear program. The overall running time of the algorithm is $O(VE + V^2 \lg V)$. The second algorithm can be applied to arbitrary two-phase clocking schemes and relies on a new technique we call *integer monotonic programming*. The second algorithm runs in $O(V^3)$ time.

Adopting the notation from [10], a circuit $G_r = \langle V, E, d, w_r, \chi_r \rangle$ is a *legal retiming* of another circuit $G = \langle V, E, d, w, \chi \rangle$ if, for some *retiming function* $r : V \to \mathbf{Z}$, we have

$$w(e) + r(v) - r(u) \geq 0 \tag{11}$$

for every edge $u \xrightarrow{e} v \in E$, and

$$\chi_r(v) = \begin{cases} \chi(v) & \text{if } r(v) \text{ is even }, \\ 1 - \chi(v) & \text{if } r(v) \text{ is odd }, \end{cases}$$

for all $v \in V$. The variable $r(v)$ counts the *lag* of vertex v, and indicates the number of latches that the retiming process moves from the output of v to the input of v. By definition, the number of latches $w_r(e)$ in the retimed circuit G_r is given by

$$w_r(e) = w(e) + r(v) - r(u) \tag{12}$$

for every edge $u \xrightarrow{e} v \in E$. If a given circuit $G = \langle V, E, d, w, \chi \rangle$ is not properly timed by its clocking scheme π, then the *retiming problem* is to either compute an appropriate retiming function r such that Inequalities (2) and (3) hold for every path in the retimed circuit G_r, or determine that no such retiming exists. If such a retiming function does exist, then π is *achievable* by G under retiming, and the retiming function r is said to *achieve* the clocking scheme.

Symmetric clocking schemes

If the clocking scheme utilized by a circuit is symmetric, *i.e.*, of the form $\pi = \langle \phi, \epsilon, \phi, \epsilon \rangle$, then Inequalities (2) and (3) can be simplified as follows. A circuit $G = \langle V, E, d, w, \chi \rangle$ is properly timed if and only if, for every path $u \overset{p}{\leadsto} v$, we have

$$d(p) \leq \pi \left(\frac{w(p)}{2} \right) + (\pi - \epsilon) . \tag{13}$$

By Definition (12), a retiming $G_r = \langle V, E, d, w_r, \chi_r \rangle$ of a given two-phase circuit $G = \langle V, E, d, w, \chi \rangle$ is properly timed if for every path $u \overset{p}{\leadsto} v$ in G_r, we have

$$\begin{aligned} d(p) &\leq \pi \left(\frac{1}{2} w_r(p) \right) + (\pi - \epsilon) \\ &= \frac{\pi}{2} (w(p) + r(v) - r(u)) + (\pi - \epsilon) . \end{aligned}$$

This last inequality can be rewritten as the following simple linear constraint over the variables $r(u)$ and $r(v)$,

$$\begin{aligned} r(u)-r(v) &\leq w(p)-\frac{2}{\pi}d(p)+\left(2-\frac{2\epsilon}{\pi}\right) \\ &= \sum_{i \xrightarrow{e} j \in p}\left(w(e)-\frac{2}{\pi}d(j)\right)+\left(2-\frac{2}{\pi}(d(u)+\epsilon)\right), \end{aligned} \tag{14}$$

which in turn can be used to prove the following theorem.

Theorem 6 *Let $G = \langle V, E, d, w, \chi\rangle$ be a circuit, and let $\pi = \langle \phi, \epsilon, \phi, \epsilon\rangle$ be a clocking scheme. Then there exists a retiming $r : V \to \mathbf{Z}$ of G such that G_r is properly timed by π if and only if there exists an assignment of a real value $R(v)$ and an integer value $r(v)$ to each vertex $v \in V$ such that the following conditions are satisfied:*

$$\begin{array}{rcll} r(u)-r(v) & \leq & w(e) & \text{for all } u \xrightarrow{e} v \in E \\ R(v)-r(v) & \leq & 0 & \text{for all } v \in V \\ R(u)-R(v) & \leq & w(e)-\frac{2}{\pi}d(v) & \text{for all } u \xrightarrow{e} v \in E \\ r(v)-R(v) & \leq & 2-\frac{2}{\pi}(d(v)+\epsilon) & \text{for all } v \in V\ . \end{array} \tag{15}$$

□

The constraints defined in Theorem 6 form a mixed-integer linear program of the special form MI from [11]. Problem MI can be solved in $O(VE + V^2 \lg V)$ time by applying Algorithm MILP from [11], which gives us an $O(VE + V^2 \lg V)$ algorithm for retiming when the clocking scheme utilized by the circuit is symmetric.

General two-phase clocking schemes

If the clocking scheme utilized by a circuit has a general form, *i.e.*, is not symmetric, the retiming problem can still be cast as a feasibility problem of a special form. There are two cases to consider for a path $u \overset{p}{\rightsquigarrow} v$ in G. First, if $\chi(u) \neq \chi(v)$, we have

$$\begin{aligned} &\left\lfloor \frac{1}{2}r(v)\right\rfloor \pi + (r(v) \bmod 2)\left(\epsilon_{\chi(u)}+\phi_{1-\chi(u)}\right)+\pi\left(\frac{1+w(p)}{2}\right)-d(p)+\phi_{\chi(u)} \\ &\geq \left\lfloor \frac{1}{2}r(u)\right\rfloor \pi + (r(u) \bmod 2)\left(\phi_{\chi(u)}+\epsilon_{\chi(u)}\right), \end{aligned} \tag{16}$$

where $w(p)$ denotes the number of latches on p. Second, a similar inequality (omitted in this extended abstract) must also hold if $\chi(u) = \chi(v)$. Following the rationale of Section 3, we can rewrite these inequalities as a set of constraints between pairs of vertices. Although there is potentially an exponential number of constraints, they can be reduced to a set of $O(V^2)$ constraints by observing that for each pair of vertices u and v, the constraint corresponding to the shortest path between u and v, with respect to the distance measure $w(e) - 2d(j)/\pi$ for each edge $i \xrightarrow{e} j$, is the strongest among all the paths from u to v. We can compute these constraints by applying Johnson's $O(VE + V^2 \lg V)$-time algorithm for all-pairs shortest-paths [1].

The problem of retiming to achieve a general, two-phase clocking scheme, as described by the set of inequalities (11) and (16), is an instance of a general feasibility problem that we call an *integer monotonic program*. This problem is defined as follows.

Problem IMP *Let S be a set of constraints in which the kth constraint has the form $f_k(x_i) \geq g_k(x_1, x_2, \ldots, x_n)$, where each f_k is monotonically increasing with respect to x_i, and each g_k is monotonically increasing with respect to each x_j, for $j = 1, 2, \ldots, n$. Find a vector $x = (x_1, x_2, \ldots, x_n)$ of nonnegative integers satisfying S, or determine that no feasible vector exists.*

In our retiming problem, the retiming variables $r(v)$ are the unknowns x_i, the left-hand sides of Inequality (16) form the functions f_k, and the right-hand sides of Inequality (16) form the functions g_k.

We can solve an integer monotonic program by setting all the unknowns x_i equal to 0 and then performing a sequence of relaxations over the set of constraints. At each relaxation step, we consider a violated constraint, and we increase the argument of f_k by as much as it is required to satisfy the constraint.

One can prove that for integer monotonic programs arising from Inequality 16, no variable $r(v)$ can attain a value larger than $3|V|$ if the constraint system is satisfiable. Consequently, if the algorithm has not terminated after $O(V^2)$ relaxations, no legal retiming exists. If the clocking scheme is achievable, however, the algorithm can be made to generate a legal retiming. The algorithm can be made to run in $O(V^3)$ time.

6 Approximation schemes for minimum-period retiming

This section presents fully polynomial-time approximation schemes for computing a "minimum-period" retiming of a given two-phase, level-clocked circuit G. A retiming G_r is a *minimum-period* retiming if it achieves a clocking scheme whose period is minimal over some specified set of achievable clocking schemes.

The first algorithm computes, for a given G and a fixed duty ratio $\rho = \phi_1/\phi_0$, a retimed circuit G_r and a clocking scheme $\pi = \langle \phi_0, \epsilon_0, \rho\phi_0, \epsilon_1 \rangle$ such that G_r is properly timed by π, and the period of π is at most $(1+e)$ times the period of any clocking scheme that can be achieved by G under retiming and whose duty ratio is ρ. The running time of this algorithm is $O(V^3 \lg(V/e))$, but the running time can be reduced to $O((VE + V^2 \lg V) \lg(V/e))$ if only symmetric clocking schemes are utilized. In addition, if all functional elements have integer propagation delays, then the algorithm can be adapted to compute retimings which achieve exactly the minimum period possible.

The second algorithm lifts the fixed duty-ratio requirement and computes in $O(V^3(1/e)\lg(1/e) + (VE + V^2 \lg V)\lg(V/e))$ time a retiming G_r and a clocking scheme $\pi = \langle \phi_0, \epsilon_0, \phi_1, \epsilon_1 \rangle$ such that G_r is properly timed by π, and the period of π is at most $(1+e)$ times the period of any clocking scheme that can be achieved by G under retiming.

The full paper will contain a complete description of the two algorithms. We now sketch each.

The basic algorithm for fixed duty-ratio retiming is a binary search over a space of possible clock periods. Since only fixed duty-ratio clocking schemes are considered, each possible clock period corresponds to a unique clocking scheme. We can check in $O(V^3)$ time whether a particular period is achievable using the general retiming algorithm of Section 5. The only difficulties are determining the space of possible clock periods and determining when the search has isolated a sufficiently accurate approximation of the true minimum. Inequality (2) specifies that the period of an achievable clocking scheme must be greater than the maximum delay $d_{\max}$ of any individual functional element. In addition, Inequalities (4), (5) and (6) guarantee that a period of $|V|d_{\max} + \epsilon_0 + \epsilon_1$ is achievable, regardless of the duty-ratio.

Therefore, the minimum period must be in the range from d_{max} to $|V|d_{max}+\epsilon_0+\epsilon_1$. Moreover, since d_{max} is a lower bound on the minimum achievable clock period, a clock period that exceeds a lower bound of the optimal clock period by at most $e\,d_{max}$ is guaranteed to be less than $(1+e)$ times the period of any fixed duty-ratio clocking scheme that is achievable under retiming. Algorithm correctness follows from the fact that the inability to achieve a clock period of π guarantees the inability to achieve any clock period $\pi' < \pi$.

As in the algorithm for optimal retiming with fixed duty-ratios, the algorithm for the general problem uses the $O(V^3)$-time algorithm from Section 5 to determine whether a particular combination of clock period and duty-ratio can be achieved by retiming.

The algorithm proceeds in two stages. Let π^* denote the actual minimum clock period. The first stage assumes that the retimed circuit must utilize a symmetric clocking scheme, and uses the minimum-period retiming algorithm for the symmetric clocking scheme case to compute an approximate minimum clock period π^+. The second stage then lifts the assumption of a symmetric clocking scheme and isolates a general scheme of approximately minimum clock-period by binary searching over the range $[\pi^+/2, \pi^+]$ of potential clock periods. The fact that $\pi^* \geq \pi^+/2$ follows from the observation that the clock of any circuit with period π can be tuned to become a symmetric clocking scheme with period at most 2π; one simply increases the shorter of the two duty cycles until it is equal to the longer duty cycle. Using $\pi^+ e$ as an increment of resolution, there are only $O(1/e)$ potential minimum clock periods, and thus, the binary search performs $O(\lg(1/e))$ iterations.

Whether a particular potential clock period π is achievable is determined by searching a set of $1/e$ potential values for ϕ_0. The search is necessary, because the retiming algorithm of Section 5 cannot be applied unless a ϕ_0 value is also supplied. The task is thus to determine whether the interval $(0, \pi/2)$ contains a ϕ_0 value such that the clocking scheme $\langle \phi_0, \epsilon_0, \pi - \phi_0 - \epsilon_0 - \epsilon_1, \epsilon_1 \rangle$ is achievable. Since "achievability" is not monotonic in ϕ_0, however, a simple binary search of the interval $(0, \pi/2)$ is not possible, and the algorithm resorts to a linear search. Fortunately, it can be shown that searching the interval at increments of $\pi^* e$ is sufficient to accurately discern the achievability of periods up to a resolution of $\pi^* e$. Thus, since each π is bounded above by π^+, and since π^* is bounded below by $\pi^+/2$, only $O(1/e)$ points in the interval $(0, \pi/2)$ need be checked.

Combining the time bounds of the various stages and searches, an approximate minimum clock period with unconstrained duty ratio can be computed in $O(V^3(1/e)\lg(1/e) + (VE + V^2 \lg V)\lg(V/e))$ time.

7 Polynomial-time algorithms for minimum-period retiming

In the full paper, this section will describe polynomial-time algorithms that yield exact solutions to the minimum-period retiming problem. When the duty ratio of the candidate clocking schemes is fixed to a given ρ, we can compute in $O(V^2E)$ time a set $\Pi(\rho)$ of $O(V^3)$ potential clock periods that is guaranteed to contain the *exact* minimum clock period π^* achievable by any retimed circuit that utilizes a clocking scheme with duty ratio ρ. A binary search over $\Pi(\rho)$ yields π^* in $O(V^2E + V^3 \lg V)$ time, where the binary search uses the $O(V^3)$ retiming algorithm from Section 5 to test particular elements of $\Pi(\rho)$. If only symmetric clocking schemes are considered, then we can use algorithm MILP for retiming, and the overall algorithm can be made to run in $O(V^2E)$ time. When the duty-ratio of the candidate clocking schemes is unrestricted, we can compute a set Π of potential clock-period/duty-cycle pairs which is guaranteed to contain a pair (π, ϕ_0) such that π is the minimum achievable clock period. The method for computing Π is similar to the method for isolating the optimal clock period by tuning. Unfortunately, the best bound we have on the size of Π is $O(V^8)$. Using the retiming algorithm from Section 5 to check each such point, (π, ϕ_0) can be found in $O(V^{11})$ time—polynomial, but impractical.

8 Multiphase clocking

Many of the results for two-phase clocking can be generalized to "simple" multiphase clocking. For example, a straightforward generalization of the algorithm for verifying the proper timing of two-phase circuits runs in $O(kVE)$ time, where k is the number of clock phases employed. Clock tuning can also be performed, but may require the solution of a general linear programming problem. The algorithms for retiming to achieve a given clocking scheme can be generalized to run in $O(VE+V^2 \lg V)$ time when the given clocking scheme is symmetric, and in $O(kV^3)$ time when the given clocking scheme is not symmetric. The fully polynomial-time approximation schemes for minimum-period retiming with fixed duty ratio can also be generalized to run in $O((VE+V^2 \lg V)\lg(kV/e))$ time and $O(kV^3 \lg(kV/e))$ time for symmetric and asymmetric clocking schemes, respectively. The approximation scheme for the simultaneous clock tuning and retiming of two-phase circuits can be generalized, but the resulting polynomial running time is impractically large. We now sketch the formal multiphase clocking framework and the generalizations of the two-phase algorithms for timing verification and optimization.

A $2k$-tuple of real numbers $\pi = \langle \phi_0, \epsilon_0, \phi_1, \epsilon_1, \ldots, \phi_{k-1}, \epsilon_{k-1} \rangle$ is a *simple k-phase clocking scheme* if it satisfies the following constraints:

- $\pi = \sum_{i=0}^{k-1} (\phi_i + \epsilon_i)$;
- $0 < \phi_i < \pi$ for $i = 0, 1, \ldots, k-1$;
- $\epsilon_i + \phi_{i+1} > 0$ for $i = 0, 1, \ldots, k-1$;
- $\phi_i + \epsilon_i + \phi_{i+1} < \pi$, for $i = 0, 1, \ldots, k-1$;

where we assume here and henceforth that addition in subscripts is performed modulo k. In this formulation, we allow overlapping phases, that is, the ϵ_i values can be negative. The fourth constraint guarantees that the circuit contains no unsynchronized loops; for two-phase circuitry, this constraint reduces to $\epsilon_i > 0$ for $i = 0, 1$. We say a simple multiphase clocking scheme is *symmetric* if $\phi_i = \phi_j$ and $\epsilon_i = \epsilon_j$ for all $0 \le i, j \le k-1$.

A level-clocked circuit is *simple k-phase* level-clocked if

1. it utilizes a simple k-phase clocking scheme,
2. it contains no latch-free cycles, and
3. the latches surrounding any purely combinational path are controlled by successive clock phases, *i.e.*, the path "begins" with a latch controlled by ϕ_i and "ends" with a latch controlled by ϕ_{i+1}.

We denote such a *simple multiphase* circuit by $G = \langle V, E, d, w, \chi \rangle$, where χ now maps each vertex to a number in $\{0, 1, \ldots, k-1\}$ corresponding to the input phase of the vertex.

Clock period constraints. For simple k-phase clocking, the timing constraints analogous to Inequalities (2) and (3) are somewhat more complicated than for two-phase clocking. Let $G = \langle V, E, d, w, \chi \rangle$ be a simple k-phase circuit which uses a clocking scheme $\pi = \langle \phi_0, \epsilon_0, \phi_1, \epsilon_1, \ldots, \phi_{k-1}, \epsilon_{k-1} \rangle$. Then G is properly timed if and only if, for every path $u \overset{p}{\rightsquigarrow} v$ in G,

$$d(p) \leq \pi \left\lfloor \frac{w(p)+1}{k} \right\rfloor + \psi(\chi(u), \chi(v)) \ , \tag{17}$$

where

$$\psi(i,j) = \phi_i + \epsilon_i + \phi_{i+1} + \cdots + \epsilon_j + \phi_{j+1} \ .$$

Verifying proper timing. To check whether a simple multiphase circuit operates properly, we can use the constraints defined by Inequality (17) to derive a set of cyclic constraints identical to those described by Inequality (6), except with a k in the denominator instead of a 2. These cyclic constraints can be checked in $O(VE)$ time as in Step 4 of Algorithm VERIFY. If the cyclic constraints are met, the path constraints (17) need only be satisfied by simple paths. The $D(v,i)$ values in Algorithm VERIFY must now be computed for $i = 0, 1, \ldots, (2k-1)|V|$, and the entire verification algorithm runs in $O(kVE)$ time.

Clock tuning. Clock tuning for simple multiphase circuits can be done in much the same way as for two-phase circuits, but general linear programming techniques must be used, because the problem no longer succumbs to a simple, special-case algorithm. Each of the ϕ_i and ϵ_i become variables in the linear program. In the two-phase tuning problem, the ϵ_i values were essentially adjusted in advance to be as small as possible. Depending on the design methodology, however, there may be different technology constraints governing what legal values the ϵ_i may take on. If these technology constraints are linear, as one would typically expect, they can be incorporated into the linear programming problem. Of course, certain special cases can be handled without resorting to a general linear program. For example, a circuit with a three-phase, nonoverlapping clock can be tuned in $O(VE)$ time using the three-dimensional linear programming algorithm of Megiddo [16].

Retiming with symmetric clocks. When clock phases are symmetric, simple multiphase circuits can be retimed to achieve a given clock period in $O(VE + V^2 \lg V)$ time. The following constraints are the natural analog of the two-phase constraints (15):

$$\begin{array}{rcll} r(u) - r(v) & \leq & w(e) & \text{for all } u \xrightarrow{e} v \in E \\ R(v) - r(v) & \leq & 0 & \text{for all } v \in V \\ R(u) - R(v) & \leq & w(e) - \frac{k}{\pi} d(v) & \text{for all } u \xrightarrow{e} v \in E \\ r(v) - R(v) & \leq & \frac{k}{\pi}(\phi - d(v)) + 1 & \text{for all } v \in V \ . \end{array}$$

As with the two-phase constraints, these constraints can be solved using the algorithm MILP from [11].

Retiming with unequal phases. Even when clock phases are not symmetric, we can retime to achieve a given simple k-phase clocking scheme in $O(kV^3)$ time. For simple multiphase circuits, inequalities analogous to Inequality (16) are easily formulated as simple summations. In particular, if one assumes that $\phi_i + \epsilon_i \geq 0$, for all $0 \leq i \leq k-1$, then Inequality (17) can be rewritten as

$$\sum_{i=\chi(v)}^{\chi(v)+r(v)} (\epsilon_i + \phi_{i+1}) - \left(\epsilon_{\chi(v)} + \phi_{\chi(v)+1}\right) + \pi \left\lfloor \frac{w(p)+1}{k} \right\rfloor + \psi(\chi(u), \chi(v)) - d(p)$$

$$\geq \sum_{i=\chi(u)-1}^{\chi(u)-1+r(u)} (\phi_i + \epsilon_i) - \left(\phi_{\chi(u)-1} + \epsilon_{\chi(u)-1}\right) \qquad (18)$$

for any retimed circuit. These constraints can be solved using the integer monotonic programming algorithm. By maintaining the two summations dynamically, the claimed running time can be obtained, with the additional k-factor stemming from the fact that some $r(v)$ can now be as high as $O(kV)$.

Minimum-period retiming. Polynomial-time approximation schemes for the minimum-period retiming of simple k-phase level-clocked circuits can be obtained in a straightforward manner. The algorithms for retiming a multiphase circuit to achieve a given clock period can be employed in practical fully-polynomial approximation schemes for fixed duty-ratio minimum-period retiming. Straightforward generalizations also yield an approximation scheme for the simultaneous clock tuning and retiming of multiphase circuits. Unless significant restrictions are placed on the extent to which clock tuning can be performed, however, the running time of the approximation scheme is likely to contain a e^{-k} factor, which is prohibitively large for most relative error bounds e. We have not yet investigated polynomial-time algorithms for the exact minimum-period retiming of simple multiphase circuits.

The results for simple k-phase level-clocked circuits are optimal for circuits whose functional elements have minimum propagation delays of 0. In particular, such circuits can only be properly timed if they contain no unsynchronized loops, *i.e.*, at every point in time, every loop contains a latch whose controlling phase is low. The formulation above guarantees this property by specifying that $\phi_i + \epsilon_i + \phi_{i+1} < \pi$, for $i = 0, 1, \ldots, k-1$. It is possible to show that this specification does not restrict optimality with respect to minimum clock period, *i.e.*, for any circuit that has no unsynchronized loops and achieves a clock period of π, there exists another circuit that meets the specification and also achieves the period π. Similarly, the assumption that $\phi_i + \epsilon_i \geq 0$, for $i = 0, 1, \ldots, k-1$, also does not restrict optimality with respect to minimum clock period. In fact, this assumption can even be lifted at the expense of complicating Inequality (18).

9 Conclusion

Our algorithms for verifying and optimizing level-clocked circuits have their limitations. For example, as is the case with most work on retiming, it is hard to incorporate data-dependent propagation delays in the framework without making most of the interesting questions NP-hard. There are, however, several topics which appear more amenable to efficient algorithmic solutions.

The first of these is the incorporation of functional elements where the propagation delay may differ for different input-output pairs (the "nonuniform propagation delay" model from [12]). Incorporating such elements changes the time complexities of our algorithms, but the essential algorithms remain unchanged.

Another generalization of the circuit model involves the incorporation of minimum propagation delays (sometimes called "contamination" delays). In this model, the output of a functional element does not become invalid until some specified minimum amount of time after an input changes. We believe that many of our algorithms can be generalized to handle such circuits. This is a topic of current research.

Three generalizations of our work which seem more problematic are the handling of gated clocks, precharged circuits, and nonsimple multiphase clocking disciplines. When logic circuits involve the clock signals themselves (so-called gated clocks), it is possible to show, in the general case, that many timing verification and optimization problems are NP-hard. Nevertheless, we suspect that by making conservative assumptions regarding the behavior of such circuits, many of these problems become tractable. In a precharged circuit, combinational paths may exist between latches controlled by the same phase. Optimizations, such as retiming and clock tuning, may have efficient algorithmic solutions, but there are many subtleties that arise in the formulation of the constraints. Nonsimple multiphase circuits also exhibit many subtleties. A path in such circuits may pass through latches in an arbitrary order, rather than the canonical order assumed in a simple multiphase circuit. Though the proper timing of such circuits can be verified using the analysis and algorithms from [7], the timing optimization of such circuits is possibly more complex than even precharged circuits. Whether these problems have efficient solutions is a topic for further research.

Most of the algorithms described in this paper have been programmed using the *C* programming language, although we have not typically used the most asymptotically efficient implementations. We are currently in the process of applying these algorithms to actual circuits in order to evaluate their performances. An interesting direction for further research in this area is to develop algorithms which provide interactive feedback to designers. For example, rather than simply reporting the minimum clock period of a circuit, an algorithm that performs a "sensitivity analysis" could report the extent to which noncritical propagation delays can be increased without affecting the clock period. We believe that the area of algorithms for the interactive analysis of circuit timing is a rich and largely unexplored field.

Acknowledgements

This research was supported in part by the Defense Advanced Research Projects Agency under Grant N00014-91-J-1698.

References

[1] T. H. Cormen, C. E. Leiserson, and R. L. Rivest. *Introduction to Algorithms*. McGraw-Hill, MIT Press, 1990.

[2] M. R. Daganais and N. C. Rumin. Automatic determination of optimal clocking parameters in synchronous mos vlsi circuits. In *Advanced Research in VLSI: Proc. of the 5th MIT Conference*, pages 19–33, 1988.

[3] G. B. Dantzig, W. Blattner, and M. R. Rao. Finding a cycle in a graph with minimum cost to times ratio with application to a ship routing problem. *Theory of Graphs*, 1967. P. Rosenstiehl, editor, Dunod, Paris, and Gordon and Breach, New York, pp. 77–84.

[4] J. P. Fishburn. Clock skew optimization. *IEEE Transactions on Computers*, 39(7):945–951, July 1990.

[5] M. Hartmann and J. Orlin. Finding minimum cost to time ratio cycles with small integral transit times. Technical Report UNC/OR/TR/91-19, University of North Carolina, Chapel Hill, October 1991.

[6] A. T. Ishii. *Timing in Level-Clocked Circuits*. PhD thesis, Massachusetts Institute of Technology, 1991. Available as MIT/LCS/TR-522.

[7] A. T. Ishii and C. E. Leiserson. A timing analysis of level-clocked circuitry. In *Advanced Research in VLSI: Proceedings of the Sixth MIT Conference*, pages 113–130, April 1990.

[8] D. B. Johnson. Efficient algorithms for shortest paths in sparse networks. *Journal of the Association for Computing Machinery*, 24(1):1–13, January 1977.

[9] C. E. Leiserson, F. M. Rose, and J. B. Saxe. Optimizing synchronous circuitry by retiming. *3rd Caltech Conference on VLSI*, 1983. R. Bryant, ed., pp. 87-116.

[10] C. E. Leiserson and J. B. Saxe. Optimizing synchronous systems. *Journal of VLSI and Computer Systems*, 1(1):41–67, 1983.

[11] C. E. Leiserson and J. B. Saxe. A mixed-integer programming problem which is efficiently solvable. *Journal of Algorithms*, 9:114–128, 1988.

[12] C. E. Leiserson and J. B. Saxe. Retiming synchronous circuitry. *Algorithmica*, 6(1), 1991. Also available as MIT/LCS/TM-372.

[13] B. Lockyear and C. Ebeling. Optimal retiming of multi-phase level-clocked circuits. Technical Report TR 91-10-01, Dept. of Comp. Sci and Eng., University of Washington, Seattle, October 1991.

[14] S. Malik, E. Sentovich, R. K. Brayton, and A. Sangiovanni-Vincentelli. Retiming and resynthesis: Optimizing sequential networks with combinational techniques. In *Proceedings of the Hawaii International Conference on System Sciences*, 1990.

[15] C. A. Mead and L. A. Conway. *Introduction to VLSI Systems*. Addison-Wesley, Reading, Massachusetts, 1980.

[16] N. Megiddo. Partitioning with two lines in the plane. *Journal of Algorithms*, 6:430 – 433, 1985.

[17] C. Papadimitriou and K. Steiglitz. *Combinatorial Optimization, Algorithms and Complexity*. Prentice-Hall, Inc., 1982.

[18] M. C. Papaefthymiou. On retiming synchronous circuitry and mixed-integer optimization. Master's thesis, Massachusetts Institute of Technology, September 1990. Available as MIT/LCS/TR-486.

[19] M. C. Papaefthymiou. Understanding retiming through maximum average-weight cycles. *3rd ACM Symposium on Parallel Algorithms and Architectures*, July 1991.

[20] N. Park and A. C. Parker. Theory of clocking for maximum execution overlap of high-speed digital systems. *IEEE Transactions on Computers*, 37(6):678–690, June 1988.

[21] K. A. Sakallah, T. N. Mudge, and O. A. Olukotun. Optimal clocking of synchronous digital circuits. In *Proceedings of the 1990 IEEE International Conference on CAD*, pages 552–555, November 1990.

[22] N. Shenoy, R. K. Brayton, and A. Sangiovanni-Vincentelli. Retiming of circuits with single phase level-sensitive latches. In *ICCD-91*, 1991. To appear.

[23] S. H. Unger and C. J. Tan. Clocking schemes for high speed digital systems. *IEEE Transactions on Computers*, C-35(10):880–895, October 1986.

Optimal Retiming of Multi-Phase, Level-Clocked Circuits

Brian Lockyear and Carl Ebeling
Department of Computer Science and Engineering
University of Washington
Seattle, Washington 98195

Abstract

Using level-sensitive latches instead of edge-triggered registers for storage elements in a synchronous system can lead to faster and less expensive circuit implementations. This advantage derives from an increased flexibility in scheduling the computations performed by the circuit. While in edge-clocked circuits the amount of time available for computation between two registers is precisely the length of the clock cycle, in circuits using level-sensitive latches a computation can borrow time across latches thus reducing the amount of dead time in the clock cycle. In either type of circuit, achieving maximum performance requires locating the storage elements in such a way as to spread the computation uniformly across a number of clock cycles.

In this paper, we extend the retiming techniques developed for edge-clocked circuits by Leiserson, Rose and Saxe to a general class of multi-phase, level-clocked circuits. We first describe this class of well-formed circuits and define what it means for a well-formed, level-clocked circuit to operate correctly. We then show that a set of constraints can be efficiently derived for a circuit which preserve its correctness under retiming. These constraints can then be used to retime a level-clocked circuit using efficient integer linear programming techniques similar to those used for edge-clocked circuits.

1 Introduction

Synchronous circuits rely on clocked storage elements to hold values while computation is performed. The most widely used storage element is the edge-triggered register which samples its input at the beginning of each clock period, holding that value for the entire clock period. Edge-triggered registers allow straightforward analysis of the minimum clock period by determining the maximum delay between any two registers. This simplified timing analysis leads to efficient retiming techniques for adjusting the placement of registers to optimize the cycle time or the number of registers [6, 7].

Level-clocked circuits use level-sensitive latches which allow inputs to flow through during the active period of the clock, latching the value during the inactive period. In level-clocked circuits the time available for computation between latches depends on when the input signal arrives and when the output signal is required. This borrowing of time between clock cycles makes the determination of the constraints on latch placement difficult. However, the flexibility in scheduling the computation provides greater opportunity for optimization of the clock period.

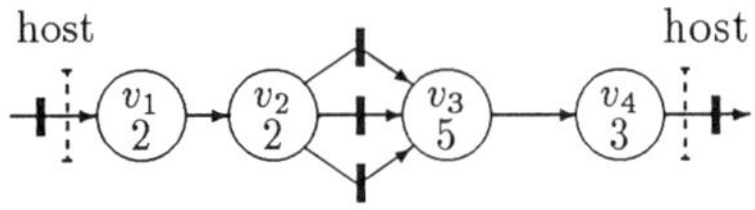

Figure 1.a: *Edge-Clocked Circuit*

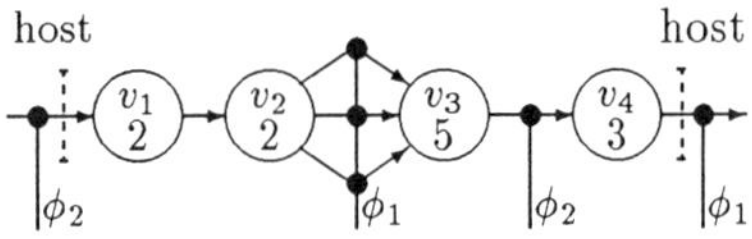

Figure 1.b: *Level-Clocked Circuit*

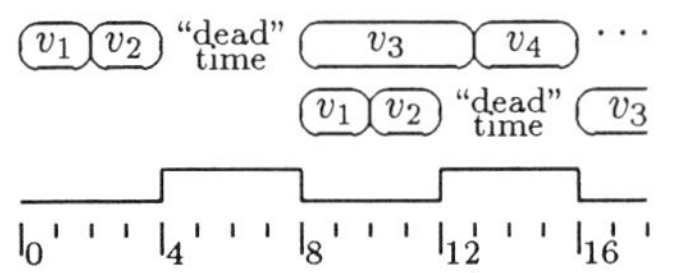

Figure 1.c: *Edge-Clocked Schedule*

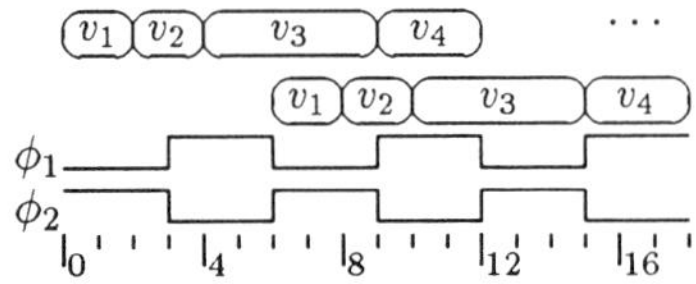

Figure 1.d: *Level-Clocked Schedule*

Figure 1: *Comparison of a simple circuit optimally timed using edge-triggered registers and level-clocked latches.*

This difference in scheduling is shown by the example of Figures 1.a and 1.b where the same computation is implemented using an edge-clocked and a level-clocked design. The circuit of Figure 1.b uses a two-equal-phase, non-overlapping clock. The edge-clocked circuit shows an optimal placement of registers which achieves a cycle time of $T_\Phi = 8$. By contrast, the level-clocked circuit shows an optimal placement of latches that achieves a cycle time of $T_\Phi = 6$. Assuming the cost of a register is R verses $\frac{R}{2}$ for a latch, this level-clocked circuit is also cheaper. This flexibility can also be used to reduce cost while meeting a particular clock period. For instance, if the circuit in Figure 1.b is to be retimed with $T_\Phi = 7$, the latches on the three edges $v_2 \rightarrow v_3$ may be moved to the single edge $v_1 \rightarrow v_2$ reducing the storage element cost to R, one-third that of the slower optimal edge-clocked circuit.

In this paper we extend the retiming techniques developed by Leiserson et. al. for edge-clocked circuits to optimize level-clocked circuits. Figure 2 shows the edge-clocked correlator example from their paper which can be retimed to operate with a clock period of 13. We convert this into an equivalent two-phase level-clocked circuit by replacing each register with a pair of ϕ_1, ϕ_2 latches. The result is also shown in Figure 2 with an optimal clock period of 10 using an equal phase clock. The retiming techniques we describe also handle more complex clock schedules with unequal phases and phase overlap or underlap. For example, retiming the two-phase correlator example using an equal-phase clock with 10% underlap between phases achieves a optimal period of 10.345.

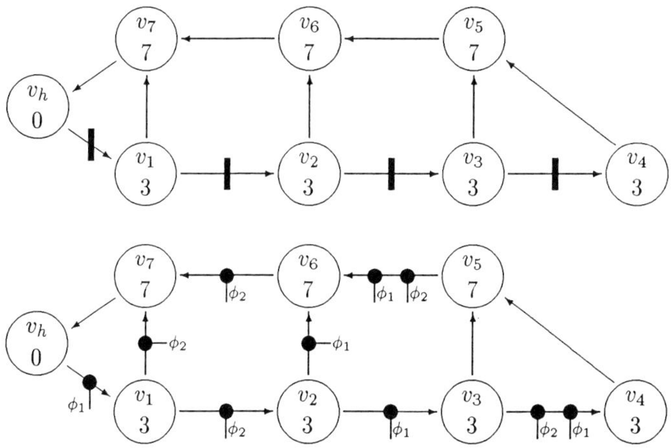

Figure 2: *The correlator circuit from Leiserson et. al. in its initial configuration with registers shown as solid bars, and a two-phase, level-clocked version optimally retimed to* $T_\Phi = 10$ *using an equal-phase clock schedule.*

1.1 Overview of the Paper

We first review the work of Leiserson et. al. [6, 7] and present the underlying circuit graph model. In Section 3 we review the clock model adopted from the work of Sakallah, Mudge and Olukotun [12]. Section 4 then describes the class of well-formed circuits to which we will be limited and defines correct circuit operation. We then use this model to derive constraints that fully specify the multi-phase, maximum delay timing restrictions of level-clocked circuits. Section 6 applies these timing restrictions to clocks using equal phases while Section 7 extends the capability to unequal phase clocks.

2 Background

In this section, we briefly review the terminology and graph model of digital circuits described in Leiserson et. al. [6] and extend it to handle level-sensitive latches. We then review the basic retiming results of their paper. The reader is encouraged to read [6, 7] for full details.

A circuit is modeled as a directed multigraph $G = \langle V, E, w, d, s\rangle$ with vertices V, the functional elements of the circuit, and edges E, the interconnections between these elements. Each vertex v has an associated delay $d(v)$. A unique host vertex v_h with $d(v_h) = 0$ represents the environment of the circuit. Each edge has weight $w(e)$ which is the number of registers (latches) along the connection. The notion of edge weight must be extended for level-clocked circuits which use latches controlled by different clock phases by associating with each edge e the sequence $s(e) = (l_1, l_2, \ldots, l_{w(e)})$ of latches along the connection.

The notation $u \xrightarrow{e} v$ represents an edge e between vertices u and v. A path p between vertices u and v is denoted $u \overset{p}{\twoheadrightarrow} v$. A *simple* path contains no vertex twice. For level-clocked circuits we also refer to paths between latches l and m using the notation $l \overset{p}{\twoheadrightarrow} m$. The weight of a path $w(p)$ is the sum of the edge weights along the path. The weight of a latch terminated path $l_0 \overset{p}{\twoheadrightarrow} l_{n+1}$ is defined to be n; that is, the initial and final latches are not included in the path weight.

The delay of a path, $d(p)$ is the sum of the delays of the vertices along the path, including the initial and final vertices. For edge-clocked circuits, correct timing requires that the following conditions be maintained:

- C1. For any path p in G, if $d(p) >$ clock period, then $w(p) \geq 1$.
- C2. For any cycle c in G, $w(c) \geq 1$.

Retiming a circuit is the process of transforming a circuit graph G into another graph G_r by relocating registers (or latches) such that the input/output behaviors of G and G_r are identical. Transforming a circuit can be viewed as assigning a *retiming* (or lag) value $r(v)$ to each of the vertices of G. This retiming value represents the number of registers (latches) removed from the output edges of vertex v and added to the input edges. More formally, for any edge $u \xrightarrow{e} v$, $w_r(e) = w(e) + r(v) - r(u)$.

In addition to the two timing correctness constraints, correct transformation of the circuit graph requires maintaining the relative difference in the weight of two paths between the same two vertices. In [6, 7] it is shown that retiming maintains a constant number of registers on any cycle, thus meeting condition C2.

The key result of [6, 7] defines the following set of constraints which must be met by a legal retiming of an edge-clocked circuit graph with clock period c. These constraints are given in terms of the maximum delay along critical paths in G, defined below.

$$
\begin{aligned}
\text{I/O:}\quad & r(v_h) = 0. \\
\text{Positive edge weight:}\quad & r(u) - r(v) \leq w(e) \text{ for all edges } u \xrightarrow{e} v. \\
\text{Maximum path delay:}\quad & r(u) - r(v) \leq W(u,v) - 1 \\
& \qquad\qquad \text{for all } u \twoheadrightarrow v \text{ where } D(u,v) > c.
\end{aligned}
$$

A critical path in an edge-clocked circuit graph is a minimum-weight path from u to v and is used to define the matrices W and D:

$$
\begin{aligned}
W(u,v) &= \min\{w(p) \mid u \overset{p}{\twoheadrightarrow} v\} \\
D(u,v) &= \max\{d(p) \mid u \overset{p}{\twoheadrightarrow} v \text{ and } w(p) = W(u,v)\}.
\end{aligned}
$$

Integer Linear Programming (ILP) techniques can be used to solve this constraint set and return a valid assignment of the retiming variables if one exists. A binary search over the set of possible optimum clock periods derived from the delay of critical paths determines the fastest possible clock period to which the circuit may be retimed.

3 Clock Model

We have adopted the clock model of Sakallah, Mudge & Olukotun [12] which provides a convenient way to describe the constraints on multi-phase clocks. A *k-phase clock* is a set of k periodic signals $\Phi = \{\phi_1 \ldots \phi_k\}$ where ϕ_i is referred to as *phase* i of the clock Φ. All ϕ_i have a common cycle time T_Φ. Each phase divides the clock cycle into two intervals as shown in Figure 3: An *active* interval of duration T_{ϕ_i} and a *passive* interval of duration $(T_\Phi - T_{\phi_i})$. The latches controlled by a clock phase are *enabled* during its active interval and *disabled* during its passive interval. The transitions *into* and *out of* the active interval are called the *enabling* and *latching* edges respectively. We refer to the clock phase controlling latch l by $P(l)$.

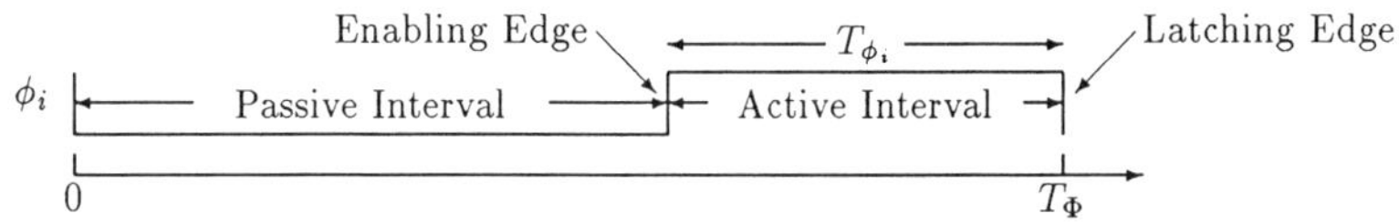

Figure 3: *Diagram from Sakallah et. al. showing a clock phase* ϕ_i *and its local time zone.*

Associated with each phase is a *local time zone* such that its passive interval starts at $t = 0$, its enabling edge occurs at $t = T_\Phi - T_{\phi_i}$, and its latching edge occurs at T_Φ. The domain of the local time zone is defined to be the interval $(0, T_\Phi]$ since the start of the current clock cycle coincides with the end of the previous cycle. Sakallah et. al. additionally introduce an arbitrary *global time reference* and the value e_i which denotes the time relative to the global time reference at which phase ϕ_i *ends.*

Phases are ordered relative to the global time reference so that $e_1 \leq e_2 \leq \cdots \leq e_{k-1} \leq e_k$. The global time reference is arbitrarily set such that $e_k \equiv T_\Phi$. The phase sequentially following ϕ_i in the clock set is referred to as ϕ_{i+1} with phase $\phi_{k+1} \equiv \phi_1$ and $\phi_{1-1} \equiv \phi_k$.

Finally a *phase shift* operator is defined:

$$E_{i,j} \equiv \begin{cases} (e_j - e_i), & \text{for } i < j \\ (T_\Phi + e_j - e_i), & \text{for } i \geq j \end{cases}$$

which takes on positive values in the range $[0, T_\Phi]$. When subtracted from a timing variable in the *current* local time zone of ϕ_i, E_{ij} changes the frame of reference to the *next* local time zone of ϕ_j, taking into account a possible cycle boundary crossing. Because the period of each clock phase is identical and $e_i \geq e_{i-1}$, the sum of the shifts between all successive phases is T_Φ:

$$\sum_{i=1}^{k} E_{i,i+1} = T_\Phi. \tag{1}$$

We assume that the setup, hold and propagation delay times of latches are zero. The timing characteristics of a given latch may vary as it is moved across combinational logic nodes and thus we treat latches as pass gates followed by a non-inverting buffer of zero delay and infinite drive capability. Refining this simplified latch model is a topic for further research.

The final parameters of interest are the values for the arrival and departure times of a latch. The arrival time of a signal at latch l is denoted by A_l and the departure time is denoted by D_l in the local time zone. The departure time is given by:

$$D_l = max\{A_l, T_\Phi - T_{P(l)}\} \tag{2}$$

and the arrival time at a latch m of a signal from latch l connected by a zero-weight path is:

$$A_m = D_l - E_{P(l),P(m)} + d(p) \tag{3}$$

Note that this clock model does not provide for clock phases with differing periods nor for gated clock signals.

4 Well-Formed Circuits and Valid Clocks

We impose a restriction on the structure of level-clocked circuits that simplifies our definition of correctness and is least restrictive of the movement of latches during retiming. The resulting "well-formed" circuits form a large and useful class of circuits including those that are easily produced by automatic synthesis tools. These restrictions can be eased by placing appropriate additional constraints, but this is not addressed in this paper.

A *well-formed* circuit is one in which the latches occur in clock phase order along any path through the circuit. More precisely, a circuit graph G is well-formed if:

W1. For every zero-weight path between latches, $l \twoheadrightarrow m$: $P(m) = P(l) + 1$.

W2. For every cycle c in G, $w(c) \geq 1$.

The first constraint simplifies equations defining the minimum weight along a circuit path by forcing any two paths which begin or end at the same vertex to have the identical ordered latch sequence. The second constraint avoids races and is the same as that required for edge-clocked graphs. Together these constraints require every cycle to contain a multiple of k latches for a k-phase clock. We define the clock phase of a vertex v, denoted $P(v)$, as the phase of the latch immediately preceding v on any path leading to v.

Retiming a level-clocked graph can now be defined similarly to retiming a edge-clocked graph. That is, for any edge, $u \xrightarrow{e} v$, the relationship $w_r(e) = w(e) + r(v) - r(u)$ still holds. In addition, $r(v)$ latches (in phase order) are appended to $s(e)$ and $r(u)$ initial latches are deleted from $s(e)$ to form $s_r(e)$. (The case where r is negative is treated symmetrically.) The following lemma assures us that well-formed graphs remain well-formed under retiming.

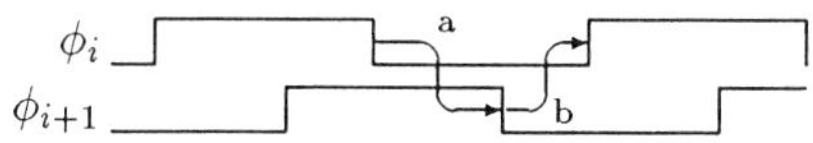

Figure 4: *Graphical representation of the constraints on the clock phases that are required for correct operation of a well-formed level-clocked circuit.*

Lemma 4.1 *A well-formed circuit graph remains well-formed under a valid retiming.*

Proof Sketch: Maintenance of constraint W1 is shown by induction on $r(v)$. That W2 is maintained for cycles follows from the retiming results for edge-clocked circuits in [6]. □

4.1 Correct Operation of Level-Clocked Circuits

Two conditions must be met to ensure the correct operation of a level-clocked circuit. The first states that along any zero-weight path between two latches, a signal departing the initial latch must arrive at the terminating latch before its next latching edge. The second that the signal departing a latch must not arrive at the terminating latch before its *previous* latching edge. More precisely, for any zero-weight path $l \overset{p}{\twoheadrightarrow} m$:

L1. Maximum delay: $A_m = D_l - E_{P(l),P(m)} + d(p) \leq T_\Phi$.

L2. Non-interference: $A_m = D_l - E_{P(l),P(m)} + d(p) > 0$.

To avoid writing two-sided delay constraints and allow retiming to locate latches without concern for retaining vertices between latches, we now remove any assumption about minimum delay in a correctly operating level-clocked circuit by defining valid clock schedules and show that any well-formed circuit operated by a valid clock schedule satisfies the non-interference constraint L2. That is, if a retiming satisfies the maximum delay constraint, then it results in a correctly operating circuit even with zero delays between latches.

A clock schedule is *valid* if it meets the following constraints:

P1. $e_{i+1} \geq e_i$, which follows from the definition of a clock schedule (constraint **a** in Figure 4).

P2. $\left\{ \begin{array}{cl} e_i + T_\Phi - T_{\phi_i} > e_{i+1} & \text{for } i \neq k \\ e_k - T_{\phi_k} > e_0 & \end{array} \right\}$ (constraint **b** in the figure).

It follows from constraint P2 that there is no time t where $e_i - T_{P(i)} < t < e_i$ for $i = 1, ..., k$. That is, not all latches can be active simultaneously and thus we avoid race conditions in cycles. Note that these constraints allow for multiple phase clock schedules with overlapping and underlapping phases. However, two-phase clocks are required to be non-overlapped.

Theorem 4.2 *Any well-formed level-clocked circuit operating with a valid clock schedule meets the non-interference constraint L2.*

Proof: By Eqn. 2, $D_l \geq T_\Phi - T_{\phi_i}$, that is, the departure time from a ϕ_i latch must occur at or after the enabling edge. By constraint P2, $E_{i,i+1} = e_{i+1} - e_i < T_\Phi - T_{\phi_i}$ and so $E_{i,i+1} < D_l$. Since $P(m) = P(l) + 1$ for any path $l \overset{p}{\twoheadrightarrow} m$ with $w(p) = 0$ in a well-formed graph, $E_{P(l),P(m)} < D_l$ and thus $D_l - E_{P(l),P(m)} > 0$. Thus constraint L2 holds for any $d(p) \geq 0$. □

5 Level-Clocked Timing Constraints

This section derives a fundamental theorem and two corollaries which will provide the basis for ILP path constraint sets which will insure a graph G_r retimed from G under the constraints will operate correctly for a given multi-phase clock schedule Φ. The theorem provides an upper bound on the delay of an n-weight simple path in a level-clocked graph in terms of the departure time of a signal at the beginning of the path and the arrival time at the end. The proof is based on the maximum delay constraint of the previous section and extends it to paths of non-zero weight. Figure 5 gives a graphical representation of this theorem.

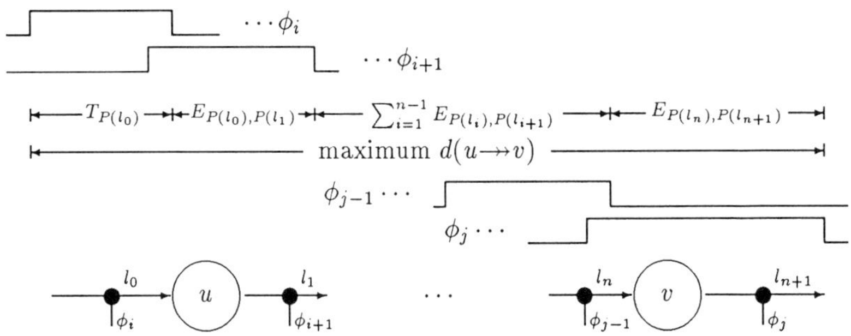

Figure 5: *Graphical representation of the constraint on the simple path delay between two latches l_0 and l_{n+1} in a correctly operating circuit.*

Theorem 5.1 *A multi-phase, level-clocked circuit graph is correctly timed using a valid clock schedule if and only if for every simple path $l_0 \overset{p}{\twoheadrightarrow} l_{n+1}$ with weight $w(p) = n$ and latch sequence $s(p) = \{l_0, l_1, \ldots l_{n+1}\}$, the path delay $d(p)$ is bounded by:*

$$d(p) \leq A_{l_{n+1}} - D_{l_0} + \sum_{i=0}^{w(p)} E_{P(l_i),P(l_{i+1})}.$$

Proof Sketch: ($\Rightarrow$) By induction on the weight of a path $w(p)$ beginning with Eqn. 3. ($\Leftarrow$) By contradiction, if the above constraint is not met then some zero-weight path must violate Eqn. 3. For a complete proof see [8]. □

The following two corollaries use the minimum departure and maximum arrival times of signals from latches to state the maximum simple path delay and maximum cycle delay in terms of the clock schedule.

Corollary 5.2 *A multi-phase, level-clocked graph using a valid clock schedule is correctly timed if and only if the delay of any simple path $l_0 \overset{p}{\twoheadrightarrow} l_{n+1}$ is bounded by:*

$$d(p) \leq T_{P(l_0)} + \sum_{i=0}^{w(p)} E_{P(l_i),P(l_{i+1})}.$$

Proof: The result follows directly from Theorem 5.1 by observing from Eqn. 2 that the minimum departure time $D_0 = T_\Phi - T_{P(l_0)}$ and from constraint L1 that the maximum arrival time $A_{l_{n+1}} = T_\Phi$. □

Corollary 5.3 *A multi-phase, level-clocked graph using a valid clock schedule is correctly timed if and only if the delay of any cycle $l_0 \overset{c}{\twoheadrightarrow} l_{n+1}$ is bounded by:*

$$d(c) \leq \sum_{i=0}^{w(c)-1} E_{P(l_i),P(l_{i+1})}.$$

Proof: The result follows directly from Theorem 5.1 by observing that $l_0 = l_{n+1}$ and thus $A_{l_{n+1}} \leq D_{l_0}$. Further, as the weight of a cycle includes all latches on it, the weight of a path $l_0 \overset{p}{\twoheadrightarrow} l_{n+1}$ is $w(c) - 1$. □

The following theorem provides a tight lower bound on possible clock periods at which the circuit will operate correctly based on the delay and weight of cycles in the circuit.

Theorem 5.4 *A well-formed graph using a k-phase clock schedule is correctly timed if and only if:*

$$\forall \textit{ cycles } c \in G: \qquad T_\Phi \geq k\left(\frac{d(c)}{w(c)}\right).$$

Proof: By Corollary 5.3: $d(c) \leq \sum_{i=0}^{w(c)-1} E_{P(l_i),P(l_{i+1})}$. In a well formed graph each cycle must contain $\frac{w(c)}{k}$ latches of each phase. By Eqn. 1, $\sum_{i=1}^{k} E_{i,i+1} = T_\Phi$. Hence:

$$d(c) \leq \sum_{i=0}^{w(c)-1} \frac{T_\Phi}{k} = w(c) \cdot \frac{T_\Phi}{k}, \quad \text{or} \quad T_\Phi \geq k\left(\frac{d(c)}{w(c)}\right). \square$$

The value of $\max\{\frac{d(c)}{w(c)}\}$ over all cycles in the graph may be found by setting the values $\alpha(u \overset{e}{\rightarrow} v) = d(v)$ and $\varepsilon(e) = w(e)$, and solving the *maximum-ratio-cycle* problem for $\frac{\alpha(c)}{\varepsilon(c)}$. Polynomial-time algorithms are available to solve this problem from Lawler [5], Megiddo [10] and Burns [2]. The resulting $\frac{d(C_R)}{w(C_R)}$ provides a fast lower bound on the cycle time of the circuit.

Although this clock cycle may not be realizable, it provides a useful starting point in searching for the optimum cycle time of the circuit.

We now redefine a *critical path* for a graph to be the path $u \xrightarrow{p} v$ such that if the minimum weight constraint is met for p, then it is met for all paths from u to v for any valid retiming. Unlike in edge-clocked circuits, in level-clocked circuits the path limiting the clock period may not have minimum weight. Instead, we identify the most constraining path from u to v for a given clock period. The following lemma provides the basis for efficiently determining a critical path.

Lemma 5.5 *A path $u \xrightarrow{p} v$ in a well-formed circuit is a critical path if*

$$\{w(p)\frac{T_\Phi}{k} - d(p)\} \leq \{w(q)\frac{T_\Phi}{k} - d(q)\} \text{ for all } u \xrightarrow{q} v.$$

Proof Sketch: By contradiction beginning with Corollary 5.2 and using the additional facts that for a well-formed graph, $w_r(p) - w_r(q) = km = w(p) - w(q)$ for some integer m, and from Eqn. 1: $\sum_{i=1}^{km} E_{l_i,l_{i+1}} = mT_\Phi$. For a complete proof see [8]. □

We now determine the values in the matrices $D(u,v)$ and $W(u,v)$ as $d(p_c)$ and $w(p_c)$ for a critical path $u \xrightarrow{p_c} v$. This in turn requires identifying the path which minimizes the quantity $\{w(p)\frac{T_\Phi}{k} - d(p)\}$ over all paths $u \xrightarrow{p} v$. We can find the paths which minimize this value by running an all-pairs shortest path algorithm on G (such as the Floyd-Warshall algorithm) using new edge weights $\tilde{w}(e) = (w(e)\frac{T_\Phi}{k} - d(v_2))$ for each $v_1 \xrightarrow{e} v_2$. This requires showing that there is no cycle c for which $\tilde{w}(c) = w(c)\frac{T_\Phi}{k} - d(c) < 0$. Theorem 5.4 places a lower bound on possible clock periods, that is: $T_\Phi \geq k\left(\frac{d(c)}{w(c)}\right)$. Thus for any clock period of interest there are no negative weight cycles in our relabeled graph. Note that unlike the case of edge-clocked circuits, D and W must be recomputed for every clock period attempted by the retiming.

6 Retiming for Equal Phase Clocks

The theorems in the previous section form the basis for a set of constraints which can be used to determine whether a retiming exists for a particular clock schedule. In this section we investigate clock schedules with equal length phases.

An *equal-phase clock schedule* is a valid k-phase clock set $\Phi = \{\phi_1, ..., \phi_k\}$ where all active phase periods T_{ϕ_i} are equal, and all phase shifts $E_{i,i+1} = \frac{T_\Phi}{k}$. We use T_ϕ to refer to the length of the active period of any phase. Note that this definition allows overlapped clock phases under the general constraints on valid clocks. Because the active periods and phase shift values are equal, the retiming process can ignore the actual phase of individual latches. Thus, in the identical manner to edge-clocked circuits, a retiming value $r(v)$ is assigned to each vertex of the circuit graph. However, a value of $r(v) = n$ now moves n latches across the vertex rather than n registers.

Proofs in [6, 7] for edge-clocked circuit graphs showing that all cycles maintain the same number of latches and that phase differences between paths with common endpoints remain constant also hold for level-clocked graphs. We now restate the maximum delay constraint as a minimum weight constraint providing a lower bound on the number of latches on a simple path in terms of the path delay.

Corollary 6.1 *A well-formed, level-clocked circuit graph G using a k-equal-phase clock schedule is correctly timed if and only if the weights of all simple paths p are bounded by:*

$$w(p) \geq \left\lceil \frac{d(p) - T_\phi}{\frac{T_\Phi}{k}} \right\rceil - 1$$

Proof: The result follows from Corollary 5.2 using the fact that for k-equal-phase clock schedule, $E_{i,i+1} = \frac{T_\Phi}{k}$ and $\forall_{i,j},\ T_{\phi_i} = T_{\phi_j} = T_\phi = T_{P(l_0)}$. □

We define $L(u,v) = \left\lceil \frac{D(u,v)-T_\phi}{\frac{T_\Phi}{k}} \right\rceil - 1$ as the minimum number of latches required on a critical path from u to v. This value forms the basis for a set of constraints for retiming well-formed circuit graphs controlled by a k-equal-phase clock schedules for a given clock period T_Φ:

$$\begin{aligned} \text{I/O:}\quad & r(v_h) = 0. \\ \text{Positive edge weight:}\quad & r(u) - r(v) \leq w(e) \text{ for all edges } u \xrightarrow{e} v. \\ \text{Maximum path delay:}\quad & r(u) - r(v) \leq W(u,v) - L(u,v) \\ & \qquad\qquad \text{for all } (u,v) \text{ such that } L(u,v) \geq 1. \end{aligned}$$

Example 1: A 2-equal-phase correlator

Retiming the two-phase correlator example using a two-equal-phase, non-overlapped clock schedule leads to the circuit graph of Figure 2 with $T_\Phi = 10$ ($d/w = 5$ for any critical cycle, which requires $T_\Phi = 10$ for a 2 phase clock). The W and D matrices are the same as in the original example except that all values in W are multiplied by two to reflect the conversion to latches.

6.1 Determining Potential Optimum Clock Periods

It is not always possible to retime level-clocked circuits to the lower bound clock period given by a critical cycle as in the previous example. In an optimal edge-clocked circuit there exists some critical path of zero weight with delay exactly the value of the clock period and thus it is a simple matter to make a list of all path delays from the D array and perform a binary search on that list to determine the optimal T_Φ. In level-clocked circuits the critical path may be of non-zero weight, and the critical path between two vertices may differ for differing clock periods. In general we cannot identify a set of potential optimum clock periods over which to search, and instead perform a binary search over a range to the desired accuracy.

We can now define a new algorithm for finding the optimal retiming of a k-equal-phase, level-clocked circuit graph:

ALGORITHM: OPTIMAL k-EQUAL-PHASE RETIMING:

1. Determine the critical cycle period $T_{C_R} = \max \left\{ k \left(\frac{d(c)}{w(c)} \right) \right\}$.
2. Attempt to retime to T_{C_R}; if successful $T_{\Phi_{opt}} = T_{C_R}$.
3. Repeatedly multiply the value of T_{C_R} by α until a legal retiming is found. Set $T_{\Phi_{opt}}+$ to the clock period of the first legal retiming. Set $T_{\Phi_{opt}}-$ to $\frac{T_{\Phi_{opt}}+}{\alpha}$.
4. Perform a binary search over $(T_{\Phi_{opt}}-, T_{\Phi_{opt}}+]$ until the desired level of accuracy is reached.

For many cases the critical paths for $T_{\Phi_{opt}}-$ and $T_{\Phi_{opt}}+$ are the same and so we can determine the set of potential optimum clock periods. Because the value of slack for a given path is linear over all clock periods, there can be at most one intersection point between the slack values for two paths. Below this intersection point one path will always have less slack than the other and above will always have greater slack. Once two clock periods $T_{\Phi_{opt}}-$ and $T_{\Phi_{opt}}+$ have been determined for which both W matrices are equivalent for all u, v, no further computation of critical paths is needed.

The following theorem uses the fact that for an optimal retiming of a level-clocked circuit graph (for a period greater than the critical cycle period) there will exist some critical path which exactly meets the minimum weight requirement. If this were not true there would exist a faster clock period for the same weighting. Assuming a range over the optimal period is found such that the critical paths do not change, we change the inequality from the minimum path weight result in Corollary 6.1 to an equality and form a set of possible optimal clock periods. Step 4 of the previous algorithm is terminated when matching critical paths are found and is followed by a binary search over the set of potential optimum clock periods.

Theorem 6.2 *The optimal cycle time $T_{\Phi_{opt}}$ of a well-formed circuit graph G clocked by a k-equal-phase clock is in the set C:*

$$C = \left\{ \left(\frac{D(u,v)}{\frac{i+1}{k} + T_d} \right) \middle| \; u, v \in V; i \in \{0, 1 \ldots n\} \right\}$$

where: T_d = duty cycle of each phase = $\frac{T_\phi}{T_\Phi}$ and n is the maximum integer value for which the resulting clock period computed is greater than the critical cycle period.

Proof: Follows from setting the left and right hand sides of Corollary 6.1 equal and solving for T_Φ. Because the value T_ϕ is proportional to T_Φ, substitute in duty cycle $(T_d \cdot T_\Phi)$ instead which remains constant for the clock schedule. □

Note that real circuit graphs have built-in error due to estimation of combinational logic delays, and thus the value of generating a set of precise possible optimum clock periods is questionable. Moreover, large, complex circuits with many combinations of possible path delays and weights have a densely populated set of possible optimum clock periods. Thus a search for the precise optimum clock period is of limited utility.

In the worst case, the algorithm stated provides an estimate of the optimal retiming of the graph to the desired level of accuracy. The complexity of Step 1 is provably $O(|E||V|^2)$ using the algorithm by Burns. Although $|E|$ may be as large as $|V|^2$, in practice circuit graphs have $|E| = O(|V|)$. Moreover, Burns' algorithm typically performs in $O(|E|)$ [2]. (Lower worst case bound algorithms may be found in Lawler [5]; however, in practice, this step is not limiting.) Thus algorithmic complexity is driven by either Step 3, the number of steps upward required to find $T_{\Phi_{opt}}+$, or Step 4, the number of steps downward required to find the desired result. Each trial step has cost $O(|V|^3)$ providing theoretical bounds of $O(|V|^3 \log \frac{T_{\Phi_{opt}}}{T_{C_R}})$ for Step 3 and $O(|V|^3 b)$ for Step 4 where b is the number of bits of accuracy desired.

Example 2: A 2-equal-phase correlator with phase underlap

A realistic clock schedule for the two-equal-phase correlator circuit might have each active period $T_\phi = (0.4 \cdot T_\Phi)$, giving an underlap between phases of $(0.1 \cdot T_\Phi)$. As in the previous example $T_{C_R} = 10$; however, in this case a retiming to that clock period cannot be found. A legal retiming is found for $T_\Phi = 20$ and the W and D matrices match at 20 and 10. The set of possible time periods C is:

$$C = \{10.0,\ 10.345,\ 10.526,\ 10.833,\ 11.053,\ 11.111,\ 11.25...\}$$

A binary search over this list finds $T_{\Phi_{opt}} = 10.345$. The resulting retiming values are: $r(v) = \langle\ 0\ \ -1\ \ -2\ \ -4\ \ -4\ \ -3\ \ -1\ \ 0\ \rangle$

7 Retiming for Unequal Phase Clocks

In this section we sketch how this work can be extended to unequal phase clocks. The reader is referred to [8] for details. In the case where clock phases are of unequal duration, the lower bound on the weight of a path depends not only on the delay of the path but also on the phase of the first vertex of the path as illustrated in Figure 6. Instead of determining a single minimum weight for a path, we define a set of minimum weights $L_i(u, v)$, $i \in \{0 \ldots k-1\}$ based on the phase of the path. Which minimum weight applies depends on how the retiming affects the phase of the path.

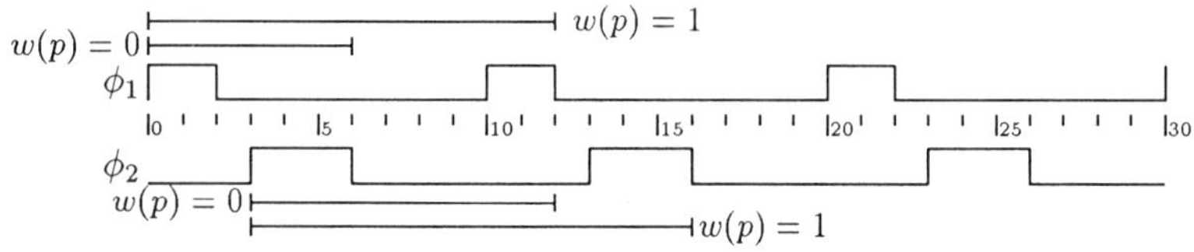

Figure 6: *A two-phase underlapped clock schedule. Distances are the maximum time period for a path of given weight beginning at a vertex preceded by the adjacent clock phase.*

The result of Corollary 5.2 may be re-written to provide an equation for the minimum weight of a path. The equation must be solved on a case by case basis.

Corollary 7.1 *The weight $w(p)$ of any path $u \overset{p}{\twoheadrightarrow} v$ in a correctly timed, well-formed circuit graph using a valid k-phase clock schedule is bounded by:*

$$w(p) \geq k \left\lfloor \frac{d(p) - T_{P(u)}}{T_\Phi} \right\rfloor + \begin{cases} -1 & d_{pm} \quad = 0 \\ 0 & d_{pm} \quad \leq E_{P(u),P(u)+1} \\ 1 & d_{pm} \quad \begin{cases} > E_{P(u),P(u)+1} \\ \leq \sum_{j=0}^{1} E_{P(u)+j,P(u)+j+1} \end{cases} \\ \vdots & \vdots \\ k-1 & d_{pm} \quad > \sum_{j=0}^{k-2} E_{P(u)+j,P(u)+j+1} \end{cases}$$

where $d_{pm} = (d(p) - T_{P(u)})$ *mod* T_Φ

Proof: For proof see [8]□

Remembering that the *initial phase* of a node u is $P(u) \in \{1 \ldots k\}$ when $r(u) = 0$, the minimum number of latches, $L_i(u, v)$, required on a critical path from u to v when $P_r(u) = P(u) + i$, $i \in \{0 \ldots (k-1)\}$ is found by substituting $D(u, v)$ for $d(p)$ and $T_{P_r(u)}$ for $T_{P(u)}$ in Corollary 7.1.

In [8] we first present a general ILP constraint set where each retiming value $r(u)$ is divided into a set of values $r_i(u)$, $i \in \{1 \ldots k\}$ which indicate how many latches of each phase $\phi_{P(u)+i}$ have been moved across vertex u. These make explicit the current phase of each node and in effect are used to select the correct constraint from the set of L_i.

This general ILP solution no longer has all equations formed as the difference of two variables on the left hand side; however, because well-formed graphs tightly constrain the relationship between node phase and values of $r(u)$, a simpler constraint set with variant weights may still be solved using a modification of the Bellman-Ford algorithm. Specifically, this can be done if constraints are formed as a difference of two variables less than or equal to a weight value which is a function of the length of the shortest path leading to the input node of the edge. A constraint set of this form is:

$$\begin{aligned} \text{I/O:} \quad & r(v_h) = 0. \\ \text{Positive edge weight:} \quad & r(u) - r(v) \leq w(e) \text{ for all edges } u \xrightarrow{e} v. \\ \text{Maximum path delay:} \quad & r(u) - r(v) \leq F(u, v, r(u)) \text{ for all } u \twoheadrightarrow v. \end{aligned}$$

where: $F(u, v, r(u)) \equiv W(u, v) - L_{r(u) \bmod k}(u, v)$.

8 Summary and Future Work

We have described an efficient method for optimally retiming the class of well-formed, multi-phase, level-clocked circuits using equal-length clock phases. This not only is a large class of circuits widely used in practice; they are also circuits that can be easily produced by current sequential synthesis tools and optimal retiming of these circuits will become increasingly important.

In independent work, Ishii, Leiserson and Papaefthymiou [4] have also obtained algorithms for 2-phase level-clocked retiming with non-overlapping clock phases of equal or unequal duration. A fast MILP solution to the equal phase case is presented. An additional efficient algorithm is provided for tuning of 2-phase clock schedules and a polynomial algorithm for simultaneous tuning with retiming.

Our next goal is to remove some of the restrictions we have placed on both circuit structure and clock schedules. Valid clock schedules can be redefined to assume a delay greater than zero between latches of specific phases. This introduces two-sided constraints and the manipulation of minimum delays as well as maximum delays. Work along these lines but in a different context has already been done by Shenoy [13] and Sakallah [12]. Extending the class of circuits beyond well-formed circuits places additional constraints on the movement of latches in the circuit. These constraints depend largely on the clock schedule itself and the implications of removing the ordering constraint on the correctness constraints.

The idea of retiming has also been used in the area of logic synthesis as a way of exposing and applying more of the functional relationships in a sequential circuit. Malik, Sentovich and Brayton describe the technique of peripheral retiming which allows registers in a sequential circuit to be moved to the periphery of the circuit, thus allowing the global resynthesis of the combinational logic as an single unit [9], and Borriello, Bartlett and Raju have explored the use of localized retiming combined with logic resynthesis to reduce the overall clock period [1]. Our techniques allow this work to be extended to level-clocked circuits.

Level-sensitive circuits have long been used for circuits where performance is important. Only recently, however, have algorithms for analyzing and manipulating these circuits become available. The potential benefits of level-sensitive circuits will make this a very fertile area of CAD research for some time to come.

Acknowledgments

We want to thank Gaetano Borriello for valuable discussions and comments on this paper and Steve Burns for pointing us to the algorithm for finding the maximum-ratio-cycle and the construction for finding critical paths in level-clocked circuits.

This research was funded in part by the Defense Advanced Research Projects Agency under Contract N00014-J-91-4041. Carl Ebeling is supported in part by an NSF Presidential Young Investigator Award with matching funds provided by IBM Corporation and Sun Microsystems.

References

[1] K. Bartlett, G. Borriello, and S. Raju. Timing optimization of multiphase sequential logic. *IEEE Trans on Comp-Aided Design*, 10(1):51–62, Jan 1991.

[2] S. Burns. *Performance Analysis and Optimization of Asynchronous Circuits.* PhD thesis, California Institute of Technology, 1991. Caltech-CS-TR-91-01.

[3] A. Ishii and C. Leiserson. A timing analysis of level-clocked circuitry. In *Advanced Research in VLSI: Proc. of the 6th MIT Conference*, 1990.

[4] A. Ishii, C. Leiserson, and M. Papaefthymiou. Optimizing two-phase, level-clocked circuitry. In *Adv Research in VLSI and Parallel Systems*, 1992.

[5] E. Lawler. *Combinatorial Optimization: Networks and Matriods.* Hold, Rinehart and Winston, New York, 1976.

[6] C. Leiserson, F. Rose, and J. Saxe. Optimizing synchronous circuitry by retiming. In *Proc. of the 3rd Caltech Conference on VLSI*, March 1983.

[7] C. Leiserson and J. Saxe. Retiming synchronous circuitry. *Algorithmica*, 6(1):5–35, 1991.

[8] B. Lockyear and C. Ebeling. Retiming of multi-phase, level clocked circuits. Technical Report 91-10-1, U of Wash, Dept. of Comp Sci, Oct 1991.

[9] S. Malik, E. Sentovich, and R. Brayton. Retiming and resynthesis: Optimizing sequential networks with combinational techniques. In *Proc. of the 23rd Hawaii Int. Conf. on System Sciences*, Kailua-Kona, HI, January 1990.

[10] N. Megiddo. Combinatorial optimization with rational objective functions. *Mathematics of Operations Research*, 4(4):414–424, 1979.

[11] G. De Micheli. Synchronous logic synthesis: Algorithms for cycle-time minimization. *IEEE Transactions on Computer-Aided Design*, 10(1):63–73, January 1991.

[12] K. Sakallah, T. Mudge, and O. Olukotun. Analysis and design of latch-controlled synchronous digital circuits. In *Proc. 27th ACM-IEEE Design Automation Conf.*, pages 111–116, January 1990.

[13] N. Shenoy, R. Brayton, and A. Sangiovanni-Vincentelli. Retiming of circuits with single phase transparent latches. In *International Workshop on Logic Synthesis*, North Carolina, May 1991. MCNC.

Event Suppression: Improving the Efficiency of Timing Simulation for Synchronous Digital Circuits

Srinivas Devadas, MIT
Kurt Keutzer, Synopsys Inc.
Sharad Malik, Princeton University
Albert Wang, Synopsys Inc.

Abstract

Timing simulation is a widely used method to verify the timing behavior of a design. In a synchronous digital system the timing property that needs to be verified is that there is no event at the outputs of the combinational parts of the circuit at or after time τ, the clock period. In this paper we first show that conventional timing simulation applied to this problem has exponential complexity. Next we demonstrate that for this problem a complete history of circuit activity before time τ is not needed. We exploit this observation and present an event suppression method that potentially leads to an exponential reduction in the number of events that need to be processed during simulation. This is backed by encouraging experimental results.

1 Introduction

Meeting delay requirements is the most important constraint imposed on a circuit and thus verifying the timing of a circuit before manufacture is one of the most important tasks of a computer-aided design system. In a synchronous digital system the final timing property to be verified is that for each input vector transition the combinational logic settles to a stable value within a given clock period τ. One approach towards ensuring this is to use timing simulation. Here a timing simulator is used to check that for each of the supplied transitions on the inputs, the combinational logic stabilizes within time τ.

In this paper we show that in simulating a vector transition the number of events at a signal in a combinational circuit is bounded above by the number of paths from the primary inputs to that signal, which may be exponential in circuit size or depth. We give a family of circuits, parameterizable by the depth of the circuit, for which this upper bound is tight. Since the computation required in event-driven timing simulation is bounded below by the number of events that must be evaluated, event-driven timing simulation can have exponential complexity and this fact is observed in practice on a widely-used timing simulator.

We make the observation that to answer the question: "Does an event occur in this circuit at any time at or after τ?" we do not need a full history of the circuit activity before time τ. In other words, an event-driven timing simulation approach is likely to compute much more information, in the form of events, than is necessary to resolve the question at hand. To fully explore the ramifications of this observation we develop a *calculus of event simulation* and then apply this calculus under a number of different models of timing behavior. Using these results we can suppress a significant number of events that need to be evaluated in timing simulation. For example on the circuit family mentioned above the number of events evaluated at a gate is reduced exponentially by our procedure.

2 Basic Definitions

A **combinational logic circuit** is an acyclic connection of gates. For simplicity, we consider circuits of only simple gates, AND , OR and NOT . Unless otherwise specified, the term circuit will henceforth refer to a combinational circuit.

If the output of a gate, g_1, is connected to an input of gate, g_2, g_1 is a **fanin** of g_2. Gate g_2 is a **fanout** of gate g_1. Primary input pins are treated as gates with no fanin and primary output pins are gates with no fanout.

A **path** in a combinational circuit is an alternating sequence of connections and gates, starting at a primary input and ending at a primary output.

With each gate, g, in the circuit is associated a real value termed its delay, $d(g)$. The semantics of this delay value are explained in Section 3. The length of a path is the sum of the gate delays along the path.

An **event** is a transition, $0 \rightarrow 1$ or $1 \rightarrow 0$, at a gate. Consider a sequence of events, $\{r_0, r_1, ..., r_n\}$ occurring at gates $\{g_0, g_1, ..., g_n\}$ along a path, such that r_i occurs as a result of event r_{i-1}. The event r_0 is said to **propagate** along the path.

We denote the value of a literal l (a variable or its complement) in a vector v as $v(l)$.

3 Delay Models

All simulation techniques as well as timing analyzers make some assumptions about the possible variations in the delay of various circuit components. We now examine some of the models used and then specify the domain of this paper in terms of these.

The most common delay model for a circuit component is one in which the delay is assumed to be a fixed number d. This is referred to as the **fixed delay** model. Thus, in this model a delay of 2 units on a gate indicates that the gate switches instantaneously to a logical 0 or 1 value but that the communication of this event to the output of the gate is delayed by 2 units.

However, in reality this number is typically an upper bound on the expected delay, so in fact the actual delay may be any number bounded above by d. This potential speedup is incorporated in the **monotone speedup** model [6], which assumes that the delay for each component lies in the range $[0, d]$.

The **bounded delay** model is more realistic about how much each gate can in fact be sped up. It specifies the delay as a range, $[d^l, d^u]$, given by the lower and upper bounds on the actual delays. There is some ambiguity in the literature as to what the bounded delay model means. It has been interpreted as either the switching delay or as the propagation delay. With the former interpretation, a gate delay of $[2, 4]$ would imply that the gate would take somewhere between 2 and 4 units to make the transition between the two logical values. Ternary algebras [7] have been used to accommodate the unknown value of the gate output (which is neither a 0 or a 1) in the interval of uncertainty. The latter assumption implies that the gate switches instantly, however it is uncertain as to when in the $[2, 4]$ interval it will switch. Note that unlike the previous case, the gate output is always a 0 or a 1 and ternary algebras are not needed. In this paper, we use the latter interpretation since it is more realistic with respect to current technologies.

We consider the fixed as well as bounded delay models. The monotone speedup model is a special case of the bounded delay model with all lower bounds being 0. It is possible that this property of the lower bounds may be exploited during analysis, however currently we do not take advantage of this.

For the purpose of developing the ideas in this paper it is sufficient that we restrict the delays in circuits to gates. (A single number in case of fixed delay and a range in case of the bounded delay model.) This is general enough to accommodate other delay quantities such as wire delays and pin to pin delays by introducing buffers with appropriate delays in the circuit. While more sophisticated delay parameters such as slope delays and separate rise and fall delays are not directly accommodated into the "delay lumped at a gate" paradigm, it can be shown with little effort that the results in this paper hold for even those models.

4 Circuit History: How Much is Enough?

Consider the following scenario: A synchronous digital circuit is being clocked at time period τ. Let v_0 be an input vector applied (to the combinational part) at the beginning of a cycle. It is possible that some gate outputs have not yet settled to their stable values under v_{-1}, the vector applied prior to v_0. Similarly, it is possible that some gates have not settled to their steady values under v_{-2}, and so on. Is there a bound on the maximum number of previous vectors that may still affect the gate outputs at any time in the circuit? In other words, what is the maximum value of i such that the gate outputs have not settled to their steady values under v_{-i} at the time when v_0 is applied.

The following theorem provides a simple bound in terms of the topology of the circuit.

Theorem 4.1 *Let C be the combinational part of a synchronous digital circuit being clocked at time period τ. Let ω be the length of the longest path in C. Then $\lceil \frac{\omega}{\tau} \rceil$ is the upper bound on the number of input vectors that can cause an event in the circuit at any time.*

Proof: This follows from the fact that in order for an event caused by v_{-i} to occur after an event caused by v_0 there must be a path of length $(i+1)\tau$ in the circuit along which the former event propagated. ∎

Thus in order for i to be more than 1 the circuit would have to be clocked at a period less than half the longest path length. For almost all circuits this is unlikely to happen. This is significant since for almost all cases prior circuit history is not needed while examining a transition from vector $v1$ to vector $v2$. This fact is exploited in developing the ideas presented in the rest of the paper. Event driven simulators are oblivious to the properties of this particular application and thus they do not exploit this observation.

The results of this paper can be extended to handle the general case, i.e. $i > 1$. Here a sequence of input changes of length i needs to be examined instead of a single transition at the inputs.

5 Analysis of Event-Driven Simulation

5.1 Computational Complexity of Event-Driven Simulation

Some of the problems we are trying to solve in this paper can also be addressed by event-driven simulation. For comparison, we would like to know the bound on the computational complexity of simulating a vector pair $< v1, v2 >$, i.e. a transition from vector $v1$ to vector $v2$. Such a question is impossible to answer without knowing the specifics of an event-driven simulator and its algorithms. Instead, we derive an upper bound on the number of transition events at a gate resulting from simulating a vector pair and then provide a circuit which generates that upper bound on the number of transitions. However efficient an event-driven simulator is, it must evaluate an event at a gate for each output event at the gate. As a result this bound is certainly a lower bound on the complexity of simulating a vector pair using event-driven simulation and will be the basis for comparison with our algorithms.

To determine the bound on the number of transitions at a particular signal when applying a vector pair to the inputs of a circuit, we make two simple assumptions about the electrical and logical properties of the gates. First, any single transition at an input of a gate can cause at most one transition at the output. Notice that such a condition may not necessarily be true, but is assumed by most gate-level event-driven simulators. Second, glitches (or spikes) are not absorbed by logic gates no matter how small they are, as long as they are of non-zero width. (This is also referred to as the transport delay model.) We will also require some additional notation:

T_x : sequence of alternating rising (R) and falling transitions ($\overline{R}$) at x, e.g., $< R(t_1), \overline{R}(t_2), \ldots >$, occurring at times t_1, t_2 etc.

Π_x : set of paths from the primary inputs to x.

The following lemma provides an upper bound on the number of transitions at a signal.

Lemma 5.1 *Let C be a combinational circuit and let g be a gate in C with output signal s. Then*

$$|T_s| \leq \sum_{t \in FANIN(g)} |T_t|$$

Proof: At any given time a set of events on the inputs of a circuit can cause at most one event on the output. Thus, if each of the input waveforms is appropriately staggered over time, and each causes an event then the number of events in the output waveform can be as much as the sum of the number of the events on the inputs waveforms, but no more. ∎

Theorem 5.1 *Let C be a combinational circuit and let g be a gate in C with output signal s. Then*

$$|T_s| \leq |\Pi_s|$$

Proof: The proof proceeds by structural induction on the depth of g. The depth of a gate is the maximum number of gates along a path over all the paths from any primary input to this gate.
Induction Hypothesis: Assume that for any gate f with an output signal r, and depth i, $0 \leq i < n$, that $|T_r| \leq |\Pi_r|$.
Induction Basis: The depth of g is 0. In other words g is a primary input. There is exactly one path to g and the number of transitions that can occur at g on the vector pair $< v1, v2 >$ is at most 1. So $0 \leq |T_s| \leq |\Pi_s| = 1$.
Induction Step: Consider gate g with depth n. We know by Lemma 5.1 that $|T_s| \leq \sum_{x \in FANIN(g)} |T_x|$. We also know that the number of paths to g is equal to the sum of the number of paths to each of the inputs to g. Thus, $\sum_{x \in FANIN(g)} |\Pi_x| = |\Pi_s|$. Finally, each of the gates that are inputs to g have depth less than n; therefore, for each such gate h with output signal r the the induction hypothesis that $|T_r| \leq |\Pi_r|$ holds. So, putting the above three observations together yields:

$$|T_s| \leq \sum_{x \in FANIN(g)} |T_x| \leq \sum_{x \in FANIN(g)} |\Pi_x| = |\Pi_s|$$

∎

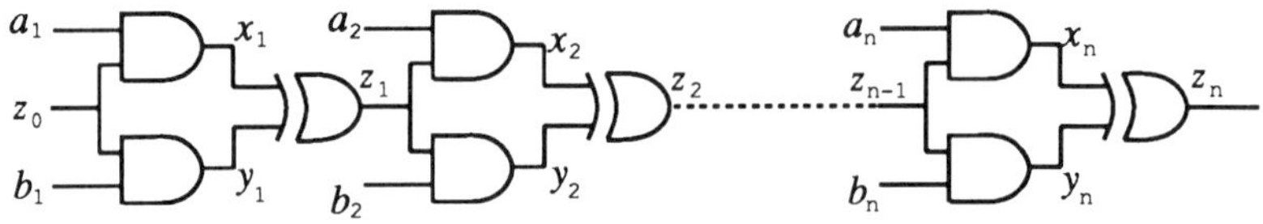

Figure 1: Circuit with exponential transitions from a single input transition. All delays are zero except $d(z_{i-1}, x_i) = 2^i$.

5.2 A Circuit Demonstrating Exponential Complexity

To justify that $|\Pi_s|$ is a tight bound on the number of transitions at s, we give a circuit in which $|T_s|$ is exactly $|\Pi_s|$ for a particular vector pair. The circuit in Figure 1 has n stages, each stage having three gates. a_i, b_i and z_{i-1} are the inputs to the ith stage. x_i, y_i and z_i are internal signals in the ith stage and z_n is the circuit output. All gate delays are zero except for the delay from z_{i-1} to x_i, $d(z_{i-1}, x_i)$, which is 2^i. (This example uses a pin to pin delay. This can be equivalently stated using only gate delays by placing a buffer between z_{i-1} and x_i with this delay value. Also, the XOR gate may be expressed in terms of simple gates.) The input vector pair $< v1, v2 >$ is defined such that $v1(a_i) = v2(a_i) = v1(b_i) = v2(b_i) = 1$ for all i and $v1(z_0) \neq v2(z_0)$.

The input vector pair $< v1, v2 >$ generates one input transition at z_0 and all other inputs remain stable at 1. It is easy to see that for every z_i, $|\Pi_{z_i}| = 2^i$.

Claim 5.1 $|T_{z_i}| = |\Pi_{z_i}|$.

Proof: Using induction on i: $|T_{z_0}| = 1$ directly from the definition of the vector pair $< v1, v2 >$. Since $d(z_{i-1}, x_i) = 2^i$ and the longest path from z_0 to z_{i-1} is of length $2(2^{i-1} - 1)$, y_i settles to its final stable value before x_i starts to change. Every change at x_i or y_i causes a change at z_i, therefore

$$
\begin{aligned}
|T_{z_i}| &= |T_{y_i}| + |T_{x_i}| \\
&= |T_{z_{i-1}}| + |T_{z_{i-1}}| && \textit{because } a_i = b_i = 1 \\
&= 2^{i-1} + 2^{i-1} && \textit{by the induction hypothesis} \\
&= 2^i \\
&= |\Pi_{z_i}|
\end{aligned}
$$

■

5.3 Signal Transitions in Time Intervals

In this section, we develop an upper bound on the maximum number of transitions a signal can have within a given time window, resulting from applying a vector pair at the circuit inputs. This bound shows that the number of transitions of a signal within a time interval can be exponentially smaller than the total number of transitions of the signal, motivating the need for algorithms to compute transitions within a time interval without computing all transitions. In particular, in order to determine if there is a transition at the circuit outputs after time τ, we just need to examine the transitions in the time window $[\tau, \omega]$, where ω is the length of the longest path in the circuit.

We will first introduce some notation:

$\Pi_x^{[\delta,\Delta]}$: set of paths from the primary inputs to x whose length is in the closed interval $[\delta, \Delta]$.

$T_x^{[\delta,\Delta]}$: sequence of alternating rising and falling transitions within the time window $[\delta, \Delta]$ at x.

Theorem 5.2 *Let C be a combinational circuit and s be the output signal of a gate g. Then*

$$|T_s^{[\delta,\Delta]}| \leq |\Pi_s^{[\delta,\Delta]}|$$

Proof: We prove the theorem by induction on a topological ordering of the signals in the circuit. The base case is trivial. If s is an input, the path length from the inputs to s is zero. If $\delta \leq 0 \leq \Delta$, then $|\Pi_s^{[\delta,\Delta]}|$ is 1 and $|T_s^{[\delta,\Delta]}|$ is at most 1. If the interval $[\delta, \Delta]$ does not include 0, then both $|\Pi_s^{[\delta,\Delta]}|$ and $|T_s^{[\delta,\Delta]}|$ are 0. For the induction step, let $d(g)$ be the delay of gate g. If a transition t_s at s in the window $[\delta, \Delta]$ is caused by a transition t_x at input x of g, t_x must be in the window $[\delta - d(g), \Delta - d(g)]$. So, we have

$$|T_s^{[\delta,\Delta]}| \leq \sum_{x \in FANIN(g)} |T_x^{[\delta-d(g),\Delta-d(g)]}|$$

. Since the topological order of s must be strictly greater than those of the inputs, by the induction hypothesis we have

$$|T_x^{[\delta-d(g),\Delta-d(g)]}| \leq |\Pi_x^{[\delta-d(g),\Delta-d(g)]}|$$

Furthermore, every path in $\Pi_x^{[\delta-d(g),\Delta-d(g)]}$ contributes to a path in $\Pi_s^{[\delta,\Delta]}$. Therefore, we have

$$|T_s^{[\delta,\Delta]}| \leq \sum_{x \in FANIN(g)} |T_x^{[\delta-d(g),\Delta-d(g)]}| \leq \sum_{x \in FANIN(g)} |\Pi_x^{[\delta-d(g),\Delta-d(g)]}| = |\Pi_s^{[\delta,\Delta]}|$$

■

It should be noted that when $\delta = \Delta = \tau$, our bound can be off by an arbitrary amount, since at the instant of time τ there can be at most one transition at a signal s and our bound, $|\Pi_s^{[\tau,\tau]}|$, counts all path of length τ from inputs to s. However, it remains open whether there is an algorithm to answer the question "Is there a transition of s at time τ?" in time better than $O(|\Pi_s^{[\tau,\tau]}|)$.

6 Transitions Under Fixed Gate Delays

We are concerned with the problem of determining if a transition occurs at an output within a time interval $[\delta, \Delta]$ when a vector pair is applied at the inputs to the circuit. Obviously, this question can be answered by simulating the vector pair and obtaining the entire waveform at the output. As we showed by the example in Section 5.2, the entire transition sequence can be very large.

In this section we consider this question under a fixed gate delay model. Our goal is to avoid a strict event-driven simulation by focusing only on those transitions which can possibly cause transitions at the output within the time interval $[\delta, \Delta]$.

6.1 Simulating Transitions

We compactly represent transitions of a signal in the circuit as a sequence of rising/falling events stamped with the time of the event. For example, the transition sequence $< R(t_1), \overline{R}(t_2) >$ represents the waveform starting with 0, rising at time t_1, falling at time t_2, and staying 0 thereafter. Given a vector pair $< v1, v2 >$ if a variable v makes a $0 \rightarrow 1$ ($1 \rightarrow 0$) transition in the vector pair, then we generate a transition sequence $< R(0) >$ ($< \overline{R}(0) >$). If a variable is a constant 0 (1) then we generate a transition sequence $< 0 >$ ($< 1 >$). Given transition sequences at the inputs of a gate we compute a resulting output transition şequence in much the same way one would operate on graphical representations of the input waveforms [1]. Determining the transitions at the output of a gate from the transitions of the inputs involves two steps: delaying the input transitions by their corresponding gate delays and combining the transitions according to the logic function of the gate. To delay a transition sequence T by d is to simply add d to each of the components of T and the result is denoted by:

$$T \rightarrow d$$

Combining two waveforms is demonstrated with an example. Given a transition sequence $T_a = < R(t_1), \overline{R}(t_2) >$ on input a of an `AND` gate converging with a transition sequence $T_b = < \overline{R}(t_3), R(t_4) >$ at the input b of an `AND` gate with delays such that $t_4 > t_2 > t_3 > t_1$, the output transition sequence, call it T_g is computed using the calculus of Table 1 as follows: First for ease of calculation we need to provide a value on each transition for each delay in the sequence $< t_1, t_3, t_2, t_4 >$. Thus transition sequence T_a becomes $< R(t_1), 1, \overline{R}(t_2), 0 >$ and transition sequence T_b becomes $< 1, \overline{R}(t_3), 0, R(t_4) >$. Since $R(t_1)$ is the first event it is paired with the 1 in transition sequence T_b and the result is $R(t_1)$. Thc rcsulting transition sequence T_g is now $< R(t_1) >$. The next event is $\overline{R}(t_3)$ and it is paired with the 1 on transition sequence T_a (that is the result of the rise $R(t_1)$) so the result is a fall and the resulting transition sequence T_g is now

$i_1 \rightarrow$ $i_2 \downarrow$	0	1	R	$\overline{R}$
0	0	0	0	0
1	0	1	R	$\overline{R}$
R	0	R	R	0
$\overline{R}$	0	$\overline{R}$	0	$\overline{R}$

Table 1: Fixed time, fixed delay calculus for an AND gate

$< R(t_1), \overline{R}(t_3) >$. The next event is $\overline{R}(t_2)$ and it is paired with the 0 transition sequence T_b resulting from the fall $\overline{R}(t_3)$ and as a result the output stays 0 and no event occurs. The final event is $R(t_4)$ and it is paired with the 0 on input a that is the result of the fall $\overline{R}(t_2)$ and again no event occurs. So the final waveform T_g is $< R(t_1), \overline{R}(t_3), 0, 0 >$ which can be abbreviated to: $< R(t_1), \overline{R}(t_3) >$. The operations of combining transition sequences at AND gates and OR gates in such a manner are denoted by: $\prod_{x \in FANIN(g)} T_x$ and $\sum_{x \in FANIN(g)} T_x$. The computation of transition sequences at a gate g is then denoted by:

$$T_g = \prod_{x \in FANIN(g)} (T_x \rightarrow d(g))$$

$$T_g = \sum_{x \in FANIN(g)} (T_x \rightarrow d(g))$$

for AND and OR gates, respectively. For NOT gates, it is sufficient to delay the transition sequence and reverse the polarity of each transition in it.

6.1.1 Event Suppression Outside Time interval $[\delta, \Delta]$

To compute transitions of the circuit output at or after time δ, we want to keep the transition sequence at intermediate signals as small as possible. Let s be an intermediate signal, ω_s be the length of the longest path from s to the primary outputs and α_s be the length of the shortest path from s to the primary outputs.

Lemma 6.1 *Any transition at s before $\delta - \omega_s$ will not contribute to a transition at the outputs after δ.*

Proof: There is no path in the circuit long enough for the transition to propagate along and appear at the outputs during $[\delta, \Delta]$. ■

So, we can suppress all the transitions before $\delta - \omega_s$ at s. Such an operation is denoted by: $T_s \downarrow (\delta - \omega_s)$.

Lemma 6.2 *Any transition at s after $\Delta - \alpha_s$ will not contribute to a transition at the output before Δ.*

Proof: There is no path in the circuit short enough for the transition to propagate along and appear at the outputs during $[\delta, \Delta]$. ■

So, we can suppress all the transitions after $\Delta - \alpha_s$ at s. Such an operation is denoted by: $T_s \uparrow (\Delta - \alpha_s)$.

The following theorem, which follows from Lemmas 6.1 and 6.2, gives the range of transitions at an intermediate signal s sufficient to compute transitions at the circuit output within interval $[\delta, \Delta]$.

Theorem 6.1 *Let s be an intermediate signal in a single-output combinational circuit C. Only when a transition at s is within $[\delta - \omega_s, \Delta - \alpha_s]$, could it be responsible for a transition at the output within $[\delta, \Delta]$.*

Operationally, we keep a time interval $[\alpha_s, \omega_s]$ for each intermediate signal s, and after computing the transition sequence of s, suppress all transitions before $\delta - \omega_s$ and after $\Delta - \alpha_s$. For example, for an AND gate:

$$T_s = \left(\prod_{x \in FANIN(g)} (T_x \rightarrow d(g)) \right) \downarrow (\delta - \omega_s) \uparrow (\Delta - \alpha_s)$$

6.2 Simulation Complexity for Example Circuit

Consider the circuit of Figure 1. Assume that we wished to check if there is a transition at the circuit output on a given input vector pair at or after time $\delta = 2^n$. We merely check if a transition can propagate down the single path of length 2^n in the circuit. The number of events we need to compute is at most one. In general, to answer questions regarding transitions at a fixed time for this circuit, the complexity of simulation is $O(n)$.

6.3 Demand-Driven Simulation

Demand-driven simulation [8] has been proposed as a means to reduce the number of unnecessary events during logic simulation. The basic concept of demand-driven simulation is straightforward: given a a request for a particular value of a signal at a particular time (SVR), the simulator proceeds backwards from the given output through the circuit, gathering the data required to determine the signal's value at the specified time. This backward search is potentially exponential since it may end up exploring all the paths in the circuit. It should be noted, however, that the pruning techniques described in the previous sections can be used to further reduce the number of events in demand-driven simulation. This has not been explored in [8].

Another problem with demand-driven simulation is that the backward evaluation of a gate is more complicated than normal forward evaluation. This is exacerbated when sophisticated gate delay models are used. In [8] the authors concede that for timing analysis, demand-driven simulation forfeits the accuracy and modeling capabilities of event-driven simulation.

7 Transitions with Bounded Gate Delays

The simulation techniques used in the bounded delay model use the same basic principles developed for the fixed delay model. However, the uncertainty in the delay introduces some complexity. It is interesting to note that for the bounded delay model even traditional simulation methods have not been well developed. Perhaps the most important work here is that of Ishiura *et al.* [5, 4]. The ideas developed in this section use ideas similar to those in [5] and sets them in the context of restricted simulation.

7.1 Waveforms and Regions

Since the gate delays are not fixed, it is necessary to use a separate variable to denote the delay of each gate. (The alternative of assuming that the output could be either a 0 or a 1 in periods of uncertainty is known to yield pessimistic results [5].)

Let d_k denote the delay of gate g_k and let $\vec{d}$ denote the vector of gate delays for all the gates in the circuit. Further, let d_k^l and d_k^u represent the minimum and maximum delays of the gate under the bounded delay model. Each distinct value of $\vec{d}$ corresponds to the delays in a particular instance of the circuit. Note that $\vec{d} \in \mathcal{R}^n$ where $\mathcal{R}$ is the set of real numbers and there are n gates in the circuit. Let D be a region of $\mathcal{R}^n$ such that D is specified in terms of linear inequality relations between the d_k's. For example, $D = \bigcup_k \{d_k^l \le d_k \le d_k^u\}$ defines the region that $\vec{d}$ is restricted to. Henceforth the letter D (possibly subscripted) will be used to denote regions of $\mathcal{R}^n$ that are specified in terms of linear inequalities of the d_k's and will be represented by the set of these inequalities.

Like the fixed delay model we consider waveforms, defined in terms of transitions, at signal lines. However, the introduction of delay variables results in a major difference from the fixed delay case. Instead of a single waveform at each signal, multiple waveforms need to be preserved which correspond to different possible combinations of delay values for the delay variables. Each waveform then denotes the signal's response under a particular combination of these delay values. A signal i in a circuit has a set of waveforms $W_i = \{w_{i,j_i}\}$ where j_i varies from 1 to $|W_i|$. Each individual waveform w_{i,j_i} has two parts: D_{i,j_i} and T_{i,j_i} and is represented as $D_{i,j_i} : T_{i,j_i}$. Here, D_{i,j_i} is the region of its validity, (note that D_{i,j_i} denotes some combination of the delay values), and T_{i,j_i} is the transition sequence in the waveform. As before the transition sequence T_{i,j_i} is given by set of rising/falling transitions denoted by $R/\bar{R}\ (t_{i,j_i,l_{j_i}})$. $R/\bar{R}\ (t_{i,j_i,l_{j_i}})$ has a time value $t_{i,j_i,l_{j_i}}$ associated with it. Since the gate delays are variables, $t_{i,j_i,l_{j_i}}$ is actually a linear sum of a subset of the d_k variables.

7.2 Simulating Transitions

Let s be a fanout of a primary input in a given multilevel circuit. Let the transition at the primary inputs from $v1$ to $v2$ occur at time 0. There will be exactly one waveform for s with the region $\mathcal{R}^n$. Let us now look at the transition sequence for this waveform. By Theorem 6.1 we know that we just need to look at transitions in the window $[\delta - \omega_s, \Delta - \alpha_s]$. If $v1(s) = v2(s)$ then the transition sequence is simply $< v1(s) >$. Let us now consider the case where there is a transition on s. If $\Delta - \alpha_s < 0$ then the value corresponding to $v1$ is the only relevant value on this signal. The waveform used for this primary input during simulation will be $< 0 >$ or $< 1 >$ depending on the value of the primary input on $v1$. If $\delta - \omega_s > 0$ then the value corresponding to $v2$ is the only relevant value. Otherwise a $R(0)(\bar{R}(0))$ is the singleton element of the transition sequence if this input changed from a 0 to a 1 (1 to a 0).

Now let us examine a simple case of the simulation at a gate given the waveforms at the inputs. Consider a two-input `AND` gate g_k with delay d_k and the following waveforms at the inputs:

$$\begin{aligned} W_1 &= \{D_{1,1} :< R(t_{1,1,1}) >\} \\ W_2 &= \{D_{2,1} :< R(t_{2,1,1}) >\} \end{aligned}$$

This is the simplest case where all the waveforms and transition sequences have a single element.

The waveform set, W, at the output is given by:

$$\begin{aligned} W = \; & \{D_{1,1} \cap D_{2,1} \cap \{d_k^l \leq d_k \leq d_k^u\} \cap \{t_{1,1,1} > t_{2,1,1}\} :< R(t_{1,1,1} + d_k) >, \\ & D_{1,1} \cap D_{2,1} \cap \{d_k^l \leq d_k \leq d_k^u\} \cap \{t_{2,1,1} \geq t_{1,1,1}\} :< R(t_{2,1,1} + d_k) >\} \end{aligned}$$

The first waveform in this set corresponds to the case when the the first input rises after the second and the second corresponds to the case when the the second input rises no later than the first. Note that for each of these the corresponding region is the intersection of three regions. The first of these is itself the intersection of the regions for each input waveform, the second is the region corresponding to the delay bounds of gate g_k and the third is the region due to the forced order among the transitions. If there exists an assignment to the d_k's such that $t_{1,1,1}$ is greater than $t_{2,1,1}$, then $t_{1,1,1} > t_{2,1,1}$ denotes the values of $t_{1,1,1}$ for which it is greater. Otherwise $t_{1,1,1} > t_{2,1,1}$ is empty and the waveform W_1 is deleted, because there is no feasible region for it.

We now extend this to the general case, i.e. given a waveform set for each input of a m-input gate we need to determine the waveform set for the output. Let W_i denote the waveform set for input i. Consider one waveform, $w_{i,j_i} \in W_i$ for each input. The time values on the transitions for w_{i,j_i} have a total order on them, *i.e.*, $t_{i,j_i,1} < t_{i,j_i,2} < \ldots < t_{i,j_i,l_{j_i}}$. The collection of these total orders for all the inputs defines a partial order for all the time values in the m waveforms for the gate inputs. Let Ω be a total order for these time values. Then this defines a total order for all the transitions seen at all the inputs to gate g_k corresponding to the input waveforms w_{i,j_i}. Given this total order we can use the simulation semantics of the fixed delay model given in Section 6.1 to determine the corresponding output transitions.

To state this more precisely, the output waveform, w, is given by:

$$w = \bigcap_i D_{i,j_i} \cap \{d_k^l \leq d_k \leq d_k^u\} \cap \Omega : \mathcal{S}(\Omega, \{T_{i,j_i}\})$$

The region of validity is determined by the intersection of the region of each input, the region due to the gate delay bounds, and the region forced by the total order Ω. $\mathcal{S}(\Omega, \{T_{i,j_i}\})$ denotes the sequence of transitions at the output under the total order Ω given the sequence of transitions T_{i,j_i} at each input.

Of course, in order to generate all possible waveforms for this combination of input waveforms, we need to consider all possible total orders Ω. This takes care of one possible combination of waveforms at the inputs. Going further, we need to examine all possible combinations of these waveforms in order to get the complete set of output waveforms.

7.3 Pruning Conditions

The potentially exponential blowup of waveforms in the previous section needs to be constantly pruned to ensure that only relevant information is being propagated through the circuit.

The first level of the pruning is done by determining empty regions and eliminating the waveforms corresponding to them. This may be done using linear programming.

The second level of pruning uses the lower and upper bounds on the path lengths from the gate, g_k, to the primary outputs, namely α_k and ω_k, to suppress transitions in non-empty regions. Let $w = D : T$ be an output waveform and let $R(t) \in T$. Let x and y be the minimum and maximum values of t in the region D (determined by linear programming). If $x+\alpha_k > \Delta$ or $y+\omega_k < \delta$ then this transition should be suppressed since there is no path which can propagate it to the output so as to be observed in the window $[\delta, \Delta]$.

While checking that a region is feasible can be done very efficiently, it is more time-consuming to compute the minimum or maximum value of x or y within a region D. Of course we can compute an upper bound for x, call it x_t^u, simply by adding up the d_j^u's corresponding to the variables in t. Similarly, we can compute a lower bound for y, y_t^l, by adding up the d_j^l's corresponding to the variables in t.

EX	EDS	100%	95%	90%	85%	80%
C17	1292	451	451	451	451	1292
C432	23192	1718	1718	5155	7114	7590
C499	75428	10141	20015	27219	32752	37395
C880	34290	753	1360	1757	2602	4349
C1355	54800	6477	12807	17169	22769	23628
C1908	110548	17382	25486	35698	42247	49004
C2670	114403	1808	2393	3500	5138	8267
C3540	191233	6187	8784	17462	29555	42693
C5315	345584	3901	8585	20895	37037	64020
C6288	3791472	4710	87021	340853	695371	1065141
C7552	551114	7813	12912	26761	49772	78896
summary	5293356	61341	181532	496920	924808	1382275

Table 2: Comparison of raw events evaluated

8 Experimental Results

To compare the effectiveness of the techniques proposed in this paper with conventional event-driven simulation for determining transitions at or after time δ, we present a set of results using the ISCAS-85 combinational logic benchmark set [2], which are medium to large sized circuits. In particular, we compare the number of events evaluated by our algorithm based on Theorem 6.1 with those by conventional event-driven simulation. This comparison is done using the fixed delay model.

Table 2 compares the number of events evaluated by a conventional event-driven simulation (EDS) with those evaluated by our algorithm in order to determine the transitions after various target delays expressed as percentage of the true floating delays [3] of the circuits, for 100 random input vector pairs [1]. For example, the last column labeled 80% represents the time window $[.8f, f]$ where f is the floating delay. Table 3 shows the reductions in event count by our algorithm over conventional event-driven simulation. These results demonstrate a significant reduction in the number of events that need to be processed, especially as very narrow windows of time are examined.

Notice that we did not report any CPU time, because that is highly implementation dependent and reporting unbiased CPU time for event-driven simulation is almost impossible. However, all techniques for speeding up event-driven simulation could also be used in our algorithm. Secondly, in reporting just the event count, we have ignored the overhead necessary to filter out events at logic gates. The first overhead is in computing the α's and

[1] The true floating delay is the delay reported by an accurate timing verifier that takes into account false paths in the floating mode assumption and is used here since it is a tighter upper bound on the delay of the circuit than the longest path.

EX	EDS	100%	95%	90%	85%	80%
C17	1	0.3491	0.3491	0.3491	0.3491	1.0000
C432	1	0.0741	0.0741	0.2223	0.3067	0.3273
C499	1	0.1344	0.2654	0.3609	0.4342	0.4958
C880	1	0.0220	0.0397	0.0512	0.0759	0.1268
C1355	1	0.1182	0.2337	0.3133	0.4155	0.4312
C1908	1	0.1572	0.2305	0.3229	0.3822	0.4433
C2670	1	0.0158	0.0209	0.0306	0.0449	0.0723
C3540	1	0.0324	0.0459	0.0913	0.1545	0.2233
C5315	1	0.0113	0.0248	0.0605	0.1072	0.1853
C6288	1	0.0012	0.0230	0.0899	0.1834	0.2809
C7552	1	0.0142	0.0234	0.0486	0.0903	0.1432
average	1	0.0845	0.1210	0.1764	0.2313	0.3390

Table 3: Ratio of events evaluated

ω's at each gate. But, this is a linear time computation and when compared with number of vectors simulated it is insignificant. The second overhead is to filter the events at each gate which we also believe is very small, because filtering involves two comparisons with the bounds, α and ω, which is relatively cheap comparing with the amount of computation needed to schedule the events.

9 Conclusions and Future Work

In this paper we have examined techniques for resolution of the question "On a given vector pair does an event occur in this combinational circuit at any time at or after τ?". We have shown that an event-driven simulation approach to resolving this question has exponential complexity. To help address this question we have introduced an event-calculus to help us in reasoning about the timing behavior of circuits. We have given an alternative approach to resolving this question on circuits which uses a mixture of static timing analysis procedures and our event-calculi.

Our approach exploits a tight relationship between timing simulation and static timing verification. Currently we are exploring an even tighter coupling between these procedures. Having discovered that even after eliminating false paths, timing verifiers may over-pessimistically compute the delay of a circuit, we envision a scenario in which by applying our event-calculi, static timing verifiers are extended to compute dynamic two-vector *transition delay* of circuits. As a by-product of this process vector pairs are generated which may then be validated by timing simulation. We hope that such a procedure would simultaneously provide the benefits of the speed of timing verification with the increased accuracy of timing simulation.

10 Acknowledgments

Thanks to Randy Bryant and Rick Rudell for interesting discussions on transition delay simulation and computation. Srinivas Devadas is supported in part by the Defense Advanced Research Projects Agency under contract N00014-91-J-1698, and in part by a grant from IBM Corporation. Sharad Malik is supported in part by an IBM Faculty Development Award.

References

[1] L. Augustin. An Algebra of Waveforms. In *IFIP International Workshop on Applied Formal Methods for Correct VLSI Design*, pages 159–168, November 1989.

[2] F. Brglez and H. Fujiwara. A neutral netlist of 10 combinational benchmark circuits and a target translator in fortran. In *International Symposium on Circuits and Systems*, 1985.

[3] S. Devadas, K. Keutzer, and S. Malik. Delay Computation in Combinational Circuits: Theory and Algorithms. In *Proceedings of the International Conference on Computer-Aided Design*, November 1991.

[4] N. Ishiura, Y. Deguchi, and S. Yajima. Coded Time-Symbolic Simulation using Shared Binary Decision Diagrams. In *The Proceedings of the Design Automation Conference*, pages 130–135, June 1990.

[5] N. Ishiura, M. Takahashi, and S. Yajima. Time-Symbolic Simulation for Accurate Timing Verification. In *The Proceedings of the Design Automation Conference*, pages 497–502, June 1989.

[6] P. C. McGeer and R. K. Brayton. *Integrating functional and temporal domains in logic design.* Kluwer Academic Publishers, 1991.

[7] C-J. Seger and R. E. Bryant. Modelling of Circuit Delays in Symbolic Simulation. In *IFIP International Workshop on Applied Formal Methods for Correct VLSI Design*, pages 625–639, November 1989.

[8] S. P. Smith, M. R. Mercer, and B. Brock. Demand Driven Simulation: BACKSIM. In *Proceedings of* 24th *Design Automation Conference*, pages 181–187, June 1987.

Network Communication

Match and Move, an Approach to Data Parallel Computing

Thomas J. Sheffler and Randal E. Bryant
Carnegie Mellon University
Pittsburgh, PA 15213

Abstract

The **match** operation, introduced by G. Sabot, is a parallel primitive describing a communication pattern. For two vectors of keys, called the "source" and "destination," a path is indicated for each element of the source to each destination element having an equal key. The result of the **match** operation is a data structure called a "mapping" encapsulating these paths. Using a **move** operation, data may be sent from the sites of a source data vector through the paths of the mapping to a destination data vector. If the mapping is many-to-one, a combining function, such as **add**, **max** or **min**, specifies how collisions are resolved at the destination sites.

The **match** and **move** operations provide a unified set of parallel primitives that at once satisfy the conflicting demands of generality and efficiency. In this paper we illustrate the generality of the approach by showing several algorithms implemented with **match**, including binary tree isomorphism and some sparse matrix kernels. By using the **match** operation, the algorithms are straightforward and simple to understand.

Because **match** and **move** may be implemented efficiently on many different parallel and vector architectures, algorithms expressed in this manner achieve architecture independence. We support this claim by presenting data collected from implementations on the CRAY Y-MP and the Connection Machine CM-2. Our programs are competitive with more traditional methods of implementation, while remaining architecture-independent and high-level.

1 Introduction

Decreasing costs of computer hardware have allowed supercomputer architectures to evolve simultaneously in two directions. While massively parallel SIMD architectures are being explored in computers such as the Connection Machine, MasPar MP-1, and DAP, other manufacturers have been developing vector multiprocessors with increasing numbers of vector units and processors. These machines include the CRAY Y-MP, IBM 3090 VF and Alliant FX/80. With the increasing complexity these machines offer the programmer, it becomes important to adopt high level parallel primitives that simplify the view of these architectures.

The parallel vector model of computing proposed by Guy Blelloch has been successfully applied to both massively parallel SIMD machines and vector multiprocessors [2, 3]. The basic data structure of this model is the "collection," represented as a vector, with nested collections (i.e., collections of collections) represented as segmented vectors. Primitive communication operations include simple permutations through **send** and **get**, with data combining and reduction through **scan** operations.

Closely related to the parallel vector model is the Paralation model proposed by Gary Sabot [10]. Sabot introduced the **match** operation as a means to describe a data-dependent communication pattern called a "mapping" in terms of two vectors of keys. A **move** operation is used to send data through the paths of a mapping. If the mapping is many-to-one, an optional combining function, such as **add**, **max** or **min**, may be specified for **move** to resolve collisions at the destination sites.

The abstraction presented by the **match/move** paradigm is defined at a high level. Because **match** and **move** specify how data values should be permuted or combined, but not the methods by which the operations are carried out, they provide an architecture independent programming model. Implementations of **match** and **move** may be tuned for an underlying target architecture, making use of special primitives available, in a manner transparent to the programmer.

In this paper we present the **match** operation and its relatives as extensions to the vector model. The ease with which algorithms may be designed in this model is illustrated by example algorithms using **match** as a primitive. The brevity of the algorithms described in this manner is evidence of the power of the **match** operation.

Implementation strategies for the **match** operations are discussed for different target architectures. We place upper bounds on the efficiency of the implementation of the basic set of operations, but also discuss how performance may be improved in certain circumstances. The choice of an implementation strategy is guided by the target architecture, the context of the application and other factors. These considerations may be deferred however, until the fine-tuning of a final program, and do not affect the functionality of a program expressed using **match**.

Finally, we present the implementations of the **match** and **move** operations on the CRAY Y-MP and CM-2 and discuss performance data collected for each. The algorithms expressed using **match** are simple, straightforward, and not machine specific. For our sparse matrix operations, we show that the performance of the **match**-based implementation on the CRAY is competitive with more traditional methods of implementation. The fact that our algorithms expressed in an architecture-independent manner can compete with an optimized, vectorized FORTRAN code for the CRAY Y-MP is evidence of the power of the **match** operations. On the Connection Machine, while matrix multiplication algorithms have been reported before, ours represent the first reported performance figures for sparse matrix multiplication.

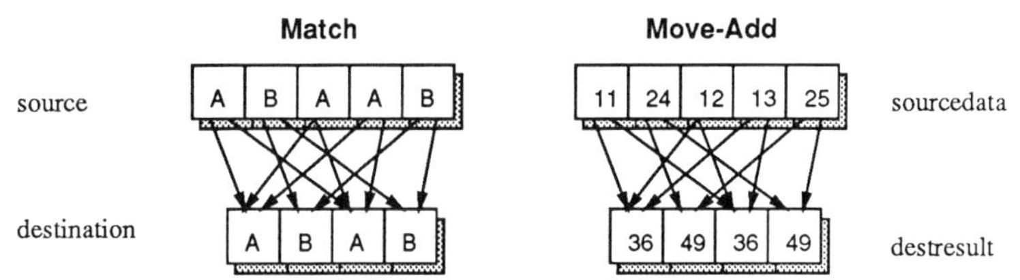

m = **match**(destination, source);
destresult = **move-add**(m, sourcedata);

Figure 1: The **match** operation constructs a mapping from a source vector to a destination vector. A **move** operation can use this mapping to send data from a source vector to a destination vector with a combining function.

2 The Match Operation

The **send** and **get** primitives of the parallel vector model described earlier provide simple many-to-one or one-to-many communications paths. These are equivalent to the concurrent-write and concurrent-read memory access methods of the CRCW PRAM model. The **match** operation along with its **move** operations provide the functionality of both **send** and **get** and allow a restricted form of many-to-many communication.

The **send** and **get** primitives use the addresses of sites as arguments. **match** is slightly different: it uses keys to show how a communication pattern should be constructed. Two vectors called the "source" and "destination" define this pattern. For each element in the source a path is indicated to each destination element with an equal key. The result returned by the **match** operation, called a "mapping," can be used with a **move** operation to send data through these paths.

The **match** operation of Figure 1 shows two vectors used to create a mapping. Arrows represent the communication paths of the mapping that are set up between equal keys. When multiple source sites map to the same destination, variants of **move**, such as **move-add**, **move-max** and **move-min**, allow combining functions to be specified. A special combining function called **move-arb** simply chooses an arbitrary value from those arriving at a destination, but also ensures that the same arbitrary value is chosen for all destination sites matching a given key. In Figure 1, the mapping is applied to a vector "sourcedata" with the **move-add** operation. The result vector contains the sums of values from source sites sharing the same key.

A mapping created with **match** expresses a data-dependent relationship. Frequently, with many algorithms designed for the Connection Machine, the same effect is achieved by performing complicated computations on site

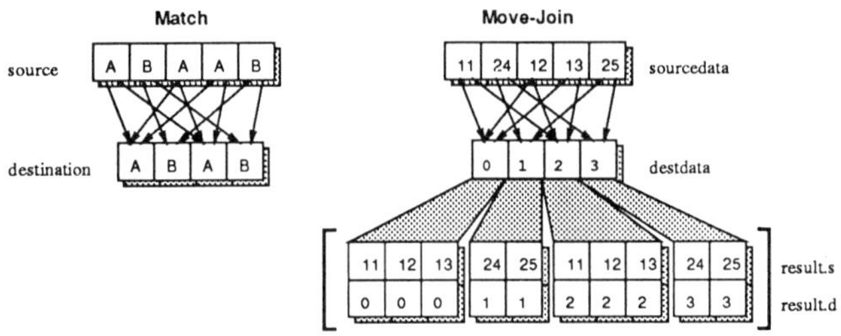

m = **match**(destination, source);
result = **move-join**(m, sourcedata, destdata);

Figure 2: The **move-join** operation provides a combining function that arranges the values colliding at a destination site into adjacent sites in a new vector. The two fields of the result vector relate each site to a source and destination vector.

addresses. Mappings make such relationships explicit, rather than a side-effect of other computations.

The basic combining functions apply a simple algebraic operation to the values sent to a destination site. Data values mapped to a destination site may also be gathered into a new vector with the **move-join** operation. With this operation, each destination site is associated with a segment in a vector containing copies of each of the data values from the source sites in the mapping.

An example **move-join** operation is shown in Figure 2. For simplicity, we define the **move-join** operation to return a vector of tuples with two fields called "s" and "d." Each of these fields contain copies of data from vectors "sourcedata" and "destdata" given as arguments to **move-join.**

The **move-join** operation takes its name from the join operation of the relational data base model [4]. In the Paralation model, this functionality is provided through **collect**. Another way of viewing **move-join** is as an operation that constructs all of the cross-products between the source and destination data vectors for the groups defined by the keys. This also has the effect of "flattening" the information encapsulated in a mapping, expressing each path of the mapping explicitly with a tuple in the result vector.

2.1 Other match operations

Certain types of mappings resulting from an application of **match** are common enough that special functions have been defined to build them. The only one used in this paper is **collapse**. This operation accepts a single vector of keys, and returns a mapping to a new vector. For each distinct key in the vector, a destination site is created in the mapping, so that all sites with the same key map to the same site in the destination vector. The length of the destination vector is exactly the number of distinct keys in the key vector. This operation forms the basis for a simple histogramming operation. By applying the mapping created by **collapse** to a vector of 1's with **move-add**, the destination vector will contain a count of the number of each key in the source.

2.2 Comparison to other parallel primitives

The functionality provided by **match** and **move** is similar to a variety of parallel primitives that exist on real and theoretical machines. The common theme among the parallel primitives to be discussed is the association of each member of a collection of data values to one and only one group. For the **match** operation, groups are identified by keys. For the other primitives discussed, groups are identified by reference to a common variable, or by their arrangement in a vector.

On the Connection Machine, a combining-**send** operation identifies groups by parallel reference to a processing site. When using the **send-add** operation, for example, the final value written in the site is the sum of all values sent to it [7]. The summing operation performed is also generalized to a number of other associative functions.

The NYU Ultracomputer provides the **fetch-and-op** primitive [5, 6]. Groups are identified by a reference to a parallel variable. When executing a **fetch-and-op** on a parallel variable, the final value of the variable is the sum of all the values sent to it, while each of the referencing sites receives one of the partial sums computed during the reduction. The order in which the partial sums is computed is unspecified.

The multiprefix operator of the Fluent abstract machine subsumes the functionality of the **fetch-and-op** primitive [9]. It provides the reduction, and also guarantees that all partial sums are computed in processor address order. This is similar to the segmented scan of the Connection Machine. Here, groups must be arranged into contiguous sites of a vector called "segments." Partial sums are computed in the site address order of the vector.

The **match** operation generalizes the specification of groups to the simple mechanism of keys. Keys are much more general than the site addresses used with **send** and **get**, the parallel variables used with the **fetch-and-op** primitives, or the segments associated with **scan** operations. **match** and **move** provide the functionality of the combining-**sends** and **get**, and also provide the reduction values associated with the **fetch-and-op** primitives and **scan** operations. However, they do not directly provide the partial sums of some of the primitives listed.

match and **move** may be implemented using any of the parallel primitives mentioned above by first associating each "group" of unique keys to the group required by the parallel primitive. The association may be through a unique variable, or a contiguous range of sites in a vector. Then, the parallel primitive supplied by the underlying hardware may be applied directly to implement a **move**. The mapping of groups to variables or sites may be achieved by sorting or hashing techniques. Thus, **match** isolates the programmer from the underlying architecture, but allows sufficient variation of implementation strategies for efficiency on a wide range of architectures. We will discuss implementation strategies in depth later.

3 Algorithms using Match

match allows the description of many parallel algorithms in just a few steps. Because mappings describe a relationship between data elements and how they should be manipulated, a programmer need not be concerned with the actual locations of data elements. This frees the programmer to focus on only the high level issues of the algorithm.

This section presents some examples that show the power of **match** for expressing parallel algorithms. The algorithms presented are not typical candidates for a data parallel style of implementation because the irregular data structures of these algorithms have no obvious mapping onto parallel hardware.

Our algorithms will be presented in a pseudo-code that makes use of vectors. Each site of a vector is identified by its index. Elements of vectors may be either simple primitive types such as integer, boolean, and float, or structured types represented through the use of tuples. Tuples have named fields, much like traditional structure types. A reference to an aggregate of tuple fields is indicated by enclosing the field names in brackets: **[field1, field2]**. The use of tuples simplifies the expression of many algorithms since it allows composite values to be treated as a whole.

In our notation, vector assignments take place on an element by element basis, and scalars are promoted to a vector of constants when appropriate. Elementwise operations are indicated by regular scalar operators applied to two vectors, and also produce their values on an element by element basis. Lastly, reduction functions such as **and-reduce** apply an associative operator to all elements of a vector to produce a scalar result.

The vector of tuples will be our basic data structure for presenting algorithms using **match**. This approach leads to an associative programming style that uses **match** to express relationships between tuples collected in vectors.

3.1 Binary Tree Isomorphism

A forest of binary trees is represented as a vector of tuples. For each vertex, there is a tuple **[id, left, right]**. The **id** of each vertex is initially unique for each vertex. The fields **left** and **right** are references to the **id** fields of other vertices. Leaf fields have the value **null** in their **left** and **right** fields. The binary tree isomorphism procedure computes new ID's for each vertex so that two vertices are given the same ID if and only if they are roots of isomorphic sub-trees.

The algorithm is a variant of the tree isomorphism algorithm of [1]. A vector called *f* contains the forest represented through tuples. In the first steps, two mappings are created. The mapping *leftm* relates left children to their parents, similarly, *rightm* is a mapping to the parents of right children. These two mappings freeze the structure of the forest as it is at the beginning of the procedure.

```
     procedure tree-isomorphism {
(1)      leftm = match(f.id, f.left);        (* from child to parent *)
(2)      rightm = match(f.id, f.right);
(3)      do{
             (* Select a unique ID for all verts whose children are isomorphic *)
(4)          m = match(f.[left,right], f.[left,right]);
(5)          oldid = f.id;
(6)          f.id = move-arb(m, f.id);
             (* Replace the references of the parent *)
(7)          f.left = move-arb(leftm, f.id, f.left);
(8)          f.right = move-arb(rightm, f.id, f.right);
(9)          done = (oldid == f.id);
(10)     } until(and-reduce(done))
     }
```

The procedure works iteratively. Initially, all vertices have different IDs. At each step, vertices that are roots of isomorphic subtrees are identified by tuples with the same **[left,right]** values. Because all IDs are unique, in the first step, only leaf vertices whose **[left,right]** value are **[null,null]** will be given the same ID. Thus, after the first step, all subtrees whose height is 1 are labeled with the same ID if and only if they are isomorphic subtrees.

Now, proceeding inductively, at step number $h+1$, all subtrees of height h have the same ID if and only if they are isomorphic subtrees. The root vertices of subtrees of height $h+1$ are isomorphic if and only if their left and right children are isomorphic. On line 4, a **match** operation on the **[left, right]** composite value of each vertex will create a mapping between the roots of all isomorphic subtrees, and the **move** on line 6 will label all isomorphic subtrees of height $h+1$ with the same ID. For a forest of trees whose maximum height is H, the algorithm terminates after H steps, after information has been propagated up the trees from the leaves to the roots.

3.2 Sparse Matrix Multiplication

Sparse matrix applications are generally considered ill-suited for parallel machines. Because sparse matrices present an irregular data structure with no direct mapping onto parallel hardware, they represent a difficult algorithm design problem. The algorithm construction using **match** is simple, and elegant.

Given two matrices A and B, the matrix product, $C = A \times B$ is defined as follows. An element of matrix A in row i and column j is denoted $A(i,j)$. Each element of the matrix product C is computed by the summation

$$C(i,j) = \sum_{1 \leq k \leq N} A(i,k)B(k,j).$$

When A and B are sparse, it is advantageous to avoid storing elements that are zero and computing products where one of the operands is zero.

Sparse matrices are represented as vectors of tuples, **[row,col,val]**. In this storage representation zero-valued elements need not be stored. Using **match**, the algorithm consists of two parts. The first phase computes all of the partial products that contribute to the sums, along with their indices in the result matrix, C. The second phase gathers partial products with the same indices in the result matrix and adds them together.

If we examine the formula above, we see that each product is of the form $A(i,k)B(k,j)$, indicating a "match" of the columns of matrix A with the rows of B. The partial products created between a column of A that matches a row of B are all of those that result when the elements from a column k of A are distributed over the elements of row k of B. A simple application of **move-join** to the mapping creates sites for all of these partial products.

```
      procedure matrix-multiply {
          (* PHASE 1 *)
(1)       m1 = match(B.row, A.col);
(2)       pp = move-join(m1, A, B);        (* distribute elements of A over B *)
          (* Elementwise calculations for each partial product *)
(3)       new.row = pp.s.row;       (* from matrix A *);
(4)       new.col = pp.d.col;       (* from matrix B *);
(5)       new.val = pp.s.val × pp.d.val;
          (* PHASE 2 *)
(6)       m2 = collapse(new.[row,col]);
(7)       C.val = move-add(m2, new.val);
(8)       C.row = move-arb(m2, new.row);
(9)       C.col = move-arb(m2, new.col);
      }
```

This application uses a very powerful feature of **move-join** in Lines 1 and 2 to move entire copies of the elements of matrices A and B to the sites of the partial products in just two steps. Lines 3, 4, and 5 illustrate the origin of the information contributing to each partial product. Finally, Line 6 creates the mapping that gathers all partial products with the same **[row,col]** address for the matrix C. The **move-add** of Line 7 calculates the sum of the partial products for each of the result elements and the final two lines simply move the row and column labels to each result sum.

3.3 Sparse Matrix-Vector Multiplication

This application is much simpler than the previous one. Because the vector is not sparse, its elements are labeled by its index vector. The procedure computes the product of a sparse matrix A, represented in tuple form, with a vector X, computing the result in vector Y. The partial products are computed by first distributing the vector elements over the elements of the matrix by columns. Then the new vector values are computed by summing the products over all of the rows.

```
procedure matrix-vector-multiply {
    (* PHASE 1 *)
    m1 = match(A.col, index(X));
    distribute = move-arb(m1, X);      (* copies of X *)
    pp = distribute × A.val;      (* partial products *)

    (* PHASE 2 *)
    m2 = match(index(X), A.row);      (* reduce by rows *)
    Y = move-add(m2, pp);
}
```

Because the vector is labeled by its index vector, both of the **match** operations of this algorithm involve an index vector. We will show later how this information can be used to optimize the **match** operation.

4 Implementation Strategies for Match

The **match** operation provides a very general mechanism for specifying groups through source and destination keys. As mentioned earlier, other parallel primitives perform combining operations on groups. However, the manner in which groups are specified for any of the combining functions mentioned earlier is much more restrictive than the key-based approach.

The two main methods for creating mappings use either sorting or hashing as a first step to assign keys to groups appropriate for the combining primitives of the underlying hardware. The choice of a method may be based on characteristics of the key vectors, or by the context in which the mapping is created or used with a **move** operation. After introducing the parallel vector model and its primitives we will show how each of these methods may be used to create a mapping.

4.1 The Parallel Vector Model of Computing

The parallel vector model will be the tool with which we express our implementation strategies for **match** and **move**. It builds on the standard RAM model by adding a separate vector memory and vector processing unit [2]. Operations on vectors have the distinction of being performed in parallel. A standard RAM would iterate through each of the elements to calculate the same result.

The vector model expresses the efficiency of algorithms by two complexity measures. The *step* complexity measures the total number of vector instructions issued. The *element* complexity measures the total number of vector elements operated on over all instructions. The *step* and *element* complexity measures are related to the PRAM model through the following relation. For an algorithm with step complexity s and element complexity e, the PRAM complexity of the algorithm is

$$t = O(e/p + s)$$

for a p-processor PRAM. This result is due to a simulation argument in which the vector algorithm is simulated on a CRCW PRAM [2].

The result of each vector primitive is a new vector. Communication operations include **send** and **get**. A simple **send** does not allow collisions. Optional extensions to the strict parallel vector model supply combining-**send** operations, as described earlier.

scan operations compute all partial sums given a data vector. Variants include different combining functions such as **add-scan**, **max-scan** and **and-scan**. A vector may also be divided into segments. Segmented **scan** operations provide partial sums for each segment individually. Reduction operations are another class of vector primitive that provide the total "sum" of a vector with respect to an operator such as **add**, **max** or **and**.

The last class of vector operations are the elementwise primitives. These include such operations as addition, or vector assignment. Each of the primitive vector instructions is unit step in the parallel vector model, with an element complexity $e = O(n)$ for a vector of length n.

4.2 Matching by Sorting

The idea behind constructing a mapping through sorting is that equal keys should be brought near each other by a sorting operation. This type of mapping associates the groups specified by keys with a segment in a vector. The combining function appropriate for this type of group is the segmented-**scan**.

The steps involved in building a mapping are illustrated in Figure 3. First, the keys of the source and destination vector are concatenated into one vector called the "agent" along with pointers back to their sites of origin. These site addresses are called the "home" pointers for the keys. After sorting the agent elements by their key value, the agents are divided into segments so that each group of equal keys is in one segment. The identification of segments is done in one parallel step by comparing each key to its left neighbor in the sorted agent vector. Finally, the indices of the new sites in the agent vector are sent back to the home site of each key. The agent

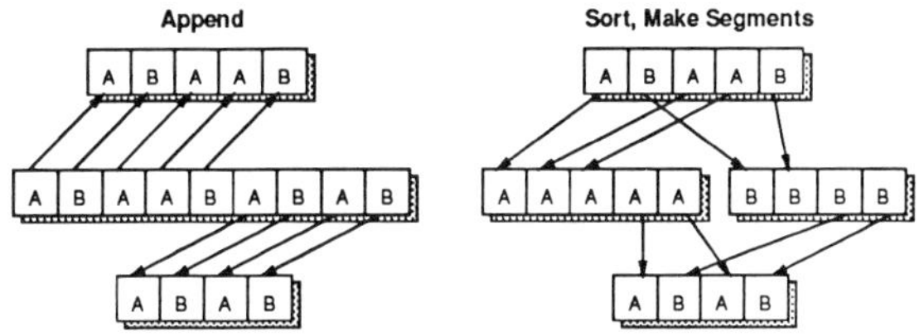

Figure 3: A mapping may be created by using sorting. In the first step, agents are appended from the source and destination vectors into a new vector. After sorting the agents, they are divided into groups by comparing each key to its neighbors.

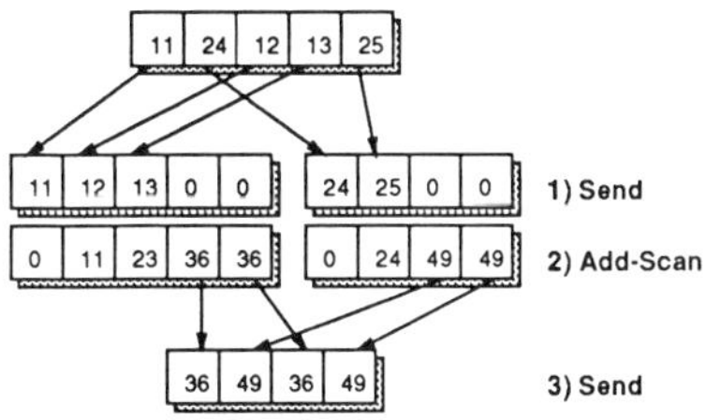

Figure 4: Using a mapping created by sorting to perform a **move-add**. The first step sends the source values to the agent vector, where they are combined with a segmented **add-scan**. The final **send** forwards the results to their destinations.

vector along with the pointers from keys to agent sites becomes the data structure of the mapping.

When creating a mapping with this technique, the complexity measures are dominated by the sort operation. For source and destination vectors of length S and D, respectively, the total number of elements involved is $n = S + D$. Thus, the complexity measures of this type of **match** are $s = O(\log n)$ and $e = O(n \log n)$.

A **move** operation using this type of mapping is illustrated in Figure 4. Three steps complete the **move**. First, the source values are sent to the agent. Next, a single **scan** combines the values of the source agents and distributes the sum across the destination elements. The last step sends the sums in the destination agents to their final sites

The operations involved in each of the three steps are unit step ($s = O(1)$) operations in the vector model. If the sum of the lengths of the source and destination vectors $n = S + D$, then the element complexity of **move** is $e = O(n)$.

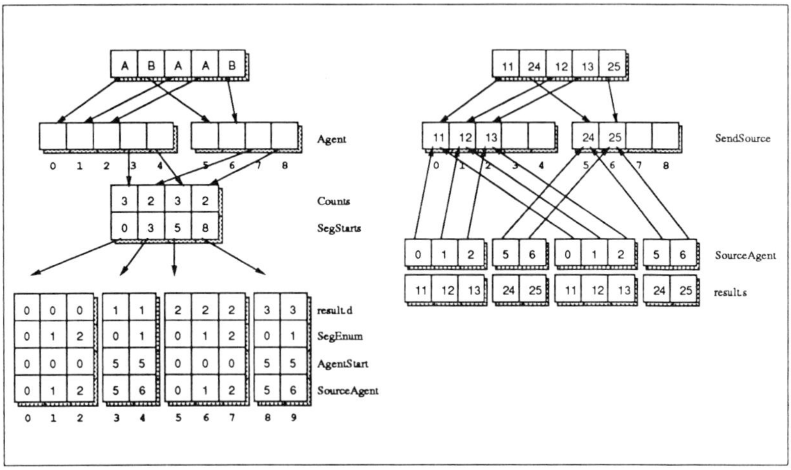

Figure 5: The implementation of **move-join** is the most complicated of the **move** operations. It is divided into two phases. The first part constructs the new segments and distributes data from each destination site over the segments. The second phase relates each site of the segments to a different source element in its group.

4.2.1 Implementing move-join

The **move-join** operation is more complicated than the other **move** operations, but uses the same agent vector to perform its work. Snapshots of the vectors involved in performing a **move-join** operation are illustrated in Figure 5. In the first step, each destination site computes the number of source sites that it matched. This vector, called *Counts*, is computed by using a **move-add** operation with a source vector of '1's.

Using vector *Counts* as the argument to an **add-scan**, the starting place in the result vector for each segment is calculated in vector *SegStarts*. Each value from the *destdata* vector is sent to the front of its segment using *SegStarts* and then copied across its segment using a segmented **copy-scan**. Assuming that an index vector was used for the *destdata* vector, the field "d" of the result is shown completed in the vector named *result.d* of Figure 5. Moving copies of data from the source sites is a more difficult problem however.

Each of the segments of the result vector are enumerated in vector *SegEnum* by using a segmented **enum** function. Next, the index of the first agent in each segment is sent to all destination sites in its group, from where it is then distributed across the segments allocated in the result vector. This value, *AgentStart*, marks the beginning of the agent segment for each group. The steps required to do this involve a few sends and scans.

By adding the *SegEnum* value of each result site to the value in *AgentStart*, each result site calculates the index of a source agent in the segment for its group. This vector is called *SourceAgent*. The join operation is completed by sending source values to the agent vector, and then using *SourceAgent* as the index for a **get** operation to copy the values to their final sites.

All operations involved are $s = O(1)$ unit step complexity operations in the vector model. However, the element complexity of this operation is $e = O(S \times D)$, where S and D are the lengths of the source and destination vectors.

4.3 Matching by Hashing

An alternative method of creating a mapping uses a hashing algorithm to associate each group with a single site in an agent vector. A **move** operation executed with this type of mapping uses a single combining-**send** to combine the values for each source, and a single **get** to copy all of the sums to the destination sites.

We will give only a brief outline of the hashing algorithm here because of space limitations. Our parallel hashing algorithm begins by creating a work vector for all of the keys that includes "home" pointers as before. In parallel, all keys compute the address of an agent site by using a hash function. Initially all agent sites are "empty," but become "filled" when they are claimed by a key. When keys find empty agent sites, they are marked "done" and the agent sites they have probed are marked filled. Other keys continue their probe steps until they find an empty agent site.

After each iteration, there are many sites in the work vector that are marked done, and hence, participate in no later iterations. A packing step compresses the remaining active keys into a smaller vector. This packing step has the effect of reducing the amount of work performed on each succeeding iteration, allowing the parallel hashing algorithm to have an expected step complexity of $s = O(\log n)$ and expected element complexity $e = O(n)$ [11].

Once the keys are assigned agent sites, these indices are sent back to the "home" site of each of the source and destination vectors. This completes the construction of this type of mapping. It is interesting to note that because the element complexity for the parallel hashing algorithm is smaller than that of parallel sorting, the hashing algorithm performs less work (fewer total operations) than the sorting algorithm for creating a mapping.

4.4 Comparison of Sorting and Hashing methods

The two different methods of implementation of **match** are each appropriate in different circumstances, most of which are governed by the context in which they are used in a program.

In some algorithms using **match**, such as the tree-isomorphism algorithm, the keys applied to **match** are pairs of integers. In other cases, they may be even longer bitstrings. Many implementations of parallel sorting are based on a radix sort. As the size of the keys increases, the speed of radix sorting decreases with each added bit. Conversely, the hashing algorithm is immune to increases in key length. For very long keys, the hashing algorithm can perform significantly better than sorting.

The mapping representation for a hash match is not well suited for the **move-join** operation and may not be used with many-to-one mappings if the target architecture does not provide combining-send operations. However, even on machines such as the CRAY Y-MP, the default **send** ensures that an arbitrary one of the values written to a location will be preserved. This is sufficient to provide a **move-arb** operation. For an algorithm such as tree-isomorphism, **move-arb** is all that is needed.

By implementing an algorithm in terms of **match**, a programmer can avoid these considerations until late in the program development cycle. After verifying an application correct, the choice of a particular implementation strategy for different instances of **match** and **move** could be the job of a compiler, or as simple as supplying optional "hints" that the **match** implementation can use at run time to make decisions about which of the available strategies to use. The latter approach reduces fine-tuning the performance of an application for a particular machine to a very simple process.

5 Implementation on the CRAY Y-MP

While the CRAY Y-MP supplies many regular vector operations, it is only with the addition of segmented scans that the full parallel vector model is supported [3]. We chose to implement the sparse matrix algorithms matrix-multiply and matrix-vector-multiplication because there were standard, optimized versions of these routines available in a FORTRAN library for comparison. Because the CRAY vector instructions do not support combining-**send** operations, the **match/move** operations all use the sort-mapping format.

The SPARSKIT library is a suite of test programs for manipulating sparse matrices. Its implementations of basic sparse matrix operations represent the standard, vectorized algorithms used by most linear algebra packages. We used SPARSKIT as a baseline from which to evaluate the performance of our approach. The routines we used from SPARSKIT include AMUB (sparse matrix multiply), AMUX (sparse matrix-vector multiplication), and MATRF2, a procedure that generates a random sparse matrix with a given density.

The representation format chosen for the **match** implementation shown earlier implies no ordering of the elements. The format used for the SPARSKIT routines were "compressed row format" (CSR), where matrix elements are ordered by rows. It is easy to write vectorized code for this format. However, for very sparse matrices, the rows are short and the vector lengths used by the program become very small.

Times measured for the comparisons included only the matrix operation, format conversions were not timed. The size indicated for each trial is the

Table 1: Execution times collected for the sparse matrix multiplication application expressed using **match** versus a traditional vectorized FORTRAN code from SPARSKIT. The size and density are a measure of the complexity of the **match** from Phase 1 of the algorithm, while the total number of partial products measures the complexity of the final **collapse**. The **match** implementation performed comparably for all trials. This is impressive considering the much higher level of abstraction presented by the **match** operations.

CRAY Matrix Multiplication				
		Time (S)		
N	ρ	SPARSKIT	MATCH	# PP
50	1.000	0.086	0.090	125K
100	0.400	0.105	0.101	169K
1000	0.010	0.074	0.074	102K
2000	0.005	0.148	0.155	202K
5000	0.001	0.110	0.116	127K
10000	0.001	0.724	0.753	1M

dimension of the square matrix generated, N, and its density, ρ, indicates a sparsity measure. A density of 1.0 is a full, dense matrix.

Results of the matrix-multiplication trials are shown in Table 1. For these trials, the matrix was multiplied by itself. Because the elements of the rows and columns of the matrix are distributed independently and each of the implementations must relate all rows of one matrix to the columns of the other, the use of the same matrix for each operand represents a fair test. This also simplified comparisons. The number of partial products generated is shown in the last column, and gives a measure of the total number of multiplications performed. In all trials the **match**-based implementation performed comparably to the SPARSKIT implementation.

Comparisons made between the implementations for sparse matrix-vector multiplication are shown in Table 2. The same matrices were used again. In these trials, the SPARSKIT implementation always out-performed the implementation using **match** for one application of the procedure. However, when repeatedly multiplying a matrix by a vector, the **match** implementation can excel.

The breakdown of the times spent in the **match** and **move** phases of the algorithm are shown. Because the time spent creating the mappings with **match** can be amortized over many applications of the **move** portion, we calculated the number of iterations required before the **match** implementation can "break even" in total time with the SPARSKIT implementation. In many real applications, AMUX is applied repeatedly to the same matrix with new vector values. This is the core of iterative system solvers such as the conjugate gradient method, where 20 or more iterations are not uncommon.

Table 2: Execution times collected for the sparse matrix-vector multiplication application using **match** versus the code from SPARSKIT. While the direct FORTRAN code outperformed the **match** implementation for a single iteration, the symbolic computation of **match** can be amortized over multiple instances of the multiplication. An application using the **match/move** approach would break-even with the SPARSKIT code after a small number of iterations. Most iterative system solvers would benefit directly as they generally require many more iterations than the break-even points shown.

CRAY Matrix-Vector Multiplication Times					
		Time (mS)			
N	ρ	SPARSKIT	MATCH	match/move	Break Even
50	1.000	0.138	1.498	1.22/0.28	
100	0.400	0.265	2.475	2.11/0.37	
1000	0.010	2.483	4.711	3.98/0.73	2.3
2000	0.005	4.958	8.247	6.93/1.31	1.9
5000	0.001	12.690	10.862	9.14/1.71	0.8
10000	0.001	25.418	48.167	42.26/5.90	2.2
20000	0.001	49.538	190.294	167.96/22.33	6.2

The uniformly low "break-even" values show that this approach could work well for sparse system solver applications.

The matrix-vector multiplication performed by AMUX in SPARSKIT is very straightforward. It's surprising that our implementation using **match** even gets close. We do because as the matrix grows sparse, the vector lengths exploited by AMUX become very short. On the other hand, the implementation of **match** makes use of very long vectors for all operations.

6 Implementation on the Connection Machine

The Connection Machine model CM-2 is an instance of a machine to which the parallel vector model applies directly. Once again we chose to implement our sparse matrix applications using **match** with a sort-based mapping. While there are published performance figures for dense matrix operations on the CM-2 [8], ours represent the first detailed results reported for sparse matrix multiplication.

The results collected from the matrix-multiplication application are shown in Table 3. All trials were run on a 16K processor Connection Machine with the same test matrices as before. The total times of the application are much higher than those for the CRAY implementation because of the great expense of communication operations on the CM-2. While our performance figures are not outstanding (especially compared to a dense matrix algorithm), the sparse matrix approach still has merit.

Table 3: Execution times for the sparse matrix multiplication application on the Connection Machine. The symbolic computations reflected in the "join and "collapse" columns dominate the execution time. The time per partial product in microseconds is given in the last column.

Connection Machine Matrix Multiplication							
		Time (S)					
N	ρ	Total	join/move	collapse/move	PP	$\rho^2 N^3$	$\frac{\mu S}{PP}$
50	1.000	0.850	.137/.235	.352/.126	125K	125K	6.8
100	0.400	1.302	.161/.260	.638/.141	169K	160K	7.7
500	0.020	0.379	.089/.089	.169/.030	52K	50K	7.3
1000	0.010	0.794	.163/.125	.442/.062	102K	100K	7.8
2000	0.005	1.645	.334/.258	.931/.120	202K	200K	8.1
5000	0.001	1.071	.293/.136	.562/.078	126K	125K	8.5
10000	0.001	7.721	1.54/1.20	4.460/.504	1002K	1000K	7.7

As illustrated in the last column, the time of our algorithm is roughly proportional to the number of partial products generated. With a randomly generated sparse matrix, the expected total number of partial products equals $\rho^2 N^3$. This estimation is very close to the actual number of partial products generated in our trials. By estimating the performance of our algorithm as a function of the number of partial products, the following equation predicts the total time.

$$t_{\text{sparse}} = K_{\text{sparse}} \rho^2 N^3$$

Because the dense matrix algorithm must compute all N^3 partial products and sum them, its total time is proportional to N^3.

$$t_{\text{dense}} = K_{\text{dense}} N^3$$

In a comparison with the results reported in [8] for trials run on a 16K Connection Machine, we have found the ratio of the constants to be $\frac{K_{\text{sparse}}}{K_{\text{dense}}} < 2500$ over a wide range of N. While this ratio may seem quite large, it indicates that for very large matrices, the sparse approach excels when

$$\rho \leq \frac{1}{\sqrt{2500}} \leq 0.02.$$

Many applications involving finite element grids and electrical circuit simulation operate on matrices of much smaller density.

Another important observation can be made from the "join" and "collapse" columns. These times include the time of the symbolic part of the join and collapse operation for PHASE 1 and PHASE 2 of the algorithm. As before, these times could be amortized over multiple invocations of the matrix multiplication algorithm applied to matrices with the same structure.

Table 4: Execution times for the sparse matrix-vector multiplication application on the Connection Machine. An enormous time savings is made by recognizing that the two **match** operations involve an index vector, and that a sort is not required.

Connection Machine Matrix-Vector Multiplication					
		Regular Match Time (S)		Optimized Match Time (S)	
N	ρ	Total	match/move	Total	match/move
50	1.000	0.227	.207/.019	0.026	.000/.026
100	0.400	0.222	.197/.025	0.029	.000/.029
500	0.020	0.223	.206/.017	0.019	.000/.019
1000	0.010	0.121	.103/.018	0.020	.000/.020
2000	0.005	0.379	.254/.126	0.044	.000/.043
5000	0.001	0.330	.305/.025	0.037	.000/.037
10000	0.001	0.990	.934/.056	0.091	.001/.090

Results collected for the matrix-vector multiplication application are presented in Table 4. Once again, the same matrices were used. We compare the straightforward **match** implementation to a version that employs a simple optimization: when either the source or destination of a **match** is an index vector, the **match** indicates a simple **send** or **get**. This simplc optimization saves a great deal of time in all cases, and is transparent to a program using **match** and **move**.

7 Conclusions

This paper presented the match operation as a parallel primitive for expressing relationships between data. Because **match** and its **move** operations operate at a very high level, the implementor of these functions has a wide latitude for tuning their performance to a particular architecture.

We have shown that not only are **match** and **move** powerful enough to express symbolic computations, but that this symbolic approach to scientific computing is competitive with more traditional vectorized FORTRAN code. We argue that algorithms expressed at this level are clear, easy to understand, and architecture independent. The **match/move** operations offer an appropriate layer of insulation from architecture specific considerations for program development, and reduce fine-tuning to supplying a set of "hints" that the **match** implementation can make use of when appropriate.

Applications written at this high level benefit transparently from better implementations of the **match** primitives. We expect to implement multiprocessor versions of the operations on the CRAY Y-MP and to use a better sort operation on the CM. Our applications written using **match** on both machines will boast improved performance with no change of high-level code.

Acknowledgments

This research was supported by the Defense Advanced Research Project Agency, ARPA Order 7597. Connection Machine time was provided by the Northeast Parallel Architectures Center, and by the Pittsburgh Supercomputing Center, grant number ESC910004P.

References

[1] A. V. Aho, J. E. Hopcroft, and J. D. Ullman. *The Design and Analysis of Computer Algorithms.* Addison-Wesley series in computer science and information processing. Addison-Wesley Publishing Co., 1974.

[2] G. E. Blelloch. *Vector models for data-parallel computing.* MIT Press, 1990.

[3] S. Chatterjee, G. E. Blelloch, and M. Zagha. Scan primitives for vector computers. In *Proceedings Supercomputing '90*, November 1990.

[4] E. Codd. A relational model of data for large shared data banks. *CACM*, 13:377–387, June 1970.

[5] J. Edler, A. Gottlieb, C. P. Kruskal, K. P. McAuliffe, and L. Rudolph. Issues related to MIMD shared-memory computers: the NYU Ultra-Computer approach. In *12th International Symposium on Computer Architecture*, pages 126–135, June 1985.

[6] A. Gottlieb, B. Lubachevsky, and L. Rudolph. Coordinating large numbers of processors. In *Proc. 1981 International Conference on Parallel Processing*, 1981.

[7] W. D. Hillis. *The Connection Machine.* The MIT Press series in artificial intelligence. MIT Press, Cambridge, Mass., 1985.

[8] S. L. Johnsson, T. Harris, and K. K. Mathur. Matrix multiplication on the Connection Machine. In *Proceedings of Supercomputing '89*, pages 326–332, 1989.

[9] A. G. Ranade, S. N. Bhatt, and S. L. Johnsson. The Fluent abstract machine. In J. Allen and F. T. Leighton, editors, *Advanced Research in VLSI, Proceedings of the Fifth MIT Conference*, pages 71–93, 1988.

[10] G. W. Sabot. *The Paralation model: architecture-independent parallel programming.* Series in Artificial Intelligence. MIT Press, 1988.

[11] T. J. Sheffler and R. E. Bryant. A work efficient solution to the naming problem using parallel hashing. Submitted to the Sixth International Parallel Processing Symposium (IPPS '92), 1992.

Scattering and Gathering Messages in Networks of Processors

Sandeep N. Bhatt
Department of Computer Science
Yale University
New Haven, CT 06520, USA

Geppino Pucci
Dipartimento di Informatica
Università di Pisa
56100 Pisa, ITALY

Abhiram Ranade
Computer Science Division
University of California
Berkeley, CA 94720, USA

Arnold L. Rosenberg
Department of Computer Science
University of Massachusetts
Amherst, MA 01003, USA

Abstract

The operations of scattering and gathering in a network of processors involve one processor of the network—call it $\mathcal{P}_0$—communicating with all other processors. In scattering, $\mathcal{P}_0$ sends (possibly) distinct messages to all other processors; in gathering, the other processors send (possibly) distinct messages to $\mathcal{P}_0$. We present algorithms for scattering and gathering messages in processor-networks based on interconnection networks of arbitrary topologies. The algorithms are:

- quite general, in that the messages transmitted can differ arbitrarily in length;
- quite strong, in that they send messages along noncolliding paths, hence do not require any buffering or queueing mechanisms in the processors;
- quite efficient: the algorithms for scattering in general networks are optimal, the algorithm for gathering in a ring is optimal, and the algorithms for gathering in general networks are nearly optimal.

1 Introduction

1.1 Communication in Parallel Computation

Communication is an essential component of parallel computation. Several modes of communication have been studied within the framework of *networks of processors*—identical processing elements (PEs) that communicate by means of an interconnection network. Most notable among these are: *(partial) permutation routing*, *broadcasting*, and *gossiping* [1, 3, 5, 6, 7, 8, 9, 10, 11]. Several of these sources mention the following additional algorithmically useful forms of communication.

- In *scattering* (or, *one-to-all personalized communication*), one PE sends (possibly) distinct messages to all other PEs.
- In *gathering*, all PEs send (possibly) distinct messages to one specific PE.

In this paper, we study the most general version of the scattering and gathering operations. We present optimal algorithms for scattering in general networks and gathering in a ring, and nearly optimal algorithms for gathering in general networks. Our study is general in three senses.

1. We allow messages to differ in length by arbitrary amounts; indeed, some messages may be null. (This contrasts with [1, 4, 6, 8], wherein all messages have the same length.)
2. We consider the operations of scattering and gathering in networks of arbitrary topologies. (This contrasts with [1, 4, 6, 8, 9], which focus on a small repertoire of familiar networks.)
3. We transmit messages along noncolliding paths in our networks, hence do not require any buffering or queueing mechanisms in the PEs. (This contrasts with virtually all other studies of message transmission in networks.)

One might rationalize our demand for unbuffered communication in terms of resource conservation: buffering requires both additional memory (each PE must be prepared to store the longest message in the system) and time (e.g., for processing addresses). However, our overriding motivation is to understand communication in networks better, by determining the cost of this strict assumption in terms of the complexity of the problems of scattering and gathering general messages in general networks.

1.2 The Computing Model

A. Networks of Processors

We focus on a *synchronous* network $\mathcal{A}$ of arbitrary topology, comprising $n+1$ identical PEs, $\mathcal{P}_0, \mathcal{P}_1, \ldots, \mathcal{P}_n$. We represent the interconnection network underlying $\mathcal{A}$ as an undirected graph whose nodes are PEs and whose edges are inter-PE communication links. (Henceforth, we use "network" to refer ambiguously to $\mathcal{A}$ and its underlying interconnection network.)

The PEs of $\mathcal{A}$ have neither message buffers nor queues. Messages must, therefore, be scheduled so as never to "collide" with one another; this scheduling is a major challenge, particularly with the gathering operation.

The networks we study use *single-port* communication: a PE can send and receive information from only *one* of its immediate neighbors during each communication step. We do, however, allow a PE to perform computations while communicating. This regimen contrasts with *multiport* communication, in which a PE can send and receive information from *each* of its immediate neighbors in one step. In the full paper, we indicate how our algorithms can be adapted to a multiport regimen.

The networks we study communicate *in rounds:* in each scattering operation, precisely one PE transmits messages to all other PEs, and in each gathering operation, precisely one PE receives messages from all other PEs. This restriction contrasts with the study in [2], wherein bufferless PEs pass messages in arbitrary ways (but only in a small repertoire of topologies).

With no loss of generality, we always let $\mathcal{P}_0$ be the "singular" PE in $\mathcal{A}$, i.e., the source of messages in the scattering operation and the destination of messages in the gathering operation.

B. Messages and Message Sequences

Each transmitted message $\mathcal{M}_i$ is a sequence of some number $L_i \geq 0$ of atomic flits: a *flit* is the largest unit of information that $\mathcal{A}$ can transmit between adjacent nodes in one communication step (i.e., in one so-called transit time). A message is treated as an indivisible unit during transmission, since addressing information appears only in the first flit (thereby lessening both the setup time for messages and the aggregate length devoted to addressing information).

Throughout this paper, we focus on a given network $\mathcal{A}$ and a given sequence $\mathcal{M} = \langle \mathcal{M}_1, \mathcal{M}_2, \ldots, \mathcal{M}_n \rangle$ of messages to be scattered or gathered in $\mathcal{A}$, where message $\mathcal{M}_i$ has length $L_i \geq 0$ and where

$$\mathbf{N}(\mathcal{M}) = \langle i_1, i_2, \ldots, i_k \rangle$$

denotes, in increasing order, the subsequence of message indices whose message are nonnull, i.e., for which $L_{i_j} > 0$.

C. The Scattering and Gathering Problems

In a scattering operation, $\mathcal{P}_0$ has a message $\mathcal{M}_i$ of length $L_i \geq 0$ for each $\mathcal{P}_i$, $i > 0$. In a gathering operation, each $\mathcal{P}_i$, $i > 0$, has a message $\mathcal{M}_i$ of length $L_i \geq 0$ for $\mathcal{P}_0$. We perform these operations in networks of arbitrary topologies, subject to the following constraints.

- Once a message has been dispatched by its originating PE, it encounters no interruption until it is received by its destination PE. In particular,

- intermediate PEs do not queue or buffer messages;
- messages are treated as indivisible units.

- For each i, message $\mathcal{M}_i$ is routed between $\mathcal{P}_0$ and $\mathcal{P}_i$ along a preassigned routing path ρ_i, of length $\delta(i)$.

Most conveniently, the paths $\{\rho_i\}$ will be derived from an "efficient" spanning tree of network $\mathcal{A}$, rooted at $\mathcal{P}_0$. The notion "efficient" depends both on the operation (scattering or gathering) and on the communication regimen (single-port or multiport). This issue is discussed in the full paper.

D. Problem Complexity

We measure the complexity of a (scattering or gathering) operation in terms of the time for delivering all relevant messages.

Scattering Time. A *schedule for scattering message sequence* $\mathcal{M}$ is a permutation σ of the index-sequence $\mathbf{N}(\mathcal{M})$, i.e., a one-to-one function

$$\sigma : \{1, 2, \dots, k\} \longrightarrow \mathbf{N}(\mathcal{M}).$$

The intended scenario is that $\mathcal{P}_0$ sends out message $\mathcal{M}_{\sigma(1)}$, then message $\mathcal{M}_{\sigma(2)}$, and so on, in that order, with no gaps. Thus, under schedule σ, if $L_i \neq 0$, then $\mathcal{P}_0$ begins transmitting message $\mathcal{M}_i$ at *dispatch time*

$$\tau_\sigma(i) = \sum_{\{j | \sigma^{-1}(j) < \sigma^{-1}(i)\}} L_j. \tag{1}$$

Message $\mathcal{M}_i$ arrives at its destination $\mathcal{P}_i$ at *arrival time*

$$\alpha_\sigma(i) = \tau_\sigma(i) + L_i + \delta(i). \tag{2}$$

The *time for scattering message sequence* $\mathcal{M}$ under scattering-schedule σ is the time it takes for every flit of $\mathcal{M}$ to reach its final destination; symbolically,

$$T^{\text{scat}}(\sigma; \mathcal{M}) = \max_{i \in \mathbf{N}(\mathcal{M})} \{\alpha_\sigma(i)\}. \tag{3}$$

Proposition 1.1 *For any scattering-schedule σ for message sequence $\mathcal{M}$,*

$$\max\left(\sum_{i \in \mathbf{N}(\mathcal{M})} L_i, \max_{i \in \mathbf{N}(\mathcal{M})} \{\delta(i)\}\right) \leq T^{\text{scat}}(\sigma; \mathcal{M}) \leq \sum_{i \in \mathbf{N}(\mathcal{M})} L_i + \max_{i \in \mathbf{N}(\mathcal{M})} \{\delta(i)\}.$$

Gathering Time. A *schedule for gathering message sequence* $\mathcal{M}$ is a sequence γ of integers; letting $\mathbf{N}(\mathcal{M}) = \{i_1, i_2, \dots, i_k\}$,

$$\gamma = \tau_\gamma(i_1), \tau_\gamma(i_2), \dots, \tau_\gamma(i_k),$$

The intended scenario is that each $\tau_\gamma(i)$ is the *dispatch time* for $\mathcal{M}_i$, i.e., the time when $\mathcal{P}_i$ begins transmitting $\mathcal{M}_i$. The last flit of $\mathcal{M}_i$ arrives at $\mathcal{P}_0$ at *arrival time*

$$\alpha_\gamma(i) = \tau_\gamma(i) + L_i + \delta(i).$$

The *time for gathering message sequence* $\mathcal{M}$ under gathering-schedule γ is the time it takes for every flit of $\mathcal{M}$ to reach $\mathcal{P}_0$; symbolically,

$$T^{\text{gath}}(\gamma; \mathcal{M}) = \max_{i \in \mathbf{N}(\mathcal{M})} \{\alpha_\gamma(i)\}.$$

The Challenges. Neither T^{scat} nor T^{gath} allows for any delay of messages at intermediate nodes. Hence, our message-scheduling algorithms cannot rely on—so the network need not provide—any mechanism for buffering or queuing messages in PEs. This lack of buffering provides a major challenge in scheduling the gathering operation: the scheduling algorithm must—in a distributed manner— have PEs dispatch messages in $\mathcal{A}$ so that they never collide on their paths to $\mathcal{P}_0$.

2 Scattering on Networks of Processors

Scattering-schedule σ is *optimal* for message sequence $\mathcal{M}$ if
$T^{\text{scat}}(\sigma; \mathcal{M}) = \min \{T^{\text{scat}}(\sigma'; \mathcal{M}) \mid \sigma' \text{ is a scattering-schedule for } \mathcal{M}\}$.

It is shown in [4] that the *unique* optimal scattering-schedule for equal-length messages on a unidirectional ring is given by the permutation $\sigma(i) = n - i + 1$, i.e., by sending out messages according to a *farthest-destination-first* (*FDF*) schedule. We now prove that the optimality of FDF schedules persists when the lengths of messages and the topology of the network are general. Specifically, we show that every FDF scattering-schedule is optimal. As one would expect, there may be optimal non-FDF schedules also: a single enormous message could dominate the message transmission time, masking the order of a collection of small messages sent out right after it.

Theorem 2.1 *Every FDF scattering-schedule is optimal.*

The following two Lemmas prove the Theorem.

Lemma 2.2 *For every scattering-schedule σ for $\mathcal{M}$, there is an FDF scattering-schedule σ' for $\mathcal{M}$ with*

$$T^{\text{scat}}(\sigma'; \mathcal{M}) \leq T^{\text{scat}}(\sigma; \mathcal{M}).$$

Moral. *Every message sequence has an optimal FDF scattering-schedule.*

Proof Sketch: Assume that there is an $\mathcal{M}$ that admits no optimal FDF scattering-schedule. Let σ_1 be any optimal (hence, non-FDF) scattering-schedule σ_1 for $\mathcal{M}$. There must exist $i, j \in \mathbf{N}(\mathcal{M})$ such that:

$$\sigma_1^{-1}(i) = \sigma_1^{-1}(j) + 1; \ \delta(i) > \delta(j); \ L_j > 0.$$

Create the scattering-schedule σ_2 from σ_1 by interchanging $\sigma_1^{-1}(i)$ and $\sigma_1^{-1}(j)$:

$$\sigma_2(k) = \begin{cases} i & \text{if } \sigma_1(k) = j \\ j & \text{if } \sigma_1(k) = i \\ \sigma_1(k) & \text{otherwise} . \end{cases}$$

One can show that $\alpha_{\sigma_2}(k) = \alpha_{\sigma_1}(k)$ for all $k \notin \{i, j\}$, while

$$\max\{\alpha_{\sigma_1}(i), \alpha_{\sigma_1}(j)\} \geq \max\{\alpha_{\sigma_2}(i), \alpha_{\sigma_2}(j)\},$$

whence $T^{\text{scat}}(\sigma_2; \mathcal{M}) \leq T^{\text{scat}}(\sigma_1; \mathcal{M})$. If σ_2 observes the FDF regimen, then we are done; otherwise, we iterate the argument. □-Lemma 2.2

Lemma 2.3 *All FDF scattering-schedules take the same time.*

Moral. *Every scattering-schedule that observes the FDF regimen is optimal.*

Proof Idea: Rearranging the transmission order of messages destined for equidistant PEs does not alter the time for a schedule. □-Lemma 2.3

When $\mathcal{A}$ is a ring, one can visualize Theorem 2.1 easily (cf. Fig. 1): each $\mathcal{M}_i$ sweeps out a parallelogram in the space-time domain. (The parallelogram associated with $\mathcal{M}_i$ has length-L_i sides parallel to the time axis, corresponding to the path traversed by $\mathcal{M}_i$'s flits, and length-i sides at a 45-degree angle, corresponding to the flits' progress along the line of PEs.) Using such parallelograms, one sees that a portion of the upper slanted side of $\mathcal{M}_i$'s parallelogram can "hide in the shadow" of the parallelogram of an earlier dispatched $\mathcal{M}_j$ (via pipelining of the intermediate PEs). Analogs of the competing dispatch orders of Figs. 1(a,b) illustrate what Theorem 2.1 verifies, namely, that the parallelogram of a message destined for a more distant PE can "hide" more of the parallelogram of a message destined for a nearby PE than vice versa.

3 Gathering on Networks of Processors

As noted earlier, the gathering operation requires a *distributed* protocol that schedules the dispatch times of messages so that they never collide in transit. A straightforward synchronization-like protocol will accomplish this scheduling. We first present a simple version of this protocol that implements the gathering operation on a ring by interlacing the synchronization and scheduling activities. This protocol yields an optimal gathering algorithm for a ring but is too simple to work on general networks. Therefore, we decouple the synchronization and scheduling activities to get a protocol that takes more time but is adaptable to general network topologies.

3.1 Gathering on a Ring of Processors

We present two algorithms. The first attempts to gather by running an FDF scattering algorithm "backwards;" by reasoning analogous to that of Theorem 2.1, this will lead to an optimal algorithm. Of course, one can not literally achieve this goal, because in scattering, PEs other than $\mathcal{P}_0$ are passive, while in gathering, they are active—they must initiate message transmissions. Our first algorithm uses "shoulder-tapping" to "wake up" PEs according to a schedule that allows them to transmit their messages with no collisions. "Shoulder-tapping" seems not to extend to multiple-successor networks, so we develop a second algorithm that modifies the "backward FDF" strategy by decoupling the synchronization of PEs from the transmission of messages, thereby trading time for adaptability to general network topologies.

A. Gathering via Shoulder Tapping

Our simple algorithm exploits the single-successor structure of the ring in an essential way, thereby precluding its extension to multiple-successor networks. The algorithm seeks a gathering-schedule for $\mathcal{M}$,

$$\gamma = \tau_\gamma(i_1), \tau_\gamma(i_2), \ldots, \tau_\gamma(i_k)$$

($i_j \in \mathbf{N}(\mathcal{M})$) which minimizes each dispatch time $\tau_\gamma(i_j)$ subject to the requirement that messages never collide, and subject to the inequalities

$$\tau_\gamma(i) \geq \begin{cases} i+2 & \text{if } i < n \\ i+1 & \text{if } i = n, \end{cases} \tag{4}$$

which reflect the following facts which hold for *all* PE indices $\{1, 2, \ldots, n\}$.

- Each PE $\mathcal{P}_i$ (save, of course, $\mathcal{P}_0$) awaits a *wakeup call* telling it when to begin transmitting its message $\mathcal{M}_i$ (assuming that $L_i > 0$).
- Each wakeup call is initiated by $\mathcal{P}_0$, hence must take at least i steps to reach $\mathcal{P}_i$.
- The single-port regimen does not allow $\mathcal{P}_i$ to overlap sending its message and relaying a wakeup call to $\mathcal{P}_{i+1}$.

Each $\mathcal{P}_i$ remains dormant until it receives its wakeup call from $\mathcal{P}_{i-1}$; the call is a (single-flit) message:

TRANSMIT AFTER $\geq s_i$ STEPS

where s_i is a positive integer. When $\mathcal{P}_i$ receives its wakeup call at time t_i, it responds by serially entering the following operational phases:

Algorithm **Shoulder-Tap**:

Phase 1: For all $i < n$: At time $t_i + 1$, $\mathcal{P}_i$ transmits to $\mathcal{P}_{i+1}$ a wakeup call

TRANSMIT AFTER $\geq s_{i+1}$ STEPS

where the integer $s_{i+1} > 0$ is computed using the following time-line. (Note the effect of the single-port regimen.)

time t_i**:** $\mathcal{P}_i$ receives its wakeup call

time $t_i + 1$**:** $\mathcal{P}_{i+1}$ receives its wakeup call from $\mathcal{P}_i$

time $t_i + \max(2, s_i)$: $\mathcal{P}_{i-1}$ receives the first flit of $\mathcal{M}_i$ from $\mathcal{P}_i$ (assuming that $L_i > 0$)

time $t_i + L_i + \max(1, s_i - 1)$: $\mathcal{P}_{i-1}$ receives the last (L_ith) flit of $\mathcal{M}_i$ from $\mathcal{P}_i$; $\mathcal{P}_i$ begins relaying messages it receives from PEs $\mathcal{P}_j$ with $j > i$.

Since $\mathcal{P}_{i+1}$ receives its wakeup call at time $t_i + 1$, $\mathcal{P}_i$ sets s_{i+1} to the *maximum* of 1 and $L_i + \max(0, s_i - 2)$.

Phase 2: For all $i \in \mathrm{N}(\mathcal{M})$: from time $t_i + \max(2, s_i)$ to time $t_i + L_i + \max(1, s_i - 1)$, $\mathcal{P}_i$ transmits $\mathcal{M}_i$ to $\mathcal{P}_0$, via $\mathcal{P}_{i-1}$, one flit at a time.

Phase 3: From time $t_i + L_i + \max(1, s_i - 1)$ on, $\mathcal{P}_i$ begins passively transmitting messages it receives from PEs $\mathcal{P}_j$ for $j > i$.

Fig. 2 depicts a "typical" instance of Algorithm **Shoulder-Tap**.

Theorem 3.1 *Algorithm* **Shoulder-Tap** *is an optimal algorithm for gathering on a ring.*

Proof Sketch: When all messages in $\mathcal{M}$ are nonnull, the algorithm delivers the messages in a *gap-free* fashion, hence is optimal. When $\mathcal{M}$ contains null messages, the only gaps are those necessary to deliver the wakeup calls to PEs with higher indices than the PE whose message is null. □

Interlacing synchronization with message passing in multiple-successor networks seems to be rather complicated: messages must have end-of-message delimiters so that each $\mathcal{P}_i$ can coordinate the message streams of its descendants. We turn now to a simpler alternative avenue toward coordination.

B. Gathering via Transmission Certificates

The algorithm we present now decouples the synchronization and message passing activities of Algorithm **Shoulder-Tap**. In its general version, this algorithm gathers within a spanning tree of $\mathcal{A}$. In preparation, we describe the algorithm using the obvious spanning tree of the ring, namely, the path having $\mathcal{P}_0$ as root, $\mathcal{P}_n$ as leaf, and $\mathcal{P}_{i+1}$ as the child of $\mathcal{P}_i$.

The algorithm operates in four phases. In *phase 1*, $\mathcal{P}_0$ "awakens" all PEs by broadcasting a *synchronization token*. In *phase 2*, PEs pass *certificates* to their parents, indicating the earliest time that they can start transmitting their descendants' messages *in a gap-free stream*. In *phase 3*, each PE, starting with $\mathcal{P}_0$, schedules the transmissions of its descendants so that it can relay their messages in a gap-free stream. In *phase 4*, the PEs follow the schedule of phase 3, transmitting messages in a gap-free stream. The gap-free transmission guarantees that this algorithm is optimal, up to the time required for the synchronization and scheduling, each of which takes time $O(n)$. We now describe phases 2 and 3 in detail.

Transmission Certificates. During phase 2, each $\mathcal{P}_i$ ($i > 0$) sends its parent a *certificate of lag time* that announces the smallest number of steps $\mathcal{P}_i$ must wait, after receiving a TRANSMIT-MESSAGE order, before it can start to transmit all messages $\mathcal{M}_j$ with $j \geq i$ *with no gaps*. $\mathcal{P}_n$, of course, needs no lag time; each other $\mathcal{P}_i$ inductively generates its certificate from that of $\mathcal{P}_{i+1}$. $\mathcal{P}_0$ then uses the certificate of $\mathcal{P}_1$ to schedule all the transmissions, using TRANSMIT-MESSAGE orders that mimic the shoulder-taps of Algorithm **Shoulder-Tap**. The transmission schedule produced by the current algorithm avoids gaps in message transmission like that in Step 8 of Fig. 2. We avoid gaps at the cost of a two-pass scheduling protocol.

Computing Transmission Certificates. Let c_i denote the transmission certificate generated by $\mathcal{P}_i$. Since $\mathcal{P}_n$ needs no lag time, it can start transmitting with no gaps one step after receiving the TRANSMIT-MESSAGE order; hence, it transmits the certificate $c_n = 1$ to $\mathcal{P}_{n-1}$. Each other $\mathcal{P}_i$ must consider how its message interacts with the message stream that will come from $\mathcal{P}_{i+1}$; it therefore computes c_i from c_{i+1}, using the following time-line. Say that $\mathcal{P}_i$ will receive the order

TRANSMIT-MESSAGE IN s_i STEPS

at time t (in the future). The following actions will ensue.

time $t+1$**:** $\mathcal{P}_i$ relays the order to $\mathcal{P}_{i+1}$, with an appropriately modified s_i.

time $t+s_i...t+s_i+L_i-1$**:** $\mathcal{P}_i$ transmits $\mathcal{M}_i$. Note that $s_i \geq 2$, because of the single-port regimen.

time $t+s_i+L_i$ **on:** $\mathcal{P}_i$ begins to transmit the stream $\mathcal{M}_{i+1}, \mathcal{M}_{i+2}, \ldots, \mathcal{M}_n$, without gaps. Note that $s_i \geq c_{i+1}$, because $\mathcal{P}_{i+1}$ begins its gap-free transmission of the stream one step before $\mathcal{P}_i$ can.

One can visualize $\mathcal{P}_i$'s computation of c_i geometrically: Consider the positive X-axis (representing time), with a line segment X of length L_i (representing $\mathcal{M}_i$) and a line segment Y of length $\sum_{j=i+1}^{n} L_j$ (representing the message stream $\mathcal{M}_{i+1}, \mathcal{M}_{i+2}, \ldots, \mathcal{M}_n$). $\mathcal{P}_i$ computes c_i by shifting X and Y to form a single line segment Z of length $\sum_{j=i}^{n} L_j$, subject to the following rules.

- The left endpoint of X can be placed at any point $p \geq t+2$; the left endpoint of Y can be placed at any point $p' \geq t+2+c_{i+1}$.
- By shifting either X or Y to the right, but neither to the left, $\mathcal{P}_i$ places the left endpoint of Z as far to the left as possible.

The final position of the left endpoint of Z is $\mathcal{P}_i$'s certificate c_i:

- If $c_{i+1} = L_i$, then both X and Y can be placed in their leftmost legal positions; accordingly, $c_i = 2$.
- If $c_{i+1} < L_i$, then X can be placed in its leftmost possible position, but Y must be shifted right $L_i - c_{i+1}$ positions,accordingly, $c_i = 2$ (but C_{i+1} is "revised" to $c'_{i+1} = L_i$).

- If $c_{i+1} > L_i$, then Y can be placed in its leftmost possible position, but X must be shifted right $c_{i+1} - L_i$ positions; accordingly, $c_i = c_{i+1} - L_i + 2$.

In summary, $\mathcal{P}_i$ transmits the certificate

$$c_i = \max(2, c_{i+1} - L_i + 2).$$

Because message streams (line segments) can be delayed (moved right) but never advanced (moved left), a single pass suffices to compute all certificates; another single pass suffices to schedule dispatch times.

Message Scheduling. One step after $\mathcal{P}_0$ receives certificate c_1 from $\mathcal{P}_1$, it transmits the order TRANSMIT AFTER $s_1 = c_1$ STEPS to $\mathcal{P}_1$. In general, when $\mathcal{P}_i$, $1 \leq i < n$, receives the order TRANSMIT AFTER s_i STEPS from $\mathcal{P}_{i-1}$ at time t_i, it computes the time s_{i+1} that it will include in the TRANSMIT-MESSAGE order to $\mathcal{P}_{i+1}$. It then transmits the order TRANSMIT AFTER s_{i+1} STEPS to $\mathcal{P}_{i+1}$ at time $t_{i+1} = t_i + 1$. s_{i+1} is computed as follows.

Computing Transmission Deadlines. $\mathcal{P}_i$ computes s_{i+1} from s_i and t_i, via the following reasoning. If $\mathcal{P}_i$ transmits the first flit of $\mathcal{M}_i$ at time $t_i + s_i$, then it transmits the last flit at time $t = t_i + s_i + L_i - 1$. Hence, it must receive the first flit of the stream $\mathcal{M}_{i+1}, \ldots, \mathcal{M}_n$ at time t if it is to transmit with no gaps. Therefore, $\mathcal{P}_{i+1}$ must start transmitting its stream at time $t_i + s_i + L_i - 1 = t_{i+1} + s_i + L_i - 2$. Accordingly, $\mathcal{P}_i$ sends $\mathcal{P}_{i+1}$ the order

TRANSMIT AFTER $s_i + L_i - 2$ STEPS.

Figs. 2 and 3 contrast shoulder-taps and transmission certificates.

Although some of the roughly $2n$ steps taken by explicit synchronization is recovered by the absence of gaps in the message stream, shoulder-taps are definitely faster than transmission certificates when gathering on a ring. However, the synchronize-certify-transmit regimen underlying gathering via transmission certificates generalizes gracefully to networks of arbitrary topologies, while shoulder-taps do not.

3.2 Gathering on a General Network

The inter-PE communication in the preceding algorithm is simplified by the fact that each PE in a ring has only one child. Since most popular networks are multi-successor networks, concern for efficiency mandates that we adapt the algorithm to run on a *spanning tree* of $\mathcal{A}$, rooted at $\mathcal{P}_0$. (This is not clear *a priori*; it follows from the fact that, even with single-port communication, a tailored synchronization protocol takes no more time than does message transmission—at least when all messages are nonnull.)

Modifying the algorithm to work efficiently on a multi-successor network demands substantive clerical modifications, many of which result from the single-port regimen. We now describe each necessary modification, keeping in mind that the basic four-phase structure of the algorithm, as described in Section 3.1.B, persists.

Assume henceforth that $\mathcal{P}_i$ has d_i children in the spanning tree, denoted $\mathcal{P}_{i1}, \mathcal{P}_{i2}, \ldots, \mathcal{P}_{id_i}$ in some arbitrary but fixed order.

Broadcasting and Receiving Messages. The single-port regimen forces us to orchestrate communication in $\mathcal{A}$. When $\mathcal{P}_i$ receives a synchronization token from its parent, it relays the token in turn to its d_i children. In order to collect transmission certificates from its children, $\mathcal{P}_i$ then enters a polling mode: it sends the order SEND-CERTIFICATE on successive steps to $\mathcal{P}_{i1}$, until it receives $\mathcal{P}_{i1}$'s transmission certificate; it then repeats this process in turn with $\mathcal{P}_{i2}$, $\mathcal{P}_{i3}$, and so on, until it has collected all d_i certificates. ($\mathcal{P}_i$ must "remember" which certificate came from which child.)

The Certificates. When $\mathcal{P}_i$ has more than one child (i.e., $d_i > 1$), it must know how long each child's message stream is, as well as when the stream can begin coming. Therefore, we must augment transmission certificates, to indicate the lengths of message streams; the certificate from $\mathcal{P}_i$ will now have the form (c_i, n_i), meaning

> c_i steps after receiving the TRANSMIT-MESSAGE order, $\mathcal{P}_i$ can start transmitting a gap-free stream of n_i flits, comprising all the messages originating at PEs in the sub-spanning tree rooted at $\mathcal{P}_i$.

Computing Certificates. Say that $\mathcal{P}_i$ has received the certificates

$$(c_{i1}, n_{i1}), (c_{i2}, n_{i2}), \ldots, (c_{id_i}, n_{id_i})$$

from its d_i children. It uses these certificates, plus the length L_i of its message, to compute its certificate (c_i, n_i), as follows.

Determining the stream-length n_i is straightforward, since $\mathcal{P}_i$ will just concatenate its message with the message streams of its children; hence,

$$n_i = L_i + \sum_{j=1}^{d_i} n_{ij}.$$

Computing the certified lag time c_i for a general network is a direct extrapolation from the computation for a ring, in Section 3.1.B: We must now schedule $d_i + 1$ line segments instead of two. $\mathcal{P}_i$ starts with its *preliminary certificate* $(d_i + 1, L_i)$ and "adjusts" each of its children's certificates: its jth child's certificate becomes $(d_i + c_{ij}, n_{ij})$, $1 \leq j \leq d_i$. $\mathcal{P}_i$'s preliminary certificate is a guarantee just for its message $\mathcal{M}_i$; both $\mathcal{P}_i$'s preliminary certificate and the "adjustments" to its childrens' certificates acknowledge the fact that no child can start transmitting its message stream until all children have received their TRANSMIT-MESSAGE orders. Next $\mathcal{P}_i$ sorts the certified lag times $\{c_{ij} \mid 1 \leq j \leq d_i\}$, to obtain a permutation π of the set $\{1, 2, \ldots, d_i\}$ which orders the children of $\mathcal{P}_i$ in *increasing* order of their certified lag times. ($\mathcal{P}_i$ will need to "remember" the permutation π, in order to transmit the TRANSMIT-MESSAGE orders to its children.) Now, $\mathcal{P}_i$ considers the following $d_i + 1$ segments X_{ij}, $0 \leq j \leq d_i$, of the X-axis.

- X_{i0} is a length-L_i segment with left endpoint $d_i + 1$.
- For $1 \leq j \leq d_i$, X_{ij} is a length-$n_{i\pi(j)}$ segment with left endpoint $d_i + c_{i\pi(j)}$.

In analogy with the two-segment analysis in Section 3.1.B, $\mathcal{P}_i$ adjusts the positions of the segments X_{ij}, moving segments rightward if necessary but never moving them leftward, to combine them into a single line segment of length n_i, with as small (i.e., leftward) a left endpoint as possible. This left endpoint is the sought quantity c_i; specifically,

$$c_i = d_i + 1 + \max(0, c_{i\pi(1)} - L_i) + \sum_{j=2}^{d_i} \max(0, c_{i\pi(j)} - c_{i\pi(j-1)} - n_{i\pi(j-1)}).$$

Computing TRANSMIT-MESSAGE Times. When $\mathcal{P}_i$ receives the order TRANSMIT AFTER s_i STEPS, it passes along updated versions of the order to its children, much as in Section 3.1.B, but altered to apprise each child of the following sources of delay to its message transmission:

- the number of children $\mathcal{P}_i$ has in the spanning tree
- the length of the message stream from sibling PEs that will precede its stream to $\mathcal{P}_i$.

Accordingly $\mathcal{P}_i$ sends out the following orders, in the indicated order.

1. To PE $\mathcal{P}_{i\pi(1)}$:
   TRANSMIT AFTER $d_i - 1 + L_i$ STEPS
2. To PE $\mathcal{P}_{i\pi(2)}$:
   TRANSMIT AFTER $d_i - 1 + L_i + n_{i\pi(1)}$ STEPS
3. In turn to PEs $\mathcal{P}_{i\pi(j)}$, $3 \leq j \leq d_i$:
   TRANSMIT AFTER $d_i - 1 + L_i + \sum_{\ell=1}^{j-1} n_{i\pi(\ell)}$ STEPS.

$\mathcal{P}_i$ then proceeds to transmit $\mathcal{M}_i$ to its parent, at its appointed time, and thence to relay the message streams originating with its children.

Acknowledgments

A portion of this research was done at the Fibonacci Institute in Trento, Italy. The research was supported in part by the following grants. S. N. Bhatt: NSF Grant CCR-88-07426, NSF/DARPA Grant CCR-89-08285, Air Force Grant AFOSR-89-0382; G. Pucci: CNR project "Sistemi Informatici e Calcolo Parallelo;" A. Ranade: AFOSR Grant F49620-87-C-0041; A. L. Rosenberg: NSF Grant CCR-90-13184.

Figures

Legend:

Processors: $\mathcal{P}_0, \mathcal{P}_1, \mathcal{P}_2, \mathcal{P}_3, \mathcal{P}_4, \mathcal{P}_5$
Message Notation: $\mathcal{M}_i = m_{i1} m_{i2} \cdots m_{iL_i}$
TRANSMIT AFTER $\geq s_i$ STEPS order: (s_i)
Synchronization Token: $[\star]$
Certificate Notation: $< c_i >$
Message Lengths for Fig. 1: $L_1 = 0$, $L_2 = 0$, $L_3 = 0$, $L_4 = 4$, $L_5 = 3$

Network traffic when $\mathcal{M}_5$ precedes $\mathcal{M}_4$:

Step	$\mathcal{P}_0$	$\mathcal{P}_1$	$\mathcal{P}_2$	$\mathcal{P}_3$	$\mathcal{P}_4$	$\mathcal{P}_5$
0	m_{51}					
1	m_{52}	m_{51}				
2	m_{53}	m_{52}	m_{51}			
3	m_{41}	m_{53}	m_{52}	m_{51}		
4	m_{42}	m_{41}	m_{53}	m_{52}	m_{51}	
5	m_{43}	m_{42}	m_{41}	m_{53}	m_{52}	m_{51}
6	m_{44}	m_{43}	m_{42}	m_{41}	m_{53}	m_{52}
7		m_{44}	m_{43}	m_{42}	m_{41}	m_{53}
8			m_{44}	m_{43}	m_{42}	
9				m_{44}	m_{43}	
10					m_{44}	

Network traffic when $\mathcal{M}_4$ precedes $\mathcal{M}_5$:

Step	$\mathcal{P}_0$	$\mathcal{P}_1$	$\mathcal{P}_2$	$\mathcal{P}_3$	$\mathcal{P}_4$	$\mathcal{P}_5$
0	m_{41}					
1	m_{42}	m_{41}				
2	m_{43}	m_{42}	m_{41}			
3	m_{44}	m_{43}	m_{42}	m_{41}		
4	m_{51}	m_{44}	m_{43}	m_{42}	m_{41}	
5	m_{52}	m_{51}	m_{44}	m_{43}	m_{42}	
6	m_{53}	m_{52}	m_{51}	m_{44}	m_{43}	
7		m_{53}	m_{52}	m_{51}	m_{44}	
8			m_{53}	m_{52}	m_{51}	
9				m_{53}	m_{52}	m_{51}
10					m_{53}	m_{52}
11						m_{53}

Figure 1: *The effect of message order on the time for scattering on a ring.*

Message Lengths for Figs. 2, 3: $L_1 = 2,\ L_2 = 3,\ L_3 = 0,\ L_4 = 2,\ L_5 = 1$

Network Traffic:

Step	$\mathcal{P}_0$	$\mathcal{P}_1$	$\mathcal{P}_2$	$\mathcal{P}_3$	$\mathcal{P}_4$	$\mathcal{P}_5$
1		(1)				
2		m_{11}	(2)			
3	m_{11}	m_{12}	m_{21}	(3)		
4	m_{12}	m_{21}	m_{22}		(1)	
5	m_{21}	m_{22}	m_{23}		m_{41}	(2)
6	m_{22}	m_{23}		m_{41}	m_{42}	m_{51}
7	m_{23}		m_{41}	m_{42}	m_{51}	
8		m_{41}	m_{42}	m_{51}		
9	m_{41}	m_{42}	m_{51}			
10	m_{42}	m_{51}				
11	m_{51}					

Figure 2: *Gathering via shoulder-tapping on a 6-node ring.*

Network Traffic:

Step	$\mathcal{P}_0$	$\mathcal{P}_1$	$\mathcal{P}_2$	$\mathcal{P}_3$	$\mathcal{P}_4$	$\mathcal{P}_5$
1		$[\star]$				
2			$[\star]$			
3				$[\star]$		
4					$[\star]$	
5						$[\star]$
6					$\langle 1 \rangle$	
7				$\langle 2 \rangle$		
8			$\langle 4 \rangle$			
9		$\langle 3 \rangle$				
10	$\langle 3 \rangle$					
11		(3)				
12			(3)			
13		m_{11}		(4)		
14	m_{11}	m_{12}	m_{21}		(2)	
15	m_{12}	m_{21}	m_{22}		m_{41}	(2)
16	m_{21}	m_{22}	m_{23}	m_{41}	m_{42}	m_{51}
17	m_{22}	m_{23}	m_{41}	m_{42}	m_{51}	
18	m_{23}	m_{41}	m_{42}	m_{51}		
19	m_{41}	m_{42}	m_{51}			
20	m_{42}	m_{51}				
21	m_{51}					

Figure 3: *Gathering via transmission certificates on a 6-node ring.*

References

[1] M. Baumslag and F.S. Annexstein. A unified approach to global permutation routing on parallel networks. *Math. Syst. Th.* 24:233–251, 1991.

[2] S.N. Bhatt, G. Bilardi, G. Pucci, A. Ranade, A.L. Rosenberg, E.J. Schwabe. General message passing in rings of processors (tentative title). In preparation, 1992.

[3] W.J. Dally and C.L. Seitz. Deadlock-free message routing in multiprocessor interconnection networks. *IEEE Trans. Comp.* C-36:547-553, 1987.

[4] P. Fraigniaud, S. Miguet, Y. Robert. Complexity of scattering on a ring of processors. *Parallel Computing* 13:377-383, 1990.

[5] J. Hromkovic, C.D. Jeschke, B. Monien. Optimal algorithms for dissemination of information in some interconnection networks. *Mathematical Foundations of Computer Science '90. Lecture Notes in Computer Science 452*, Springer-Verlag, N.Y., 1990.

[6] S.L. Johnsson and C.-T. Ho. Optimal broadcasting and personalized communication in hypercubes. *IEEE Trans. Comp.* 38:1249-1268, 1989.

[7] D. Nassimi and S. Sahni. Data broadcasting in SIMD computers. *IEEE Trans. Comp.* C-30:101-107, 1981.

[8] Y. Saad and M.H. Schultz. Data communication in parallel architectures. *Parallel Computing* 11:131-150, 1989.

[9] Q.F. Stout and B. Wagar. Intensive hypercube communication, I: Prearranged communication in link-bound machines. *J. Parallel Distr. Comput.* 10:167-181, 1990.

[10] D.M. Topkis. All-to-all broadcast by flooding in communication networks. *IEEE Trans. Comp.* 38:1330-1333, 1989.

[11] L.G. Valiant and G.J. Brebner. Universal schemes for parallel computation. *13th ACM Symp. on Theory of Computing*, 263-277, 1981.

Mesh and Torus Chaotic Routing

Kevin Bolding and Lawrence Snyder
Department of Computer Science and Engineering
University of Washington, Seattle

Abstract

The chaos router is an adaptive nonminimal message router for multicomputers that is simple enough to compete with the fast, oblivious routers now in use in commercial machines. It improves on previous adaptive routers by using randomization, which eliminates the need for complex livelock protection and speeds the router.

The two-dimensional chaos router is shown to be theoretically sound and physically realizable. Extensive simulation studies compare chaos routing with oblivious and deflection routing in mesh and torus networks. Chaos routing is shown to be competitive for mesh networks and superior for torus networks. This high performance is, perhaps, unexpected for the mesh since there is no finite bound on the delivery time of any message.

1 Introduction

Chaotic routing is a randomizing, adaptive, message routing technique that has previously been shown (in simulations) to be effective for the binary n-cube (hypercube) topology [11]. The technique is *nonminimal*, i.e. messages do not necessarily take minimal paths to their destinations, and randomization plays a critical role in preventing *livelock*, i.e. in preventing messages from continually circulating in the network without being delivered [10]. Though the principles apply as well to batched message communication, chaotic routing assumes a continuous workload where messages are presented at the nodes for injection into the network at random real (or fine-grain discrete) times. Routing decisions are made locally in the routing nodes based on the destination address stored in the headers of the messages and the availability of outgoing channels. Messages can "cut-through" nodes if an outgoing channel is immediately available, but they may also be stored in the node if all outgoing channels are blocked, motivating short, e.g. 20 flit, messages.

Chaotic routing's success on the hypercube suggests that it might be effective for other topologies. Networks of low dimension are important because the trend in parallel computer design is towards mesh and torus based communication structures such as in the Intel Paragon, the Tera computer, and the Caltech Mosaic. But applying "chaos" in two dimensions poses several problems. First, chaotic routing relies on theoretical foundations, the theorems of which have only been proved for hypercubes. This is easily

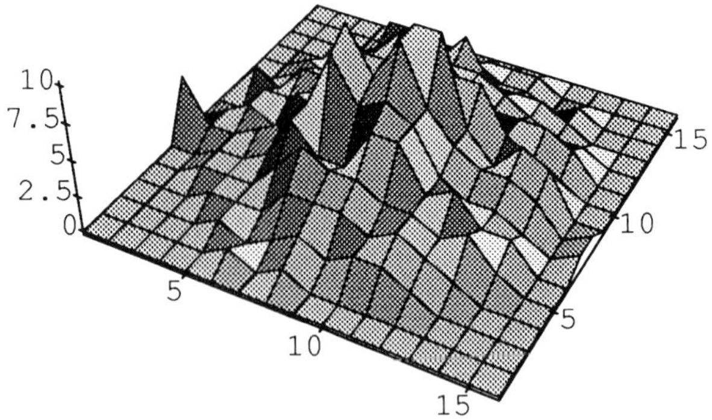

Figure 1: Average injection delay for a 256-node mesh.

remedied by the results in the [1]. The second problem is subtle and applies to any nonminimal adaptive router.

In a mesh with uniform random traffic, there is a "hot spot" in the center of the mesh. That is, the shortest message paths between two points tend to cross the center of the mesh, causing the resources in the center of the network (wires, buffers, etc.) to be more heavily used (see Figure 1). All adaptive routers will try to use these paths, but nonminimal adaptive routers, when they encounter congestion, will try to "deroute" a message away from the congestion. In such cases the hot spot can act as a barrier off of which messages can "bounce": That is, the forward paths are all congested, the message is derouted "backwards", and starts forward again, not having moved (or been able to move) away from the congestion. Though all nonminimal adaptive routers are subject to this type of behavior, the affect on performance varies depending on the type of router.

Priority adaptive routers time stamp each message and routing is governed by the rule: oldest message first. Thus, messages bouncing off the hot spot will eventually age enough to be routed through it. Matters are not so certain for the chaos router, however. The primary advancement of chaotic routers [10] is that they eliminate the time stamping and the time consuming prioritization, replacing it with a reliance on randomization. But there is no mechanism that can assure the delivery of a message in a fixed finite time. A message can continue bouncing off the hot spot for an arbitrarily long period of time. Because of the probabilistic livelock freedom proved in the [1] we know that the probability that the message has not been delivered in t steps goes to zero as t increases. So we can be confident that the message will be delivered eventually. But it could take a very very long time, leaving us with the question: Does chaotic routing work for the mesh?

In this paper we present simulation results comparing chaotic routers with oblivious routers and deflection, or "hot potato," routers on the mesh and torus topologies of sizes 64, 256 and 1024 nodes for both uniform random and hot spot loads. Three highlights are worth noting:

- On the mesh, chaotic routing performs as well as oblivious and deflection routing in throughput and latency for uniform traffic.

Thus, chaotic routing does work on a mesh, and in fact works about as well as other routers when the traffic is uniform.

- On the mesh chaotic routing performs better than oblivious and deflection routing in throughput and latency for nonuniform hot spot traffic.

Since it is likely that programs exhibit nonuniform traffic patterns the performance in such cases is perhaps more significant.

- On the torus chaotic routing is decidely superior to oblivious and deflection routing in throughput and latency.

The torus has better bisection bandwidth and better worst case path length than a mesh of similar size at the cost of a few extra wires. Its *vertex transitive* property aids the chaos router in giving it significantly better performance. The chaotic torus router is the best two dimensional packet router to our knowledge and thus a candidate for the next generation parallel computers.

2 Relationship to Previous Research

Borodin and Hopcroft [2] use the term *oblivious* to refer to routers for which the path of any message is completely determined by its [source, destination] pair. They proved that oblivious routers in an N node, d degree topology require $\sqrt{N}/d^{3/2}$ steps to route some permutations. The poor worst case performance and their fault intolerance would doom oblivious routers for use in multicomputers were it not for the fact that they are extremely simple, and thus fast. Accordingly, oblivious routers are the state-of-the-art for MIMD multicomputers such as those built by Intel, Ametech and NCUBE. Dally, Seitz and Flaig introduce oblivious routers of the type considered here for the mesh [8] and torus [4] topologies.

Randomization was first applied in the context of message routing by Valiant and Brebner [16], though in a way quite different from the chaotic approach. Their technique — select a random intermediate destination for every message, route the message to that destination and then on to the true destination — was applied to batched routing in a hypercube. It could obviously be applied continuously [3] and in two-dimensional topologies. The main difficulty with this type of randomization is that it doubles the expected path length of any message.

An adaptive mesh router was proposed by Ngai and Seitz [13], but it differs from the chaotic approach by using timestamps and prioritization

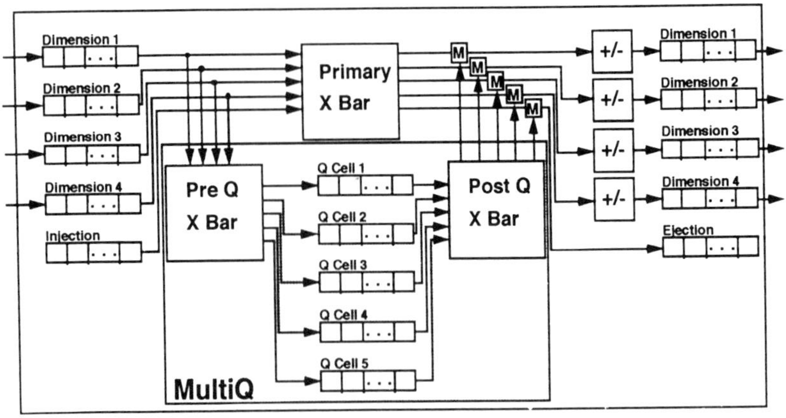

Figure 2: Two-dimensional chaos router diagram.

to prevent against livelock. Comparisons between prioritized and chaotic routers have been performed [11]. Adaptive wormhole routing using virtual channels has been studied by Duato [6].

"Hot potato" or deflection routing is another scheme capable of adaptive routing [14, 12, 7, 15]. The approach is synchronous and the time step is long enough to transmit an entire packet. At each step the incoming messages are paired with outgoing channels and are transmitted in the next step. The pairing is done in a variety of ways: Certain deterministic algorithms attempt to maximize the number of messages sent out productive channels, while others use a greedy algorithm with random selection. Those messages not receiving a productive channel are "deflected," *i.e.* derouted, out any available channel. Deflection routing differs from chaotic routing in several ways: Chaotic routing is not batched, *i.e.* does not require all headers to be present at once, thus permitting fast self-timed or high clock rate implementations, and better utilization of channels since messages can cut through, i.e. messages can be "in" multiple routers at once. Chaotic routing permits messages to wait for forward traffic to clear, thus reducing time consuming deroutes which necessarily delay the packet at least two "message times". Pausing for traffic to clear cushions the affects of bursts. Finally, chaotic routing resorts to derouting only under conditions of high load, when slower performance is inevitable.

3 Chaos Router Design

The chaotic router studied here is a two dimensional variant of the hypercube chaotic router [11]. The basic design of the chaos router is similar to a typical oblivious virtual cut-through router, with input and output frames connected by a crossbar switch, and hardware to increment or decrement the headers of messages as they pass through (see Figure 2). Two primary distinctions exist, though. The first is that the routing relation no longer specifies a single channel to traverse next, but instead a set of equally profitable channels. The first available profitable channel will be chosen for routing. The second distinction is the addition of a small (5 message) buffer, the *MultiQueue*, which holds messages for which no profitable channels are immediately available. Since the buffer space is off of the critical resource path, messages in the queue do not block messages behind them. Messages enter the queue along a separate crossbar whenever they have been denied access to any profitable channel long enough for the entire message body to have arrived in the input frame. Also, in order to prevent deadlock, when a message is read from the queue into the output frame for channel i and the input frame for channel i is full, the message in the input frame is immediately read into the queue. Messages cannot enter the queue from the injection frame and messages which are awaiting the availability of the ejection buffer do not enter the queue, as they will be consumed by the processor. Whenever an output channel that is profitable for a message in the queue becomes available, the first message in the queue which can use that channel is sent from the queue through another crossbar to the output frame for that channel. When several messages can profitably use a channel at the same time, priority is given to messages in the queue (in FIFO order); among competing input frames messages are chosen randomly.

A critical situation occurs when a message is specified to be sent to the queue, but the queue is completely full. In such a situation, a message is randomly selected from the queue to be *derouted* along the first available channel so that room will be created in the queue for the newly arriving message. Derouting provides an additional factor of adaptivity to the chaos router and allows the use of a packet-exchange protocol for deadlock prevention [13].

4 The Network Model

The performance of different routing schemes varies much according to the model of the network being studied. For the studies of chaos routing, we use the following network model:

The network is a regular two-dimensional network of bi-connected nodes. Between each pair of adjacent nodes in the network there is a *channel* consisting of control wires and a single data bus. The data bus is shared between the two directions, with arbitration occurring between message boundaries. The messages are fixed-size packets consisting of a header and several data words. The width of the data bus determines the size of a *flit* which is the amount of data which can be transmitted over the data bus in one cycle. We parameterize the packet size in our studies in terms of the number of flits per packet, L. Thus, for a 16 bit wide bus, a 20-flit message would contain a 16-bit header and 304 bits of data. We constrain our experiments to messages of size 20 flits, which is consistent with existing multicomputer designs.

We study the two-dimensional mesh and the two-dimensional torus in this investigation. To judge changes in performance with network size, we compare networks of 64, 256, and 1024 nodes.

5 Routers Studied

We study three routers in this paper: an oblivious router, the chaos router, and a deflection router. Most current multicomputers use some variant of oblivious routing. We chose a virtual cut-through oblivious router with input and output queueing to provide a baseline for current routing techniques. We provide results from a deflection router based on [7] to provide another baseline for comparison. Finally, we study the chaos router as presented in Section 3.

5.1 Oblivious Router

The oblivious router studied here is based upon the Kermani and Kleinrock [9] virtual cut-through router. Specifically, the router consists of a set of input and output frames and a crossbar switch which connects each input frame to every output frame. Each channel has one input frame and one output frame, each capable of holding exactly one fixed-size message[1]. The injection and delivery channels also have an input frame and an output frame, respectively, which are connected to the crossbar as well. Operation of the router proceeds in virtual cut-through fashion: whenever a message arrives in an input frame, it is immediately routed to the output frame for the next channel on its path to its destination,[2] if that output frame is available. It is not necessary to receive the entire message in an input frame before the header is sent to the output frame. If the output frame is not immediately available, the message will wait in the input frame until it becomes available, blocking any messages behind it if necessary. Operation of the channels proceeds in a similar demand-driven fashion.

5.2 Deflection Router

Deflection routing is an adaptive routing scheme in which messages arriving at a node are guaranteed to leave the node in the next routing cycle. An attempt to assign each message to a channel which reduces it distance to its

[1]For the oblivious torus router, virtual channels are implemented by giving each physical channel two input and output frames.

[2]Since this is an oblivious router, there will be only one possible output channel at each routing step. In order to prevent deadlock, the channels must be traversed in order of dimension.

destination is made, giving preference to messages with only a single profitable direction, followed by randomly assigning any remaining messages to the remaining free outgoing channels. The scheme does not quite fit the network model presented in Section 4, as channels must always be available and, thus, cannot be shared. We compensate for this by dividing each deflection routing channel into two uni-directional channels of one half the width of the chaos and oblivious routers. Also, the deflection protocol requires that all the headers of incoming messages arrive at the same moment, which is generally accomplished in their network models by using very high bandwidth channels capable of transmitting an entire message in each flit. Since our model includes multiple-flit messages, a routing decision may occur only once an entire message arrives at a node, resulting in a store-and-forward technique without virtual cut-through. Finally, in the analytical models presented in [7], all messages which arrive at a single destination node are removed from the network at once. In our simulations, we limit the delivery capability to the bandwidth of a standard network channel (one flit per cycle), as would be required in a realistic implementation.

6 Traffic Models

In order to compare the relative performance of the different routing schemes, a synthetic workload is applied to the simulated network and performance measurements are taken. The choice of the workload is critical when trying to compare the schemes. We provide simulation results for two workloads: *uniform random* and *hot spot* traffic.

6.1 Uniform Random Traffic

For uniform random traffic, each node presents a message to the network with a destination chosen uniformly randomly from each of the nodes in the network. The time between the presentation of messages is chosen randomly with a mean time based on the simulated applied load. The load is presented as a fraction of the maximum load the network could handle if there were no resource conflicts. This is computed as the point at which the utilization of channels cut by a bisection of the network reaches 100% assuming each message crosses this bisection with probability 0.5. If all channels of the network were utilized 100% of the time and all messages traveled on the shortest paths available, this maximum throughput would be obtained under uniform random traffic. The maximum applied load is then computed as the minimum inter-injection period for each network.

For the network model presented in Section 4, where one flit can be transmitted across a channel in one cycle, the minimum inter-injection period per node is $\frac{1}{2}\sqrt{N}L$ cycles for N-node meshes and $\frac{1}{4}\sqrt{N}L$ for N-node tori with messages of length L flits.

6.2 Hot Spot Traffic

Although uniform random traffic is a natural model of network traffic, many applications used on multicomputers create message traffic which has several *hot spot* nodes that receive considerably more traffic than the rest of the network. We attempt to model an abstract system by a synthetic load consisting of the same injection load as uniform random traffic, but with the destination distribution skewed in the following manner: ten "hot" nodes are chosen at the beginning of the simulation, each being four times as likely as the other N-10 nodes to be the destination of a message. Thus, these nodes become hot spots which could represent nodes that are used for synchronization or locking in a multicomputer application. The total loading of the network is the same as for uniform random traffic.

7 Simulations

Simulations for the networks and routers studied were conducted using a flit-based simulator written in C. The simulations were based on the *cycle* time unit. One cycle is the time necessary to transmit a single flit across a channel. Routing decisions can be made in a single cycle. Thus, if a message header enters a router at cycle t, it may enter the next router as early as cycle $t + 1$[3].

Simulations were run by applying the simulated load to the network in a continuous manner. Statistics were computed in intervals in which each node of the network has injected at least 50 messages. Average throughput and average latency were computed for each statistics interval and convergence was determined when the standard deviation of both of the measures over the most recent five intervals were less than 3%. The results presented here represent the averages and standard deviations of 3 to 5 runs.

The statistics reported here are the average throughput of the network normalized to the maximum throughput under uniform random load and the average latency of messages in cycles. We define latency as the time from presentation of a message to the network until the message has been completely removed from the network at its destination (notice that this does *not* include source queueing time).

8 Simulation Results

As described earlier, simulations were performed on mesh- and torus-connected networks of 64, 256, and 1024 nodes using random traffic and "hot spot" traffic for each of the three routing schemes studied. The average throughputs and average latencies are reported here.

[3]For the deflection router, since the entire message must be received before transmission can begin, a message entering a router at cycle t will not leave until cycle $t + L$.

To gauge performance, we concentrate on the high-load throughput and medium-load latencies. For low loads, all routing schemes are able to deliver the entire applied load without difficulty. The point at which the network saturates and the network is not able to keep up with the applied load is the interesting point in this case. Also, the shape of the throughput curve above saturation is important – *i.e.* does throughput ever decrease with increasing applied load? Latency is a more critical issue during lower load periods. At loads above saturation, since the network cannot keep up with the load applied, the latency of messages which do get through becomes of only peripheral interest. However, when the network is operating below saturation, it is latency that is the critical figure of merit. Thus, we will consider throughput saturation points and below-saturation latencies as the figures of merit for the networks studied.

We present full throughput and latency curves for 256-node networks with chaos and oblivious routing. We graph only throughput data for deflection routing because the store-and-forward nature of the router results in especially high latency figures. Since the shapes of the curves do not differ appreciably over different network sizes, we present only the 100% load throughput and 50% load latency points for other size networks. The raw data is given for all networks in [1].

8.1 Mesh networks

For mesh networks under uniform random traffic, all three routing schemes give similar throughput results (Figures 3 and 4). The throughput reaches 80–90% of the maximum throughput in each case and there is no decline in throughput as the load approaches 100%. The latencies for the oblivious and chaos routers remain very close throughout all load ranges, with the chaos router giving slightly better values. The performance of the adaptive schemes is actually lower than would be expected: the additional hardware gives little or no benefit under random loads. This is the result of the large hot spot inherently present in the center of mesh-connected networks (Figure 1) which creates a substantial barrier to cross-network traffic. While the oblivious router sends messages straight through the hot spot, even if slowly, the adaptive routers attempt to route messages around the congested center. However, since the area is so large, messages tend to bounce around the periphery for long periods of time, resulting in very long paths from source to destination.

When hot spots are added to the mesh, chaos routing becomes distinctly better than oblivious and deflection routing for small networks, with the benefit declining as network size decreases (Figures 3 and 4)[4]. This can be seen as the oblivious throughput increases with network size while the chaos throughput remains relatively stable. For small networks the oblivious throughput is especially low, resulting in high latencies from the additional

[4] Data is not currently available for the 64-node deflection router

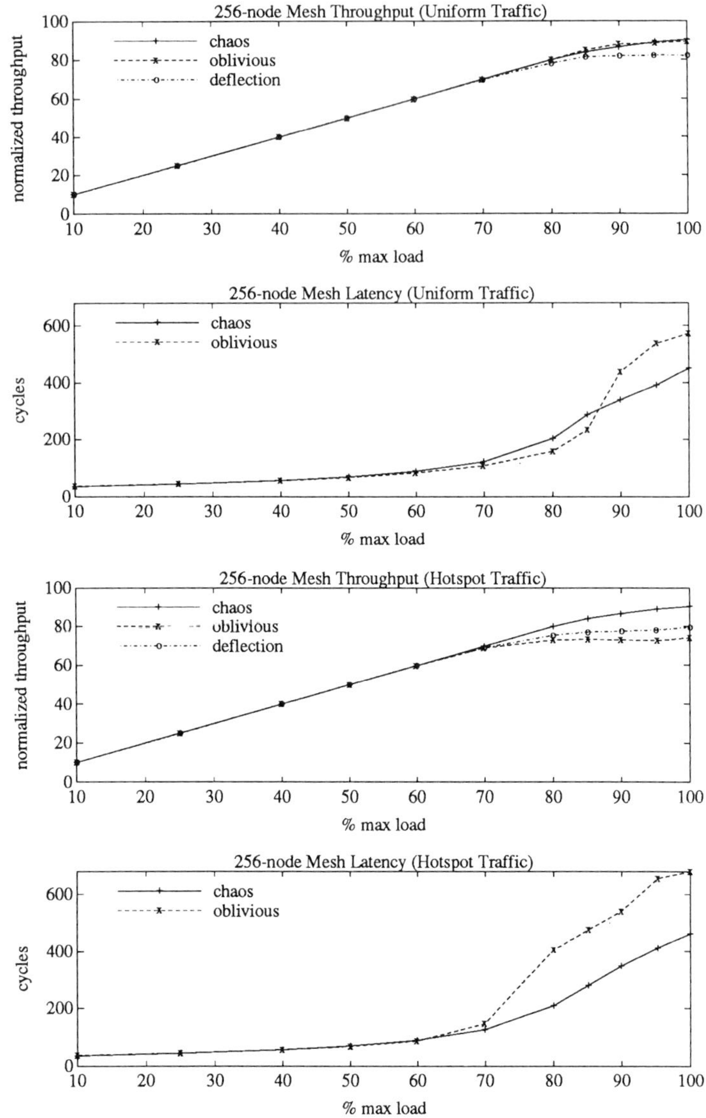

Figure 3: 256-node mesh results

congestion. This behavior is due to the fact that the central hot spot presents a more formidable barrier in larger networks – for small networks the ten hot spots influence the traffic greatly and the chaos router performs better, but as the network grows, the central hot spot dominates the traffic flow

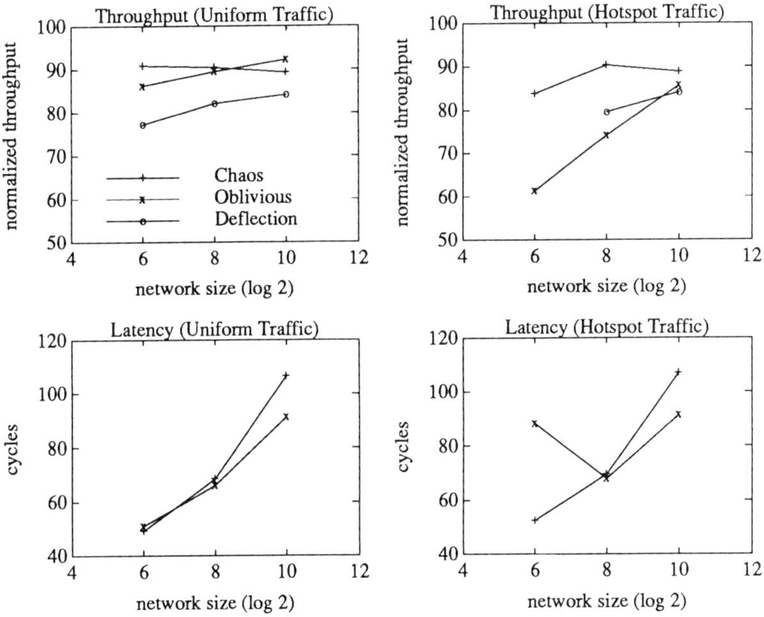

Figure 4: Mesh throughput (100% applied load) and latency (50% applied load) vs. network size

and the oblivious router performance improves. Thus, for smaller network sizes, the adaptivity of the chaos router proves useful in the mesh, but this advantage diminishes with increasing network size.

8.2 Torus networks

For torus-connected networks, the chaos router performs significantly better than both the oblivious and deflection routers in all respects (Figures 5 and 6). Since a torus is *vertex-transitive*, *i.e.* the network appears the same to every node, traffic is uniformly distributed throughout the network, unlike the mesh. This translates into a performance advantage for the chaos router, which allows messages to use the entire network without the constraints of oblivious dimension-order routing. The chaos router achieves near-maximum throughput under random traffic for all network sizes considered, while the oblivious and deflection routers top out at 55–70% performance. Again, the latencies remain low for low to medium loads, indicating very superior performance.

A disturbing property of the torus oblivious router is that the maximum throughput is achieved at less than the maximum load. This is due to a disturbance of the vertex-transitivity of the network introduced by the addi-

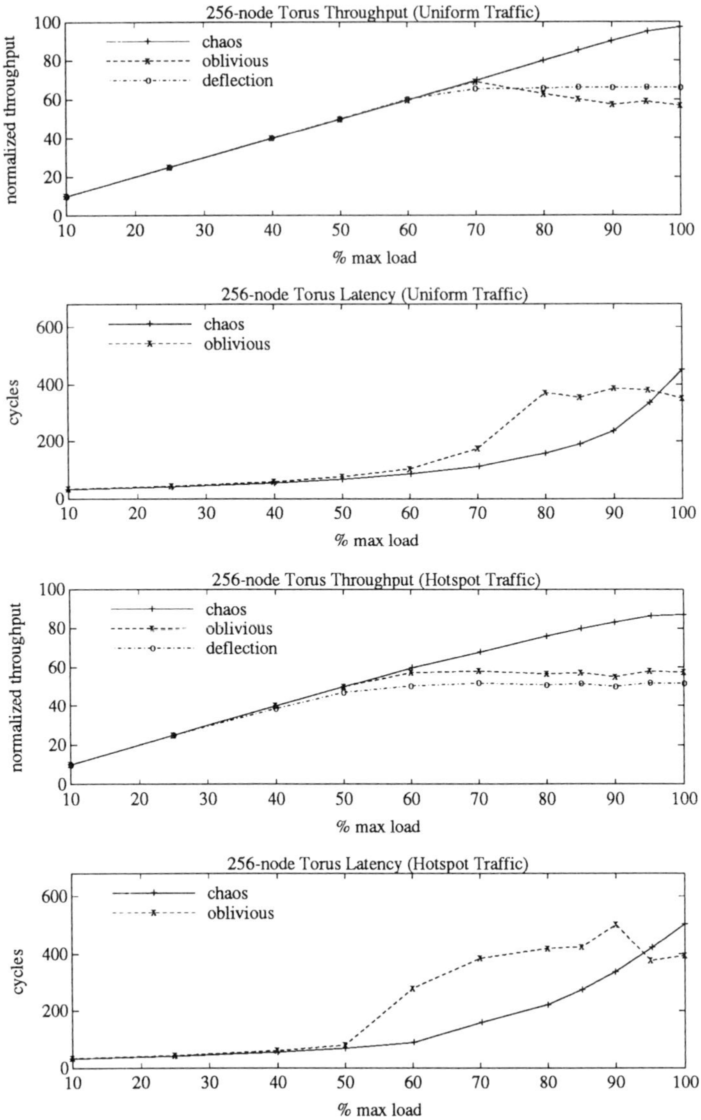

Figure 5: 256-node torus results

tion of deadlock prevention. Since the virtual-channel deadlock prevention scheme applied [5] distinguishes certain nodes as "special" in order to break cycles, the uniformity of the network is broken and hot spots are introduced at high loads. This results in the degradation of throughput as load is in-

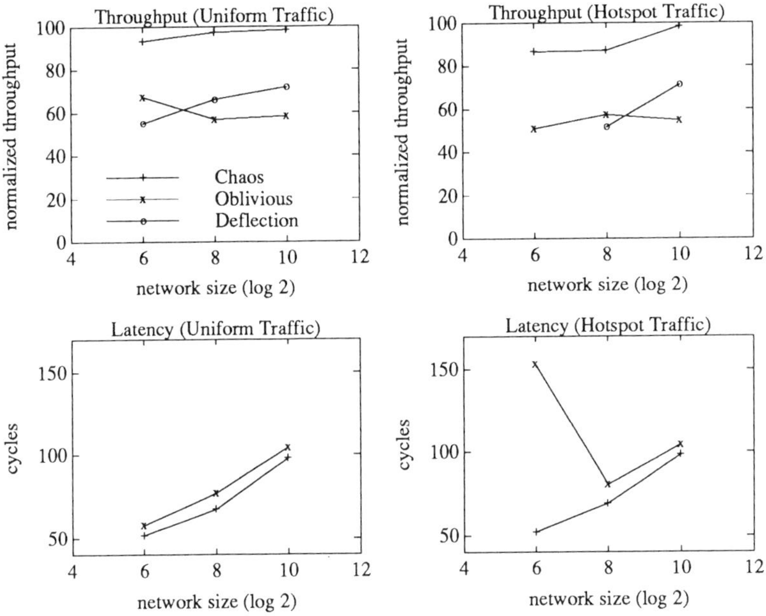

Figure 6: Mesh throughput (100% applied load) and latency (50% applied load) vs. network size

creased. The chaos and deflection routers preserve the uniformity of the network and do not exhibit this behavior.

For hot spot traffic, the adaptive routers perform well and the oblivious router suffers an earlier leveling of throughput than with random traffic[5]. The advantage of chaos routing is clearly apparent here, as throughput and latency are only minimally affected by the non-uniform traffic load. Overall, the chaos router is clearly superior to the oblivious and deflection routers for the torus network.

9 Conclusions

We have presented a two-dimensional variant of the hypercube chaos router and shown it to be a viable router. The theoretical foundations of the two-dimensional router have been presented. A working design of the chaos router has been given which is capable of competing with oblivious routers for critical-path complexity. The performance of the chaos router is comparable to oblivious routers for meshes with random traffic and better with hot spot

[5]The delivery capability of the routers was increased to 4X for the 64-node oblivious and chaos torus networks with hot spot traffic. Data is not currently available for the 64-node deflection router

traffic. For torus networks, the chaos router performs much better than the oblivious and deflection routers.

Acknowledgments

We wish to thank Sen-ching Cheung for running hundreds of simulations of the routers studied, Carl Ebeling for his help in designing the router, and Smaragda Konstantinidou for her original chaos design work.

This work was supported in part by the Office of Naval Research under Grant N00014-89-J-1368, the National Science Foundation under Grant MIP-9013274, and an Intel Foundation Graduate Fellowship.

References

[1] Kevin Bolding and Lawrence Snyder. Performance study of two-dimensional chaotic routers. Technical Report 91-04-04, University of Washington, Seattle, WA, 1991.

[2] A. Borodin and J. E. Hopcroft. Routing, merging and sorting on parallel models of computation. *Journal of Computer and System Sciences*, 30:130–145, 1985.

[3] Y. Chang and J. Simon. Continuous routing and batch routing on the hypercube. In *Proceedings of Principles of Distributed Computing*, pages 272–281. ACM, 1986.

[4] W. Dally and C. Seitz. The torus routing chip. *Journal of Distributed Computing*, 1(3), 1986.

[5] W. Dally and C. Seitz. Deadlock-free message routing in multiprocessor interconnection networks. *IEEE Transactions on Computers*, C-36(5):547–553, May 1987.

[6] J. Duato. Deadlock-free adaptive routing algorithms for multicomputers. *Technique et Science Informatiques*, 10(4):275–285, 1991.

[7] Chien Fang and Ted Szymanski. An analysis of deflection routing in multi-dimensional regular mesh networks. In *Proceedings of IEEE INFOCOM '91*, pages 859–868. IEEE, April 1991.

[8] C. Flaig. VLSI mesh routing systems. Master's thesis, California Institute of Technology, May 1987.

[9] P. Kermani and L. Kleinrock. Virtual cut-through: A new computer communication switching technique. *Computer Networks*, 3:267–286, 1979.

[10] Smaragda Konstantinidou and Lawrence Snyder. The chaos router: A practical application of randomization in network routing. In *Proceedings of the 2nd Symposium on Parallel Algorithms and Architectures*, pages 21–30. ACM, 1990.

[11] Smaragda Konstantinidou and Lawrence Snyder. Chaos router: Architecture and performance. In *Proceedings of the 18th International Symposium on Computer Architecture*, pages 212–221. IEEE, May 1991.

[12] N. F. Maxemchuk. Comparison of deflection and store-and-forward techniques in the manhattan street and shuffle-exchange networks. In *Proceedings of IEEE INFOCOM '89*, pages 800–809. IEEE, 1989.

[13] J. Y. Ngai and C. L. Seitz. A framework for adaptive routing in multicomputer networks. In *Proceedings of the Symposium of Parallel Algorithms and Architectures*, pages 1–9. ACM, 1989.

[14] B. J. Smith. Architecture and applications of the HEP multiprocessor computer system. In *Proceedings of SPIE*, pages 241–248, 1981.

[15] David Smitley. Design tradeoffs for a high speed network node. Technical Report SRC-TR-89-007, Supercomputing Research Center Institute for Defense Analysis, Bowie, Maryland, July 1989.

[16] L. G. Valiant and G.J. Brebner. Universal schemes for parallel communication. In *Proceedings of the 13th ACM Symposium on Theory of Computing*, pages 263–277, 1981.

Efficient Routing of a Class of Permutations in VLSI

Hussein M. Alnuweiri
Department of Electrical Engineering
University of British Columbia
Vancouver, B.C., Canada V6T 1Z4
email: hussein@ee.ubc.ca

Sadiq M. Sait and **Mohammed Darwish**
Computer Engineering Department
King Fahd University of Petroleum & Minerals
Dhahran 31261, Saudi Arabia

Abstract

Permutation routing plays a major role in determining the complexity of VLSI designs, particularly those implementing communication-intensive algorithms. This paper presents a scheme for implementing a wide class of permutations efficiently in VLSI. The scheme is used to derive permutation networks with N/Q input/outputs which route permutations on N elements, assuming that the elements are received in Q consecutive phases. The outputs are also produced in Q consecutive phases. A wide range of area-time tradeoffs can be achieved by selecting different values for Q. The derived networks have simple regular layouts with optimal AT^2 performance in most cases. The paper concludes by presenting an implementation of a CMOS VLSI permuter chip.

1 Introduction

Routing permutations is a problem of fundamental importance in VLSI, particularly since wire area plays a major role in determining the complexity of a VLSI chip. Bounds on the area and time of VLSI permutation networks was addressed by several researchers. For example, in his paper on sorting, Leighton [11], proved that array transpose (K−shuffle) networks on N elements have VLSI area $A = \Omega(N^2)$ in the bit-model of parallel computation. Bilardi and Sarrafzadeh [5, 6] provided an explicit construction of such networks and used them in the design of AT^2−optimal architectures for DFT. Alternative constructions were proposed by the first author of this paper for use in multidimensional DFT [1], and sorting in a constant number of rounds [2].

Motivated by the above work, this paper proposes an *index mapping* scheme for deriving reduced-area (and reduced I/O) networks to implement a wide class of permutations, particularly those of the bit-permute-complement (BPC) type defined in [12]. The proposed scheme exploits the relationship between the input and output indices of a permutation to derive a family of reduced-area networks which realize the permutation under different area-time tradeoffs. The area and time-delay of the proposed networks will be expressed in terms of certain design parameters; mainly, the number of elements to be permuted (N), the word length of each element (w), and the I/O reduction factor (Q). A family of networks, with various area-time tradeoffs, can be derived for the same permutation by varying the factor Q over a certain range. The resulting designs receive the inputs in Q phases, such that N/Q elements are received in parallel in each phase (however, each element is input bit-serially) The outputs are produced in a similar manner. The input (output) bits are received (produced) in consecutive clock cycles without additional inserted delays.

While the full version of this work provides a general scheme for implementing BPC permutations [3], this paper considers only two permutation types: $K-$shuffle and bit-reversal. These permutations appear in a wide number of applications including signal processing, computer arithmetic, and data alignment in parallel processing systems. The index mapping scheme yields reduced-area versions of the above networks with optimal or near optimal AT^2 performance. Moreover, the derived networks have simple regular layouts, simple control, and are easily partitionable among several chips to further reduce I/O pin requirements. A CMOS VLSI implementation of an 8−shuffle network is presented in the last section of the paper to demonstrate the advantages of the proposed scheme. Our results show a marked improvement over previous implementations such as RAMAT [7] and the E-network [8].

2 Routing Fixed Permutations in VLSI

This paper assumes a VLSI model [14, 15], in which the following assumptions hold: the chip is synchronous (propagation delay is independent of wire length), the chip input/output schedule is semilective-unilocal (each input is received only once, at a prespecified input port) and *when* and *where* determinate (the time and location of receiving an input or producing an output is prespecified, so as not to depend on the data or the running of the algorithm). It is further assumed that I/O pads are placed on the border of the chip. Also, a bit-serial, word-local model is assumed (the w bits representing one element are input or output bit-serially through the same I/O port).

The performance of a VLSI model of computation can be measured by two parameters, area (A) and total execution time (T). Information transfer arguments [15], have been used to establish lower bounds on the AT^2 product for several computations. Any permutation on N elements can be implemented trivially using $O(N^2)$ area (assuming bit-serial operation), although many among these may require smaller area. The scheme presented in the following section derives reduced-area versions of some fundamental permutation networks which are used in parallel processing applications, computer arithmetic, and signal processing. In particular, the proposed scheme yields designs of area $O(N^2/Q^2)$ and time delay $O(wQ)$, where w is the word length of the permuted items and Q is the *I/O reduction factor*. It must be emphasized that the area-complexity of such designs is valid only for $w \leq c \log N$, for a constant $c > 0$. It will be assumed that a reduced-area network has at most N/Q (bit-serial) inputs and outputs. This is a crucial assumption since it reduces the I/O requirements by a factor Q for a VLSI network whose area is reduced by a factor Q^2. Thus, the N input elements are presented to the network in Q phases, such that the elements with indices $0, 1, \cdots, \frac{N}{Q} - 1$ are given in parallel (but each element bit-serially) in the first phase, followed by elements with indices $\frac{N}{Q}, \cdots, 2\frac{N}{Q} - 1$ in the second phase and so on. The N output elements are produced in a similar manner.

The index-mapping scheme is a direct method for deriving reduced-area and reduce I/O versions of permutation networks, especially those in the BPC class [12]. The scheme works by decomposing the given permutation into a sequence of permutations which can be implemented more efficiently by the chip input schedule (e.g. inputs arriving in several packets at different times). BPC permutations are further decomposed into bit-complement permutations and bit-permute permutations. The details are given in [3], and will be not be discussed here due to space limitations. However, the efficiency of the method will be demonstrated for two fundamental types of permutations: the $K-$shuffle and bit-reversal. Since permutation networks operate on the indices of elements and not on their data content, in the following discussion elements will be represented by their indices only. Therefore, all subsequent operations and permutations will be performed on arrays of indices.

3 Reduced-Area $K-$Shuffle Networks

$K-$shuffle permutations have appeared in a wide number of applications such as index-rotation in multidimensional transforms [1, 5, 6, 7] and array transposition (or pseudo transposition) in multi-round sorting [2, 11]. Establishing a lower bound on the AT^2 complexity of a $K-$shuffle network on N words is not straightforward. For example, while a perfect-shuffle ($K = 1$) network requires area $\Omega(N^2/\log^2 N)$ [10], constructions for sorting networks by Leighton [11] and Alnuweiri [2], prove that $\Omega(N^2)$ area is required by $K-$shuffle networks when $K = o(N^{1/\varepsilon})$, for any constant $\varepsilon > 1$. The reduced-area $K-$shuffle presented in this paper achieve the $AT^2 = O(N^2w^2)$ and are, therefore, optimal only for the shuffle values indicated above.

In the following, the index-mapping scheme is used to develop a class of networks for the K−shuffle permutation. Each network in this class permutes N elements (indexed 0 through $N-1$), of length w bits each, in $O(wQ)$ time and $O(N^2/Q^2 + Nw)$ area. The AT^2 product for the proposed networks is $O(N^2w^2 + NQ^2w^3)$, which is optimal (under the word-local assumption) for all $1 \leq Q \leq \sqrt{N/w}$. For simplicity, it will be assumed that $N = 2^n$, $Q = 2^q$, and $K = 2^k$. The technique can be also applied when $\sqrt{N/w} \leq Q \leq \sqrt{N}$. However, it will yield networks which are optimal to within an $O(wQ^2/N)$ factor. When $Q = \sqrt{N}$, the area of the layout will be $O(wN)$ which is the same result achieved in [8].

Let $x_{n-1}, x_{n-2}, \cdots, x_0$ be an n−bit number representing the index of an element. The K−shuffle function on $N = 2^n$ inputs is a bijection $P_K : \{0,1\}^n \mapsto \{0,1\}^n$, defined as follows:

$$\begin{aligned} P_K(x_{n-1}, \cdots, \; x_{n-k} \;, x_{n-k-1}, \cdots, x_0) = \\ (x_{n-k-1}, \; \cdots \;, x_0, x_{n-1}, \cdots, x_{n-k}). \end{aligned} \tag{1}$$

It will be assumed that N input elements are received in Q columns of N/Q elements each, where column i, $0 \leq i \leq Q-1$, contains elements with indices $i\frac{N}{Q}, i\frac{N}{Q}+1, \cdots, (i+1)\frac{N}{Q}-1$. The N shuffled outputs are produced in a similar fashion. For the purpose of designing a reduced-area shuffle network, the Q input columns will be represented by an $(N/Q) \times Q$ matrix of indices arranged in increasing order from top to bottom and from right to left. Similarly, the Q output columns will be represented by an $(N/Q) \times Q$ matrix of indices. Each index in the output matrix is obtained by performing a K−shuffle on the bit representation of the index at the same position within the input matrix. The following example shows the input and output matrices for $N = 64$, $Q = 8$, $K = 4$.

56	48	40	32	24	16	8	0		14	12	10	8	6	4	2	0
57	49	41	33	25	17	9	1		30	28	26	24	22	20	18	16
58	50	42	34	26	18	10	2		46	44	42	40	38	36	34	32
59	51	43	35	27	19	11	3	K-Shuffle $\longmapsto$	62	60	58	56	54	52	50	48
60	52	44	36	28	20	12	4		15	13	11	9	7	5	3	1
61	53	45	37	29	21	13	5		31	29	27	25	23	21	19	17
62	54	46	38	30	22	14	6		47	45	43	41	39	37	35	33
63	55	47	39	31	23	15	7		63	61	59	57	55	53	51	49

Note that an output column may have elements from several input rows. This implies that a reduced-area permutation network must have register cells to store the input elements which compose a forthcoming output column. The proposed index-mapping technique will be used to derive such networks. Application of the method results in two types of shuffle networks corresponding to the cases $Q \geq K$ (or $q \geq k$) and $Q \leq K$ (or $q \leq k$). The construction of networks for each case is presented in the following two sections. We should emphasize here that realizations of reduced-area shuffle networks have been proposed in [5]. However, the index-mapping method is direct and results in uniform networks with a minimal amount of storage cells.

Networks for Case I: $(q \geq k)$

Following [2], the notation R_i^j will be used to denote a string of j bits starting at position i and ending at position $i + j - 1$ of some index string. Using this notation, the K-shuffle permutation specified by (1) can be expressed as follows:

$$R_{n-k}^{k} R_{n-q}^{q-k} R_{n-k-q}^{k} R_{0}^{n-k-q} \overset{P_K}{\longmapsto} R_{n-q}^{q-k} R_{n-k-q}^{k} R_{0}^{n-k-q} R_{n-k}^{k}, \tag{2}$$

with $R_{n-k}^{k} = x_{n-1} \cdots x_{n-k}$, $R_{n-q}^{q-k} = x_{n-k-1} \cdots x_{n-q}$, $R_{n-k-q}^{k} = x_{n-q-1} \cdots x_{n-k-q}$, and $R_0^{n-k-q} = x_{n-k-q-1} \cdots x_0$.

Note that since the inputs and outputs are arranged as a $(N/Q) \times Q$ matrix, the n-bit index of each element will be substituted by a q-bit index representing a column number and an $(n-q)$-bit index representing a row number. In this respect, the string $R_{n-k}^{k} R_{n-q}^{q-k}$, in the left-hand side of (2), constitutes the column number of each index in the input matrix while the remaining bits (in $R_{n-k-q}^{k} R_0^{n-k-q}$) constitute the row number. Similarly, the string $R_{n-q}^{q-k} R_{n-k-q}^{k}$, in the right-hand side of (2), denotes the column number of each index in the output matrix while the remaining bits denote the row number.

To further simplify our presentation, let $a = R_{n-k}^{k}$, $b = R_{n-q}^{q-k}$, $c = R_{n-k-q}^{k}$, and $d = R_0^{n-k-q}$. Following the notation given in [2], a binary string will be denoted by a small letter or by the concatenation of several small letters, and the number of all possible values of a string x will be denoted by $|x|$, for example, $|a| = K$, $|b| = Q/K$, and $|cd| = N/Q$. A fixed (constant) value of a binary string will be denoted by a capital letter, e.g. $a = W$ where $W \epsilon \{0, 1, \cdots, K-1\}$. The mapping given in (2) can be rewritten as follows:

$$ab, cd \overset{P_K}{\longmapsto} bc, da.$$

The above notation places a comma after the index which represents the column number in the input and output matrices. The method proposed below decomposes the above permutation into two permutations, as follows:

$$ab, cd \overset{P_1}{\longmapsto} bc, ad \tag{3}$$

$$bc, ad \overset{P_2}{\longmapsto} bc, da \tag{4}$$

where each permutation is to be implemented by a different network.

The design of two networks, called S-I and S-II, is presented below to implement the permutations expressed in (3) and (4), respectively. Let $M_0(ab, cd)$, denote the $(N/Q) \times Q$ matrix of input elements. Similarly, let

$M_1(bc, ad)$ and $M_2(bc, da)$ denote, respectively, the matrix of permuted elements after applying P_1 to $M_0(ab, cd)$, and the matrix of permuted elements after applying P_2 to $M_1(bc, ad)$. Any of these matrices can be partitioned by fixing one or more of the indices a, b, c, and d, as shown in figure 1. For example, $M_0(WX, cd)$ denotes the set of N/Q elements in column $ab = WX$ of matrix M_0 (i.e. the set $\{M_0(a, bcd)\}$ such that $a = W$, $b = X$, and $0 \leq c \leq K-1$, $0 \leq d \leq N/KQ - 1$). The following partitions will assist in specifying network S-I.

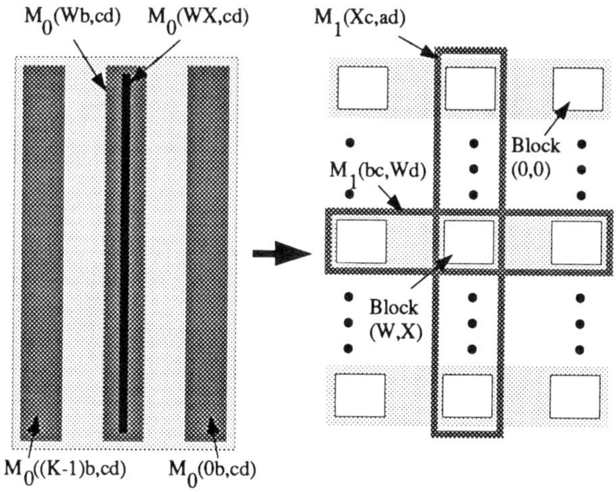

Figure 1: Index mapping Scheme for case I ($K \leq Q$)

- The index $a = W$ for a constant $W \epsilon \{0, 1, \cdots, K-1\}$. The matrix $M_0(ab, cd)$ can be partitioned into K *column-bands*, namely $M_0(0b, cd), M_0(1b, cd), \cdots, M_0((K-1)b, cd)$. Note that each column-band $M_0(Wb, cd)$ contains the elements of Q/K consecutive columns, namely columns $W\frac{Q}{K}, W\frac{Q}{K}+1, \cdots, (W+1)\frac{Q}{K}-1$.

  By the same token, the matrix $M_1(bc, ad)$ can be partitioned into K *row-bands*, $M_1(bc, 0d), M_1(bc, 1d), \cdots, M_1(bc, (K-1)d)$. Row-band $M_1(bc, Wd)$ contains N/KQ consecutive rows, namely rows $W\frac{N}{KQ}, W\frac{N}{KQ}+1, \cdots, (W+1)\frac{N}{KQ}-1$.

- The index $a = W$ and the index $b = X$ for a constant $W \epsilon \{0, 1, \cdots, k-1\}$ and a constant $X \epsilon \{0, 1, \cdots, \frac{Q}{K}-1\}$. Note that $0 \leq WX \leq Q-1$. The matrix $M_0(ab, cd)$ can be partitioned into Q *columns* $M_0(WX, cd)$.

By the same token, the matrix $M_1(bc, ad)$ can be partitioned into Q *blocks* $M_1(0c, 0d), \cdots, M_1((Q/K-1)c, (K-1)d)$. Note that each block is a $(N/KQ) \times K$ array of elements, and that the blocks are arranged in K rows and Q/K columns. The block in row W and column X of the partition will be be referred to as block (W, X).

The first partition above implies that network S-I must realize the map $M_0(Wb, cd) \mapsto M_1(bc, Wd)$ for every value of W, i.e. map elements in column-band W of M_0 onto row-band W of M_1. Also, the network must realize the map $M_0(WX, cd) \mapsto M_1(Xc, Wd)$ for every value of WX, i.e. map the elements in column WX of M_0 onto block (W, X) of M_1. The mapping $M_0(WX, cd) \mapsto M_1(Xc, Wd)$ can be realized by a network which "folds" column WX of M_0 into block (W, X) of M_1. Such a network was referred to as a *folding-network* (abbreviated to FN) in [2]. Two layouts for a folding network are shown in figure 2. The network shown in figure

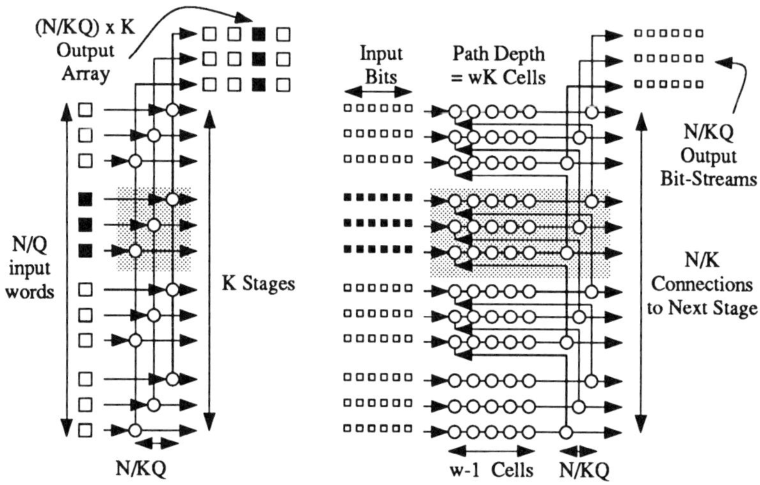

Figure 2: Folding networks for case I $(K \leq Q)$

2(a) assumes that each cell receives all w bits representing an element before transmitting it upwards (i.e. assumes a word-model). It is presented only for illustrative purposes. In the bit-model network (adopted in this paper), a node can only store a constant number of bits. Figure 2(b), shows a modified network, in which each cell is allowed to store only one bit. Note that the mapping involves indices c and d only, since a and b have fixed values in this case. Therefore, the elements of block W, X of M_1 have Xc as their column index, and Wd as their row index.

A permutation network that performs $M_0(Wb, cd) \mapsto M_1(bc, Wd)$, will be called a *band-permuter*. Note that the Q/K input columns (of each column-band W) are folded into Q/K blocks in row-band W of the output matrix. The Q/K blocks in a row-band can be produced serially using one folding network. Therefore, a band-permuter can be constructed using a single FN preceded by $w(Q/K - 1)$ stages of 1-bit register cells (to store the bits of the remaining $Q/K - 1$ columns of the input column-band). The organization of a band-permuter is shown in figure 3.

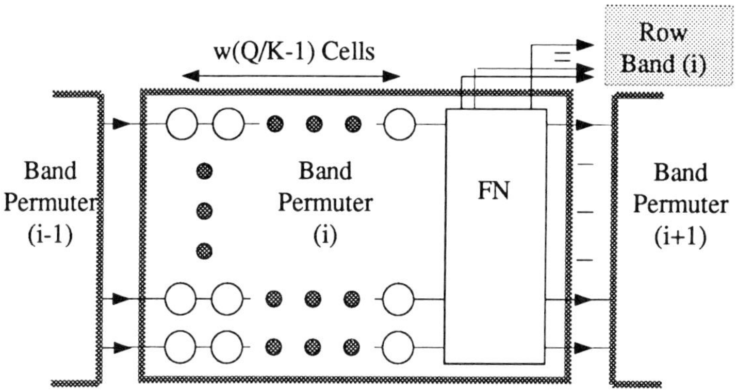

Figure 3: Layout of network S-I

Network S-I can be constructed using a horizontal cascade of K band permuters, one permuter to transform one column-band of the input, as shown in figure 3. The permutation is performed in two phases. The first phase shifts the input columns bit-serially through the band permuters using horizontal connections only. Following wQ time units, column-band i will be in permuter i. The second phase, consists of Q/K steps. Each step first shifts out the columns in the FNs (using vertical connections), then shifts the elements of the next available column into each FN. It is clear that each step requires wK time units, and thus, the second phase can be completed in wQ time units. The total delay of S-I is, therefore, $2wQ$ time units. Figures 2 and 3 show that the area of a band permuter is $O(\frac{N^2}{KQ^2} + \frac{wN}{K})$. Since S-I consists of K band permuters, its total area is $O(N^2/Q^2 + wN)$, which is $O(N^2/Q^2)$ for $Q \leq \sqrt{N/w}$.

Network S-II performs the mapping $M_1(bc, ad) \mapsto M_2(bc, da)$ (specified by (4)) which affects only the row indices of $M_1(bc, ad)$, as it maps row ad of M_1 into row da of M_2. Since $|a| = K$, the map $ad \mapsto da$ (among the rows of the matrices) is a $K-$shuffle permutation (on N/Q rows). Thus,

S-II is simply a $K-$shuffle network on N/Q elements, which can be laid out in $O(N^2/Q^2)$ area and has $O(w)$ time delay. Networks S-I and S-II thus have a total area of $O(N^2/Q^2)$ and have a total delay of $O(wQ)$ time, as desired. An example of a complete reduced-area shuffle network is shown in figure 4 for $N = 64$, $Q = 8$, and $K = 4$. Again, for illustrative purposes, a word-model (each node can store all w bits representing an element) is assumed. Note also that Q is not less than $\sqrt{N/w}$.

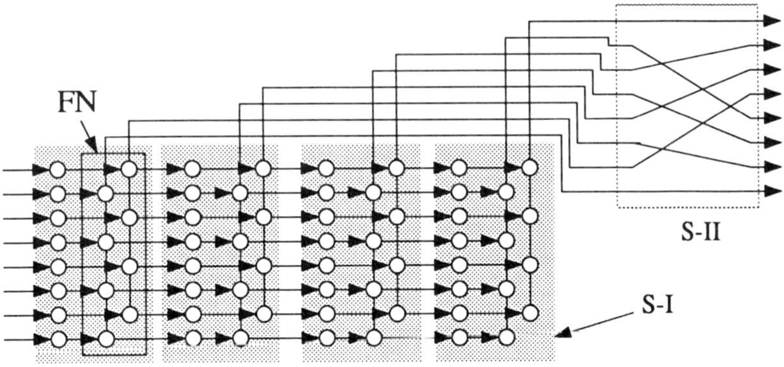

Figure 4: A reduced-area shuffle network for $N = 64$, $Q = 8$, $K = 4$.

Networks for Case II: $(1 \leq q \leq k)$

VLSI networks for this case have been developed in an earlier work by the first author [2] for implementing rotate-sort and column-sort. For completeness, a brief description of the network construction is included below. For $q \leq k$, the $K-$shuffle permutation specified by (1) can be expressed as follows:

$$R^q_{n-q} R^{k-q}_{n-k} R^q_{n-k-q} R^{n-k-q}_0 \overset{P_K}{\longmapsto} R^q_{n-k-q} R^{n-k-q}_0 R^{k-q}_{n-q} R^q_{n-q}. \quad (5)$$

Let $a = R^q_{n-q}, b = R^{k-q}_{n-k}, c = R^q_{n-k-q}$, and $d = R^{n-k-q}_0$. The above mapping can be rewritten as follows:

$$a, bcd \overset{P_K}{\longmapsto} c, dab.$$

The above permutation can be decomposed into two permutations, as follows:

$$a, bcd \overset{P_1}{\longmapsto} c, abd \quad (6)$$

$$c, abd \overset{P_2}{\longmapsto} c, dab \quad (7)$$

Two networks, U-I and U-II which realize (6) and (7), respectively, can be derived using an index mapping scheme similar to that of the previous section. The details of the derivation can be found in [2].

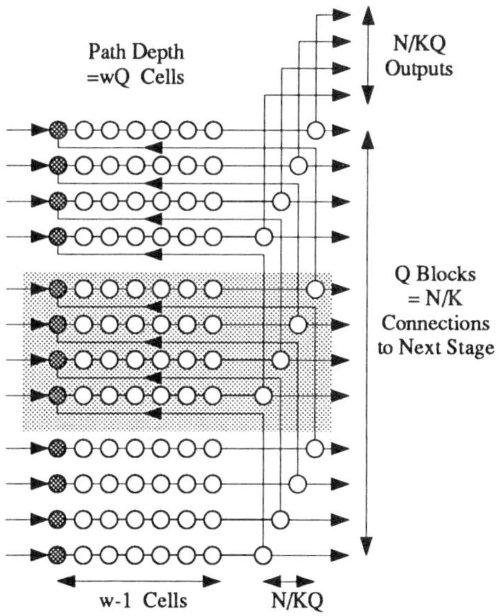

Figure 5: Folding network for case II ($K \geq Q$)

Network U-I consists of Q *column permuters* (one permuter for each column of N/Q elements) cascaded horizontally as shown in figure 6. Each column permuter is constructed from K/Q FNs as shown in figure 6. The permutation is performed in two phases (horizontal and vertical) as in the previous case. The total delay of the two phases is $2wQ$ time units, and the total area of the network is $O(Q[\frac{N^2}{Q^3} + \frac{Nw}{Q}])$, which is $O(N^2/Q^2)$ for $Q \leq \sqrt{N/w}$.

Network U-II performs the mapping specified by (4) which affects only the row indices elements, since it maps elements in row *abd* of the input into row *dab* of the output. Since $|ab| = K$, the map $abd \mapsto dab$ is a K−shuffle permutation (on N/Q row-indices). Thus, U-II is simply a K−shuffle network on N/Q elements, which can be laid out in $O(N^2/Q^2)$ area and has $O(w)$ time delay. The complete network for case II has therefore $O(N^2/Q^2)$ total area and $O(wQ)$ total delay. A CMOS VLSI implementation of networks U-I and U-II is presented in section 5.

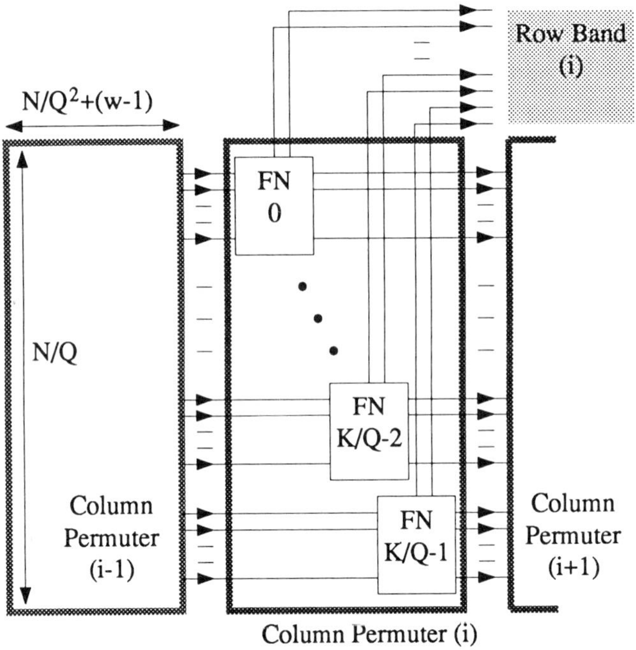

Figure 6: Layout of network U-I

4 Bit-Reversal Permutations

In the Fast Fourier transform (FFT) computations, the elements of the output transform vector are arranged by the bit-reversed order of their indices (provided that the elements of the input order were in the normal order). A bit-reversal permutation is needed to place the elements in the normal increasing order. The bit-reversal permutation is defined by the following mapping:

$$\beta(x_{n-1}x_{n-2} \quad \cdots \quad x_1x_0) = (x_0x_1 \quad \cdots \quad x_{n-2}x_{n-1}). \tag{8}$$

In compliance with the notation introduced in section 3, the permutation β can be rewritten as

$$abc \overset{\beta}{\longmapsto} \overline{c}\overline{b}\overline{a} \tag{9}$$

where $a = R^q_{n-q}$, $b = R^{n-2q}_q$, $c = R^q_0$. $\overline{a}$ denotes the string obtained by bit-reversing string a; $\overline{b}$ and $\overline{c}$ have a similar connotation. Permutation (9) can be decomposed into the following sequence of permutations.

$$abc \xmapsto{\beta_1} ab\overline{c} \tag{10}$$
$$ab\overline{c} \xmapsto{\beta_2} \overline{c}ba \tag{11}$$
$$\overline{c}ba \xmapsto{\beta_3} \overline{c}b\overline{a} \tag{12}$$
$$\overline{c}b\overline{a} \xmapsto{\beta_4} \overline{c}\overline{b}\overline{a} \tag{13}$$

As before, a is the column index and bc is the row index of the input elements, while $\overline{c}$ is the column index and $\overline{b}\overline{a}$ is the row index of the final output elements. Permutation β_2 (the block-transpose permutation) is crucial for the reduced-network design since it involves swapping the indices a and $\overline{c}$. The remaining permutations of the group perform bit-reversal on each index separately. The index map for permutation β_2 is shown in figure 7. The

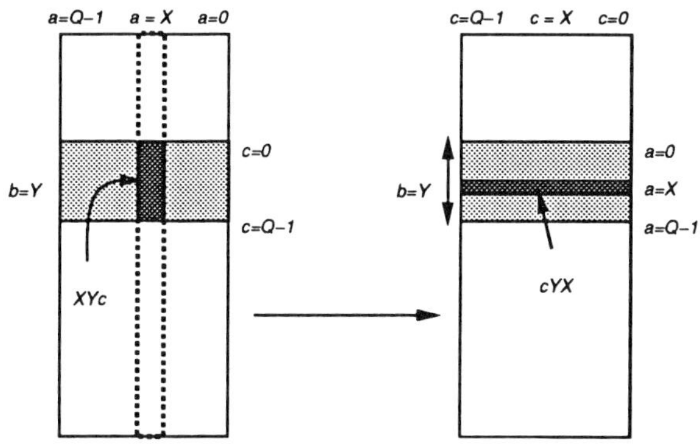

Figure 7: Index map for permutation β_2

map shows that if the input and output matrices are partitioned into $Q \times Q$ blocks, then the ith input block is mapped onto the ith output block, with the Q columns of input block mapped onto the rows of the corresponding output block. Thus, realizing permutation β_2 requires a simple $Q \times Q$ transpose network for each input block. Figure 8 shows a transpose network and its bit-serial version (with area $O(wQ^2)$).

Few derivational details details have been omitted due to space limitations. However, it can be noted that β_1 performs bit reversal on the column indices, β_2 transposes each $Q \times Q$ block of input elements, β_3 performs bit-reversal on the row-indices of each block, and β_4 performs bit reversal on the block indices. The complete structure of the reduce-area bit-reversal network is illustrated by the example of figure 9. The total area can be easily calculated as $O(N^2/Q^2)$, and the total delay as $O(wQ)$ for elements of

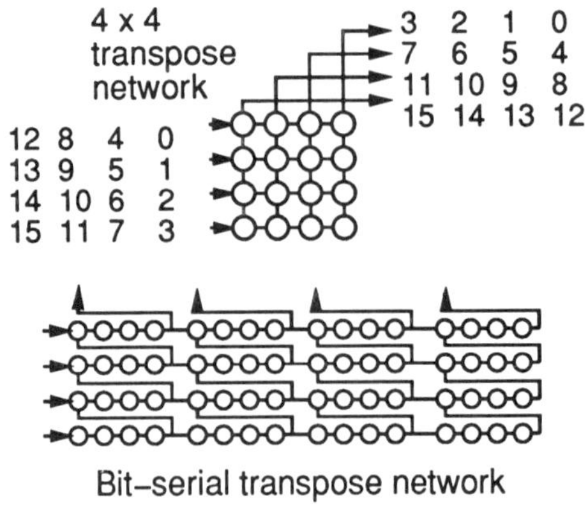

Figure 8: A network for permutation β_2 ($N = 4, w = 4$)

length w bits. To simplify presentation, figure 9 shows the block-transpose network in the word-model. In the actual design, the bit-serial network of figure 8 is used.

Network Partitioning

The reduced-area permutation networks developed in the previous sections can be partitioned among several chips so as to keep the number of I/O pins at a low count. Network S-I can be used to exemplify the partitioning strategy. Each folding network of S-I (figure 5) has N/K inputs and N/KQ outputs. Thus, each group of Q FNs (one from each column-permuter) which are connected horizontally, can be integrated on one chip with N/K I/O pins. In this case, K/Q chips can implement the the entire permuter of figure 6. For $N = 2^{16}$, $K = 2^9$, and $Q = 2^5$, this means that a single chip permuter with $N/Q = 2048$ I/O pins, can be replaced with $K/Q = 16$ chips each having $N/K = 128$ I/O pins. Note that this type of partitioning does not introduce any off chip communication delays (since the outputs of each chip are not used by other chips). Further reduction in I/O pin count per chip can be achieved by partitioning a folding network among several chips. However, off chip communication delays will be introduced in this case. The amenability of the permuter network to simple partitioning schemes is essential for practical implementations. Other advantages of the partitioning schemes include easier testing, higher fault-tolerance, and significant reduction in silicon area since part of the wiring will be routed in free space. For example, each of the K/Q chips described above will have area $O(N^2/K^2 + NQw/K)$ and the total area of all K/Q chips is $O(N^2/KQ + Nw)$.

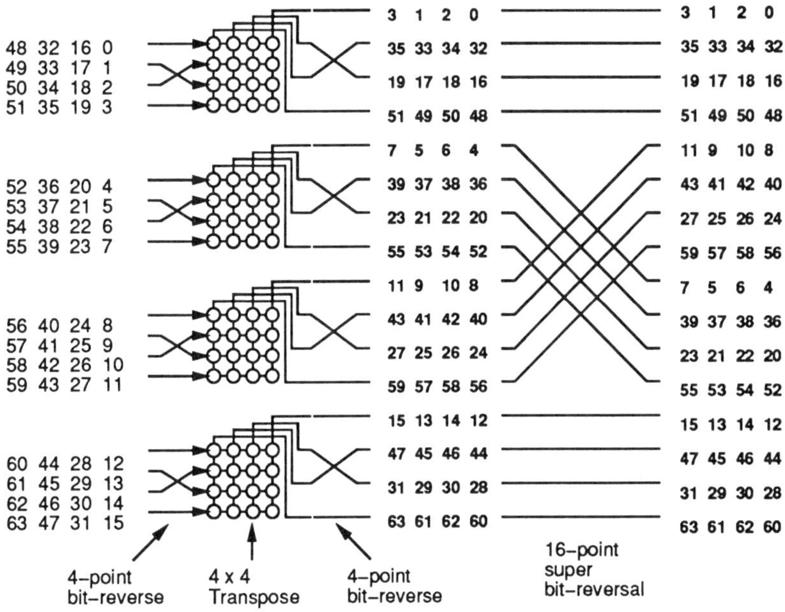

Figure 9: A bit-reversal network for $N = 64$ and $Q = 4$

5 CMOS VLSI Implementation

This section presents a CMOS VLSI layout of an 8–shuffle network (i.e. $K = 8$) on $N = 64$ inputs for $Q = 4$. Since $K > Q$, this implementation of the network specified in figure 6. For VLSI design purposes, the FN shown in figure 6 will viewed as a stack of $Q(= 4)$ row blocks, each consisting of $N/KQ(= 2)$ rows of $w(= 10)$ storage cell each. The serial outputs from the rows of one row block are fed to the inputs of the corresponding rows in the next row block. With every clock pulse the data is shifted from one cell to the next along the data path. The solid circles in the FN of figure 6 represent storage cells with two multiplexed inputs. Each such cell selects either the input from the left or from below depending on whether the network is operating in a horizontal-shift mode or vertical-shift mode, respectively. As explained in section 3, the duration of each mode is $wQ = 40$ clock pulses for this particular case. A simple control circuit is used to generate the multiplexer select control for the appropriate duration.

Logic Design

As mentioned above two types of cells are used to layout the data part of the FN; one is a standard D-latch, the other is a multiplexed two-input D-latch with a select line S. Logic diagrams of both cells are shown in figure 10. The multiplexer select control (S) is generated by a simple

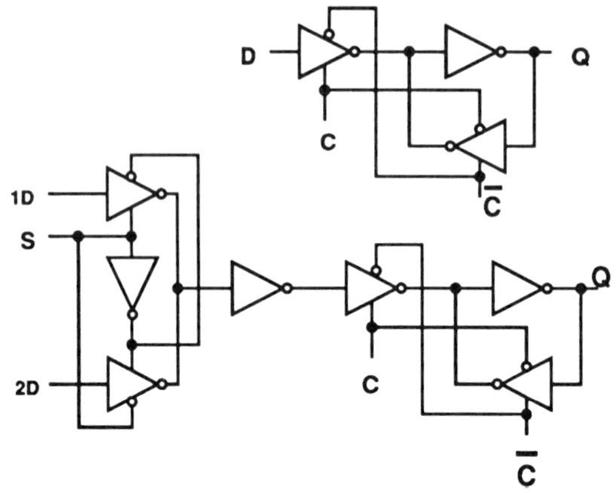

Figure 10: Logic diagrams of the basic FN cells

T-flipflop counter which counts from 0 to $(wQ - 1) = 39$ (the duration of horizontal-shift or vertical-shift mode). The counter consists $\lceil \log(wQ) \rceil = 6$ T-flipflops and additional logic. The counter is set using an external signal **mset** (synchronizing pulse) or by an internal terminal-count signal **set**.

Layout Design

The layout of the above specified shuffle network is derived using a hierarchical scheme. At the lowest level, CMOS-PW cells are obtained from a cell library [4]. These cells are used to generate the required basic cells (D-latches). The basic cells are laid out iteratively to form row-blocks. Row-blocks are then stacked to form FNs which are, in turn, stacked to form the permuter network. Owing to the regularity and uniformity (both in terms of size and function of cells) of the above architecture, the layout generation can be automated in a manner similar to that used with other structured devices such as PLAs and ROMs. A layout generation software called CFL [4] was used in our implementations. A module generator program for MOSIS 3-micron CMOS implementation of $K-$shuffle networks is presented in [3]. The reported program takes N, K, Q, and w as input parameters and automatically generates the data-part layout of an FN. The layouts of the

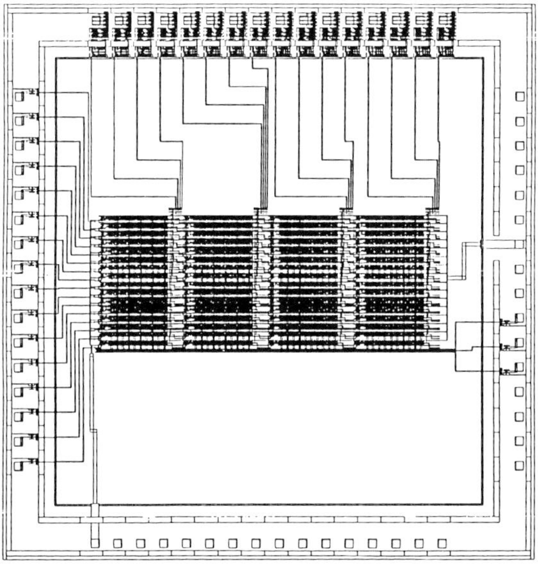

Figure 11: Chip layout for a 4–bit 8–shuffle ($N = 64$, $Q = 4$, $w = 4$)

basic cells (D-latches) are produced in *Ceaser* format [4]. Other tools used in the implementation include Mextra (Manhattan circuit extractor), Crystal (timing Analyzer), Lyra (design rule checker), and Pads (a pad frame generator). The control logic is laid out in a similar way [13]. The layout of the data part of the 8–shuffle network is shown in figure 11. Additionally, the sizes of the layouts produced for different values of the parameters N, K, Q, and w are listed in table 1.

N	K	Q	AREA (in mm^2) $\lambda = 1.5\mu m$	
			$w = 10$	$w = 1$
64	2	2	27.44	7.56
64	4	4	16.78	6.03
64	8	4	17.23	11.85
64	8	8	17.13	5.92
256	16	8	65.93	46.33
256	16	16	65.06	22.48
1024	32	32	253.33	87.55

Table 1: Area of shuffle networks for various parameter values

Conclusions and Comparisons

Index mapping is a simple method for deriving efficient permutations networks in VLSI over a wide range of the AT^2 tradeoff. Determining a suitable decomposition of the original permutation into a sequence of permutations is the most crucial step of the mapping, as the resulting permutations must have efficient layouts under the chip I/O schedule. For permutations in the BPC class [12], the general methodology is relatively simple and can be automated to a certain extent [3]. Although this paper considered permutations of the bit-permute type only, bit-complement permutations can be easily incorporated by adding one or more stages of exchange edges among the involved indices.

Implementations based on the index mapping technique show a marked improvement over previous designs. For example, the E-network proposed in [8] for 2-D transposition uses a RAM between shuffle stages and has $AT^2 = O(N^2 \log^3 N)$ for permuting elements of length $O(\log N)$ bits. In addition to the complex control and RAM addressing overheads, the E-network works only on square matrices and does not support area-time tradeoffs. Several enhancements are provided by the networks proposed in this paper. First, the implementation given in this section uses only shift registers (D-latches) with very simple control. Second, the proposed network can perform index rotation on multidimensional arrays or pseudo transposition of non-square matrices. Finally, the proposed network has $AT^2 = O(N^2w^2 + NQ^2w^3)$ for elements of any word length (w). Thus , $AT^2 = O(N^2w^2)$ for $Q \leq \sqrt{N/w}$ or $AT^2 = O(N^2w^3)$ for $\sqrt{N/w} \leq Q \leq \sqrt{N}$.

Acknowledgements

The work of first author has been supported in part by the Department of Electrical Engineering and the Centre for Integrated Computer Systems Research (CICSR) at the University of British Columbia. The work of the second and third authors has been supported in part by King Fahd University of Petroleum & Minerals.

References

[1] H. M. Alnuweiri, "A New Class of Optimal VLSI Networks for Multidimensional Transforms", Proc. International Conference on Parallel Processing, 1990.

[2] H. M. Alnuweiri, "Optimal Bounded-Degree VLSI Networks for Sorting in a Constant Number of Rounds", Proc. Supercomputing '91, Albuquerque, NM, 1991.

[3] H. M. Alnuweiri, "Efficient Permutation Routing in VLSI", Tech. Report, UBC, 1992.

[4] W. Beckett, "CFL - Coordinate Free Lab", Ref. Manual (version 1.2), Northwest LIS Release 3.1, University of Washington.

[5] G. Bilardi and M. Sarrafzadeh, "Optimal VLSI Circuits for Discrete Fourier Transform in VLSI", in *Advances in Computing Research*, Ed. F. P. Preparata, JAI Press, Greenwich, CT, 1987, pp. 87-101.

[6] G. Bilardi, S. W. Hornick, M. Sarrafzadeh, "Optimal VLSI Architectures for Multidimensional DFT", ACM Symposium on Parallel Algorithms and Architectures, Santa Fe, NM, June 1989, pp.365-272.

[7] I. Gertner and M. Shamash,"Architectures for Multidimensional Fourier Transform Processing," *IEEE Transactions on Computers*, c-36, Nov. 1987.

[8] B. T. Kelley and V. K. Madisetti, "Optimal Concurrent VLSI Architectures for 2-D Transposition", Proc. Advanced Research in VLSI, C. Sequin Ed., MIT Press, March 1991.

[9] J. Kilian, S. Kipnis, C. E. Leiserson, "The Organization of Permutation Architectures with Bused Interconnections", *IEEE Transactions on Computers*, c-39, 11, 1990, pp. 1346-1358.

[10] D. Kleitman, F. T. Leighton, M. Lepley, G. L. Miller, "New Layouts for the Shuffle-Exchange Graph" Proc. ACM Symp. on Theory of Computation, pp. 278-292, 1981.

[11] F. T. Leighton, "Tight Bounds on the Complexity of Parallel Sorting", *IEEE Transactions on Computers*, c-34, 4, April 1985, pp. 344-354.

[12] D. Nassimi and S. Sahni, "A Self Routing Benes Network and Parallel Permutation Algorithms", *IEEE Transactions on Computers*, c-30, pp. 332-340, 1981.

[13] S. M. Sait and M. A. Kulaib, "CMOS Cell for Parallel Loadable Counters", International Journal of Electronics, pp. 867-871, 1987.

[14] C. D. Thomson,"Area-Time Complexity for VLSI," Proc. ACM Symposium on Theory of Computation, May 1979.

[15] J. Ullman, *Computational Aspects of VLSI*, Computer Science Press, 1984.

Author Index

The MIT Press, with Peter Denning as general consulting editor, publishes computer science books in the following series:

ACL-MIT Press Series in Natural Language Processing
Aravind K. Joshi, Karen Sparck Jones, and Mark Y. Liberman, editors

ACM Doctoral Dissertation Award and Distinguished Dissertation Series

Artificial Intelligence
Patrick Winston, founding editor
J. Michael Brady, Daniel G. Bobrow, and Randall Davis, editors

Charles Babbage Institute Reprint Series for the History of Computing
Martin Campbell-Kelly, editor

Computer Systems
Herb Schwetman, editor

Explorations with Logo
E. Paul Goldenberg, editor

Foundations of Computing
Michael Garey and Albert Meyer, editors

History of Computing
I. Bernard Cohen and William Aspray, editors

Logic Programming
Ehud Shapiro, editor; Fernando Pereira, Koichi Furukawa, Jean-Louis Lassez, and David H. D. Warren, associate editors

The MIT Press Electrical Engineering and Computer Science Series

Research Monographs in Parallel and Distributed Processing
Christopher Jesshope and David Klappholz, editors

Scientific and Engineering Computation
Janusz Kowalik, editor

Technical Communication and Information Systems
Ed Barrett, editor